T0260225

Artificial Intelligence and Cognitive Science

Conceptual Issues

Series Editors

Andy Clark and Josefa Toribio
Washington University

Series Contents

Consciousness and Emotion in Cognitive Science

Conceptual and Empirical Issues

Edited with an introduction by

Josefa Toribio and Andy Clark
Washington University

Routledge
Taylor & Francis Group

NEW YORK AND LONDON

First published 1998 by Garland Publishing, Inc.

This edition published 2014 by Routledge
605 Third Avenue, New York, NY 10017
4 Park Square, Milton Park, Abingdon, Oxon OX14 4RN

Routledge is an imprint of the Taylor & Francis Group, an informa business

Introduction copyright © 1998 Josefa Toribio and Andy Clark.

Library of Congress Cataloging-in-Publication Data

Consciousness and emotion in cognitive science : conceptual and
    empirical issues / edited with introductions by Andy Clark and
    Josefa Toribio.
        p.   cm. — (Artificial intelligence and cognitive science ; 3)
        Includes bibliographical references.
        ISBN 0-8153-2770-6 (alk. paper)
        1. Consciousness. 2. Emotions. 3. Cognitive science. I. Clark,
    Andy, 1957– . II. Toribio, Josefa. III. Series.
    BF311.C6446 1998
    153—dc21   98-27059
                                                    CIP

ISBN 13 : 978-0-8153-2770-7 (hbk)

Contents

Introduction

Qualia Control

[We don't know], even to a first glimmer, how a brain (or anything else that is physical) could manage to be a locus of conscious experience. This . . . is, surely, among the ultimate metaphysical mysteries; don't bet on anybody ever solving it. Jerry Fodor (1995), p. 5

It is now possible, for the first time, to formulate testable hypotheses about how activities in the brain can add up to the phenomenon of consciousness. Daniel Dennett (this volume) (original ms. p. 87)

The study of consciousness is almost unique in its capacity to generate diametrically opposed diagnoses of the state of the art. Are we getting closer and closer to solving the puzzle? Or are we, as Jerry Fodor asserts bleakly, searching in conceptual and empirical darkness, without even the first glimmer of a real clue? And just what is the 'puzzle of consciousness' anyway? Is the explanation of consciousness intimately linked to the explanation of thought and reason, or is it largely orthogonal to that (somewhat better understood) domain? The papers gathered in this volume offer a broad set of perspectives upon these topics, and range from out-and-out philosophical analysis to empirically based research and conjecture. The closing papers deal with emotion — a closely related area that is only just beginning to receive the attention it deserves.

In this brief introduction, we shall comment on three central and recurring themes and issues in the study of consciousness. First, does 'consciousness' name a single or in some way unitary phenomenon? Second, what is the relation (if any) between the project of explaining consciousness and that of explaining intentionality? And third, is there really a special problem about 'qualia' or subjective feel? The questions are deeply interlinked, however, and so we shall allow our discussion to move freely between them.

Let us begin, however, with the first: Does 'consciousness' name a unitary phenomenon? Here, the general (but by no means universal) consensus seems to be that it does not. Thus Nelkin nicely distinguishes three features that a theorist might have in mind — phenomenality, intentionality, and introspectability. Thus, to follow Nelkin's example, when I look at my watch I (i) have a distinctive subjective experience in the visual modality — there is "something it is like" (Nagel 1974) to look at the watch, and that something feels very different to e.g. feeling a Braille display — (ii) exhibit intentionality, in that my mental state is *about* the watch (or perhaps the time, or both) (iii) am capable of introspectively *noticing* that I am e.g. attending to my watch. Nelkin argues that although many of the experiences we call 'conscious' exhibit all three of these features, the features themselves are conceptually and empirically distinct, and are capable of obtaining alone or in (most) other combinations. They are *not* "three features of a single noncomposite state" (p. 419 in original). These same three features are discussed by Rosenthal, who tries, nonetheless, to achieve a more unitary vision. Rosenthal suggests that a mental state is conscious if it causes us to have a higher-order thought whose content is that we are in that (lower order) mental state. Such a move would *unify* all conscious phenomena since they would all involve this higher-order awareness. It would also (to jump temporarily to our second question) pull the phenomena of consciousness into the broad gravitational field of intentionality — though it does not identify the two, since it is this special higher-order aboutness, not aboutness per se, that then emerges as the mark of conscious mentality. Lycan adopts a similar line, but he depicts the higher-order relation as less cognitive — more like inner attention and perception than like thoughts-about-thoughts. Hence the slogan 'consciousness as internal monitoring.' Lycan is explicitly concerned, however, to argue that "the mind has no special properties that are not exhausted by its representational properties" (original ms. p. 2). And in this sense he too wants to account for the qualitative and subjective dimensions under the more tractable umbrella of understanding representational role.

It is, by contrast (and jumping to our third issue), typically those who (unlike Lycan and Rosenthal) believe that there is a *special* problem about explaining subjective feel (qualia) who also defend a 'mixed bag' vision of consciousness. For the mixed bag can contain lots of phenomena that will indeed be explicable using the kind of apparatus (internal representations, internal monitoring, and the like) we use to explain other aspects of intelligence. But it will also — and problematically — contain phenomena more resistant to such analysis and explanation. In this vein, Ned Block (1995) argues that the concept of consciousness is a 'mongrel' concept denoting a number of significantly different phenomena, with the deepest difference that between 'phenomenal consciousness' and 'access-consciousness.' Phenomenal consciousness has to do with experienced feel, "what it's like," etc., whereas access consciousness has to do with the various ways in which information can be used in thought and in the control of action.

This kind of distinction is heartily promoted by David Chalmers, who distinguishes two kinds of questions about conscious phenomena. One (the "easy question") concerns functional capacities: How can a physical device access such-and-such a stored memory, achieve such-and-such a feat of perceptual recognition, and so on. We know, broadly speaking, what an explanation of such capacities might look like (witness, e.g., the work discussed in volumes one and two of this series). Not so,

Chalmers suggests, when we confront the "hard problem" of explaining not (mere) behavior but the feelings and experiences that sometimes accompany it. The deep redness of a Hawaiian sunset, the rousing blue of an exotic cocktail, the feel of the warm wind on your face, or the sweet and salty taste of the cocktail, and so on. The hard question really has two parts: Why does such-and-such a physical event feel like anything *at all*? And why does it feel like *this* rather than like *that*? Chalmers suggests that the hard question(s) cannot, in principle, be answered by telling familiar kinds of functional/computational/representational stories. For it is always imaginable that the functional story might apply and yet there be no experience at all (or a *different* experience, or a shifting dance of experiences . . .). At this point, some philosophers (e.g. McGinn 1989) diagnose a kind of human cognitive blind spot. The hard problem is simply insoluble by beings with brains like ours. Frank Jackson makes some similar comments at the end of his paper (this volume), where he too asserts the existence of a kind of 'explanatory gap' (Chalmers) such that no amount of physical storytelling can fully explain the phenomenal facts. Chalmers himself is, in a sense, more optimistic than Jackson. Chalmers asserts that no functional/computational story can explain the phenomenal facts, but thinks that some *other* kind of (still scientific) story might yet do so. He suggests that we might, for example, need to recognize a kind of fundamental 'double aspect' to physically embodied information, such that information and some degree of phenomenal content often (perhaps always — Chalmers leaves this open) go together. Positing such a link between information and experience is effectively allowing experiential properties to figure as new fundamental properties of the universe.

   This is a bold proposal. But it is not immediately clear that it really solves the problem. Instead, it seems rather like a (perhaps premature) assertion that no real explanation is to be had — that that is just how things are. While it is true that the kinds of mechanistic explanations that really make us feel that 'yes, we *see* how that works' have to stop somewhere (some properties of matter and space-time are thus simply brute facts), it seems way too soon to conclude that phenomenal consciousness is thus irreducible and basic. The more so since, unlike space-time and subatomic physics, consciousness looks (superficially at least) to be a property that is ushered into being by some specific *causal organizations* of matter. But if it is the product of a certain type of organization, then it is that organization that we need to understand.

   The paper by John Searle adopts a characteristically hard line, arguing that *all* mental phenomena are tied, ultimately, to the presence (or potential presence) of some kind of conscious, first-person awareness. Where Rosenthal, Lycan, and others thus depict consciousness as mental representation, Searle goes the other way, depicting all mental representations in terms of its potential accessibility to consciousness. And the scientific explanation of consciousness itself, Searle believes, will involve understanding its neurobiological (not functional or computational) roots. Alongside this apparent scientific optimism, however, Searle asserts (see, e.g., Searle 1992, chapter five) that consciousness is *in a sense* irreducible, in that any third-person, scientific account must indeed 'leave out' the subjective, first-person dimensions.

   Paul Churchland argues strongly against the claim (Chalmers, Jackson, and to some extent Searle) that consciousness (phenomenal consciousness) is "ontologically distinct and physically irreducible" (original ms., p. 211). The "explanatory gap" meant to characterize the hard problem is, Churchland suggests, no larger in the case of consciousness than it was in, for example, the case of light. The gap between our folk understanding of light and the scientific image of light as electromagnetic radiation might incline us to construct some (misleading) thought experiments designed to show that light cannot be fully explained as electromagnetic radiation. But such arguments are invalid and do nothing to establish the scientific story as genuinely incomplete. Churchland himself is optimistic. The march of neuroscience and the "conceptual framework of vector coding and parallel distributed processing in large-scale recurrent neural networks" (Churchland 1995, p. 226) holds out real hope, he thinks, for a satisfying scientific account of all aspects of conscious experience.

Daniel Dennett is likewise optimistic, although he tells a rather different story. Consciousness — at least the kind distinctive of much human experience — is, Dennett argues, caused by the brain's implementation of a virtual serial processor. This serial device (the 'Joycean machine,' following James Joyce's literary depiction of the stream-of-consciousness) arose as a result of a kind of (partial) software-level restructuring initiated by simple self-stimulation (via cries, exclamations, etc.) and fed and nurtured by the emergence of ever more complex language. The activity of this (virtual) machine strikes us as conscious because it encourages a kind of 'user illusion' in which only certain aspects of the overall computational process are accessible, reportable, and so on. It is this pattern of informational access that makes us look — to ourselves and others — like single subjects of experience. Elsewhere (e.g., Dennett 1991, chapter 12), Dennett extends this story to the vexed issue of qualia and phenomenal feel, insisting that this, too, is a kind of illusion created by the particular ways in which we can and cannot access the inner workings of our own cognitive and computational systems. Dennett thus rejects the idea of a 'hard problem,' insisting that once you have resolved all the 'easy,' function-and-access oriented issues, there is simply no residue left over to explain. The apparent residue — the qualitative nature of the experience — is just "an imaginary dazzle in the eye of a Cartesian homunculus" (Dennett 1995, p. 34).

Dennett and Churchland are thus confident that empirical research, carried out within broadly familiar frameworks, will prove sufficient to lay the ghost of consciousness, 'hard' aspects and all. But where both Dennett and Churchland expect (different but) largely *unified* scientific accounts, Owen Flanagan suggests we may need to work piecemeal, building theories for one domain at a time. Flanagan depicts 'consciousness' as a "superordinate term for a heterogeneous array of mental state types" (original ms., p. 1103) that share the property of having some associated phenomenal feel. And he further advocates the use of the 'natural method' of theory building, the use of multiple types of evidence and resources, and the investigation of multiple domains and competences. Flanagan thus discusses visual awareness, conscious event memory, and dreaming — all from the multiple perspectives of phenomenology, cognitive science, neuroscience, and evolutionary biology. Crick and Koch continue the focus on visual awareness, and explicitly describe their project as seeking the 'neural correlates' of conscious (visual) experience. Damasio and Damasio speak of a 'neurobiology of consciousness' and address the question of how the brain creates conscious visual imagery and (intriguingly) what it is that makes us treat such imagery as our own, what ties it to the kind of 'consistent perspective' on the world that is associated with the point of view of an individual person? It is by addressing this latter kind of issue, perhaps, that stories about neural correlates may yet hope to bridge the apparent gap between isolating crucial inner states and explaining why various kinds of inner activity present themselves as specific, felt experiences.

The papers on emotion take a more traditionally computational perspective, depicting emotion as (in part) a device for goal ordering and speedy decision-making in real-time agents (Sloman), and for efficient social coordination (Johnson-Laird and Oatley). O'Rorke and Ortony focus on reasoning *about* emotions and offer a theory of "the cognitive antecedents of emotion" (original ms., p. 321). Here too, as in the papers on visual and imagistic consciousness, the threat of the explanatory gap looms large. Such theories could clearly explain *something* — and something of major importance at that. But can they really help us understand why we have experiences *at all*, or why those experiences have the *specific characters* they have?

The proof of these puddings is indeed in the eating, and the jury is still out. If, for example, the representationalist (Rosenthal, Lycan, Tye 1996 and Dretske 1995) is right to insist that (roughly) the representational facts exhaust the phenomenal facts, then a full account of certain neural or computational correlates might indeed constitute a complete account of phenomenal experience. There need be no 'hard problem' after all, no 'phenomenal residue' to explain. But if this is the case, we at least need to understand more fully why it *seems as if* there is such a residue, and why it *seems as if* no amount of functional, neuroscientific, or computational theorizing could explain conscious awareness in all its manifest glory. The issue, thus reconceived, is about why some explanations make us so much happier than others. This is what Mark Price (1995) calls the "psychology of the hard problem." The hard problem I end on is therefore this: Is there *really* a hard problem about phenomenal consciousness, or are we just imagining a wall we can't jump over? And how are we to tell?

Josefa Toribio and Andy Clark

References

Block, N. (1995) "On a Confusion about a Function of Consciousness." *Behavioral and Brain Sciences* 18: 227–47.
Dennett, D. (1991) *Consciousness Explained*. Boston: Little Brown.
Dennett, D. (1995) "Facing Backwards on the Problem of Consciousness." In J. Shear (ed.), *Explaining Consciousness: The Hard Problem*. Cambridge: MIT Press, pp. 33–36.
Dretske, F. (1995) *Naturalizing the Mind*. Cambridge: MIT Press.
Fodor, J. (1995) "West Coast Fuzzy: Why We Don't Know How Minds Work" (Review of Paul Churchland, *The Engine of Reason, The Seat of the Soul*, Cambridge: MIT Press, 1995). *Times Literary Supplement*, August 25, 1995, pp. 5–6.
Koch, C. and Crick, F. (1994) "Some Further Ideas Regarding the Neuronal Basis of Awareness." In C. Koch and J. Davis (eds.), *Large-Scale Neuronal Theories of the Brain*. Cambridge: MIT Press, pp. 93–110.
McGinn, C. (1989) "Can We Solve the Mind-Body Problem." *Mind* 98: 349–66.
Searle, J. (1992) *The Rediscovery of the Mind*. Cambridge: MIT Press.
Tye, M. (1996) *Ten Problems of Consciousness: A Representational Theory of the Phenomenal Mind*. Cambridge: MIT Press.

DAVID M. ROSENTHAL

TWO CONCEPTS OF CONSCIOUSNESS

(Received 8 July, 1985)

No mental phenomenon is more central than consciousness to an adequate understanding of the mind. Nor does any mental phenomenon seem more stubbornly to resist theoretical treatment.

Consciousness is so basic to the way we think about the mind that it can be tempting to suppose that no mental states exist that are not conscious states. Indeed, it may even seem mysterious what sort of thing a mental state might be if it is not a conscious state. On this way of looking at things, if any mental states do lack consciousness, they are exceptional cases that call for special explanation or qualification. Perhaps dispositional or cognitive states exist that are not conscious, but nonetheless count as mental states. But if so, such states would be derivatively mental, owing their mental status solely to their connection with conscious states. And perhaps it makes sense to postulate nonconscious versions of ordinary mental states, as some psychological theories do. But if consciousness is central to mentality in the way this picture insists, any such states are at best degenerate examples of mentality, and thus peripheral to our concept of mind.

This picture is both inviting and familiar. But there are other features of the way we normally think about mind which result in a rather different conception of the relation between consciousness and mentality. We often know, without being told, what another person is thinking or feeling. And we sometimes know this even when that person actually is not aware, at least at first, of having those feelings or thoughts. There is nothing anomalous or puzzling about such cases. Even if it is only seldom that we know the mental states of others better than they do, when we do, the mental states in question are not degenerate or derivative examples of mentality. Moreover, conscious states are simply mental states we are conscious of being in. So when we are aware that somebody thinks or feels something that that person is initially unaware of thinking or feeling, those thoughts and feelings are at first mental states that are not also conscious states. These considerations suggest a way of

Philosophical Studies **49** (1986) 329–359.
© 1986 *by D. Reidel Publishing Company*

looking at things on which we have no more reason to identify being a mental state with being a conscious state than we have to identify physical objects with physical objects that somebody sees. Consciousness is a feature of many mental states but, on this picture, it is not necessary or even central to a state's being a mental state. Consciousness seems central to mentality only because it is so basic to how we know our own mental states. But how we know about things is often an unreliable guide to their nature.

These two alternative pictures of the connection between consciousness and mentality have different implications about what sort of explanation is possible of what it is for a mental state to be a conscious state. If we take the view that consciousness is not a necessary feature of mental states, then we cannot define mental states as conscious states. Accordingly, we must seek some other account of what makes a state a mental state. But once we have an account of mentality that does not appeal to consciousness, we can then try to explain what conscious states are by building upon that very account of mentality. In particular, it then makes sense to try to formulate nontrivial necessary and sufficient conditions for a mental state to be a conscious state. On this conception of mentality and consciousness, it is open for us to proceed sequentially in this way, first defining mentality and then consciousness.

No such procedure, however, is possible if instead we adopt the view that being a mental state is at bottom the same as being a conscious state. For we cannot then explain what makes conscious states conscious by appeal to a prior account of mentality, since on that view mentality presupposes consciousness itself. Any attempt to explain consciousness by formulating necessary and sufficient conditions for a mental state to be conscious will thus automatically fail. If consciousness is already built into mentality, any such explanation will be uninformative. If not, then, on the present view, the conception of mind on which our explanation of consciousness is based will unavoidably be radically defective. It is plain that there is no third way; nothing that is not mental can help to explain consciousness. So, if consciousness is essential to mentality, no informative, nontrivial explanation of consciousness is possible at all. Moreover, since we cannot then proceed sequentially, explaining mentality first and then consciousness, the gulf that seems to separate mind and consciousness from the rest of reality will appear impossible to bridge. Thomas Nagel succinctly expresses this view when he writes that "[c]onsciousness is what makes the mind-body problem really intractable."[1]

Although it seems effectively to preclude our giving any informative explanation of consciousness, the view that consciousness is essential to all mental states does have apparent advantages. For one thing, that view, which has strong affinities with the Cartesian view of mind, fits well with many of our common-sense intuitions about the mental. And perhaps that view even does greater justice to those intuitions than a view of the mind on which not all mental states are conscious. These two competing pictures of mind and consciousness seem to present us, therefore, with a difficult choice. We can opt to save our presystematic intuitions at the cost of being unable to explain consciousness. Or we can hold open the possibility of giving a satisfactory explanation, but risk being less faithful to our common-sense intuitions about what mental states are.

One reaction to this quandary is simply to accept the more Cartesian of the two pictures, and accept that an illuminating explanation of consciousness will simply prove impossible. This traditional response is now hard to credit. Mind and consciousness are continuous with other natural phenomena of which we can give impressively powerful explanations. And it is difficult to believe that a singularity in nature could exist that would utterly and permanently resist all attempts to explain it. For these reasons, some more recent writers have chosen, instead, simply to abandon common-sense intuitions about mind when they conflict with our explanatory goals. Physics does not aspire to reconstruct all our presystematic intuitions about the things around us. Why, proponents of this eliminativist approach ask, should the science of mind proceed differently?[2]

But we should, wherever possible, seek to explain our common-sense intuitions rather than just explain them away. And we should hesitate to jettison our presystematic conceptions of things, whether mental or physical, unless efforts to do justice to them have decisively failed. Indeed, even physical theories must square as much as possible with our common-sense picture of physical reality. In what follows, I argue that we need embrace neither the Cartesian nor the eliminativist stance toward the consciousness of mental states. Instead, we can both be faithful to our presystematic intuitions about consciousness and mind and, at the same time, construct useful, informative explanations of those phenomena. In section I, I develop

3

the two pictures sketched above. In particular, I articulate the two different definitions of mentality itself that comprise the core of those two pictures. I then use the non-Cartesian concept of mind and consciousness to construct a systematic and theoretically satisfying explanation of what it is for a mental state to be conscious – an explanation, that is, of what it is that distinguishes conscious from nonconscious mental states. I show further how the definition of mentality central to each of the two pictures determines a distinct conception of consciousness, and how the Cartesian concept of consciousness makes any informative explanation of consciousness impossible. In sections II and III, then, I go on to argue that the non-Cartesian explanation can save the phenomenological appearances and explain the data of consciousness as well as the more familiar Cartesian picture. And I argue there that the standard considerations that favor the Cartesian view are baseless. And in section IV I conclude with some observations about consciousness and our knowledge of the mental, and about the actual significance of the insights that underlie the Cartesian picture.

I

All mental states, of whatever sort, exhibit properties of one of two types: intentional properties and phenomenal, or sensory, properties. Something has an intentional property if it has propositional content, or if it is about something. Sensory properties, by contrast, are less homogeneous. Examples are the redness of a visual sensation and the sharp painful quality of certain bodily sensations. Some mental states may have both intentional and phenomenal properties. But whatever else is true of mental states, it is plain that we would not count a state as a mental state at all unless it had some intentional property or some phenomenal property.

Something close to the converse holds as well. For one thing, only mental states can have phenomenal properties. Although we use words such as 'red' and 'round' to refer to properties of physical objects as well as to properties of mental states, we refer to different properties in the two cases. The introspetible redness of a visual sensation is not the same property as the perceptible redness of a tomato, for example, since each can occur in the absence of the other. Moreover, mental states are not objects at all, and therefore cannot have the same properties of shape and color that physical physical objects have. Indeed, we do not even use quality words the same way when

4

we talk about mental states and about physical objects. We speak interchange-ably about red sensations and sensations of red, but it makes no nonmeta-phorical sense to talk about tomatoes of red. Similar considerations apply to properties that are special to bodily sensations. Knives and aches may both be dull, but the dullness of a knife, unlike that of an ache, has to do with the shape of its edge. Phenomenal properties, properly so called, are unique to mental states.[3]

Things are slightly less straightforward with intentional properties, since items other than metal states can exhibit in intentionality. Speech acts and works of art, for example, can be about things and can have propositional con-tent. But except for mental states themselves, nothing has intentional proper-ties other than those modes and products of behavior which express inten-tional mental states. So it is reasonable to hold that these modes and products of behavior derive their intentionality from the mental states they express. As Roderick M. Chisholm puts it, "thoughts are a 'source of intentionality' – i.e., nothing would be intentional were it not for the fact that thoughts are intentional."[4] So, even though intentional properties belong to things other than mental states, they do so only derivatively. Accordingly, all mental states have intentional or sensory properties, and sensory properties belong only to mental states and intentional properties nonderivatively to mental states alone. We thus have a compelling basis for defining mental states as just those states which have either intentionality or phenomenal quality.

There are, however, objections to this way of delineating the distinctively mental which seem to favor a mark of the mental based on consciousness instead. For one thing, a mark of the mental that relies solely on intentional and phenomenal properties may seem to underplay the special access we have to our own mental states. Even if all mental states do exhibit intentional or sensory features, one might urge that the more revealing mark of the mental would somehow appeal, instead, to that special access. On such a mark, what is essential to mental states would not be intentional or sensory character, but consciousness itself. Moreover, if we take the possession of either sensory or intentional properties to be definitive of the mental, we must then explain why we regard this disjunctive mark as determining the mental. Why do we construe as a single category the class of states that have one or the other of these two kinds of properties? It does not help to note that that some mental states, for example, perceptual states, have both sorts of characteristic. Despite the existence of such mongrel cases, it seems unlikely that pure

5

phenomenal states, such as pains, have anything interesting in common with pure intentional states, such as beliefs. And we can avoid this difficulty if, instead, we take consciousness to be what makes a state a mental state. Finally, the characteristically mental differences among kinds of mental states are all differences in what intentional or sensory properties those states have. So those properties may seem to figure more naturally in an account of how we distinguish among types of mental state than in an account of how mental states differ from everything else. These various considerations all suggest that an account of mind in terms of consciousness may be preferable to an account that appeals to intentionality and phenomenal character.

Moreover, defining mentality in terms of consciousness need not involve any circularity. We can say what it is to be a conscious state in a way that does not explicitly mention being mental. A state is conscious if whoever is in it is to some degree aware of being in it in a way that does not rely on inference, as that is ordinarily conceived, or on some sort of sensory input. Conscious states are simply those states to which we have noninferential and nonsensory access.

People do, of course, have many more beliefs and preferences, at any given time, than occur in their stream of consciousness. And the nonconscious beliefs and preferences must always have intentional properties. But this need not be decisive against taking consciousness as our mark of the mental. For we can construe beliefs and preferences as actual mental states only when they are conscious. On other occasions we could regard them to be merely dispositions for actual mental states to occur; in those cases, we can say, one is simply disposed to have occurrent thoughts and desires.

Consciousness is intuitively far more crucial for sensory states than for intentional states. This disparity is something we must explain if we take consciousness as the mark of all mental states. Construing nonconscious beliefs and preferences not as actual mental states but as mere dispositions to be in such states helps us give a suitable explanation. Sensory states normally result from short-term stimulations; so we have little reason to talk about our being disposed to be in particular types of sensory state. By contrast, we are often disposed to be in intentional states of various kinds. Since we are typically not conscious of being thus disposed, the tie between consciousness and mentality may at first sight seem less strong with intentional than with sensory states. But that tie may apply equally to both sorts of state if we count only nondispositional states as mental states, properly speaking. For when we focus on short-term, episodic intentional states, the common-sense intuition that mental states must be conscious is no less compelling than it is in the case of phenomenal states.

The two marks of the mental just sketched are independent of each other, and both lay claim to long and well-established histories. Thus writers with Cartesian leanings have generally favored some mark based on consciousness, while those in a more naturalist, Aristotelian tradition have tended to rely instead on some such mark as intentionality or sensory character. For it is a roughly Aristotelian idea that the mental is somehow dependent on highly organized forms of life, in something like the way in which life itself emerges in highly organized forms of material existence. And this idea suggests that one should try to delimit the mental in terms of the various distinctively mental kinds of functioning and, thus, by reference to the intentional and phenomenal characteristics of mental states. To the Aristotelian, such a mark has the advantage of inviting one to conceive of the mental as continuous with other natural phenomena. Thus Aristotle's own account of psychological phenomena gives great prominence to sense perception, thereby stressing the continuity between the mental and the biological.

The Cartesian tradition, by contrast, conceives of the mental as one of the two jointly exhaustive categories of existence, standing in stark opposition to everything physical. And on this view it is tempting to select some single essential feature, such as consciousness, to be the mark of the mental. For this kind of mark will stress the sharp contrast between mental and physical, and play down the differences among types of mental states compared to how different all mental states are from everything else. In this spirit, Descartes takes nonperceptual, propositional states to be the paradigm of the mental and, notoriously, has great difficulty in explaining how perception can involve both mental and bodily states.

Although both marks of the mental have enjoyed widespread acceptance, it is crucial which mark we adopt if our goal is to give an explanation of consciousness. Conscious states are simply mental states we are conscious of being in. And, in general, our being conscious of something is just a matter of our having a thought of some sort about it. Accordingly, it is natural to identify a mental state's being conscious with one's having a roughly contemporaneous thought that one is in that mental state. When a mental state is conscious, one's awareness of it is, intuitively, immediate in some way. So we can stipulate that the contemporaneous thought one has is not mediated

7

by any inference or perceptual input. We are then in a position to advance a useful, informative explanation of what makes conscious states conscious. Since a mental state is conscious if it is accompanied by a suitable higher-order thought, we can explain a mental state's being conscious by hypothesizing that the mental state itself causes that higher-order thought to occur.

At first sight it may seem that counterexamples to this explanation are rife. Although we are usually, when awake, in some conscious mental state or other, we rarely notice having any higher-order thoughts of the sort this explanation postulates. Typically, mental states occur in our stream of consciousness without our also having any evident thought that we are in those states. But such cases are not counterexamples unless we presuppose, contrary to the present explanation, that all mental states are conscious states. For otherwise, there will be no reason to assume that the higher-order thoughts that our explanation posits would, in general, be conscious thoughts. On this explanation, a mental state is conscious if one has a suitable second-order thought. So that second-order thought would itself be a conscious thought only if one also had a third-order thought that one had the second-order thought. And it begs the question against that account to assume that that these higher-order thoughts are usually, or even often, conscious thoughts. If a mental state's being conscious does consist in one's having a suitable higher-order thought, there is no reason to expect that this thought would ordinarily be a conscious thought. Indeed, we would expect, instead, that the third-order thoughts that confer consciousness on such second-order thoughts would be relatively rare; it is hard to hold in mind a thought about a thought that is in turn about a thought. So the present account correctly predicts that we would seldom be aware of our second-order thoughts, and this actually helps confirm the account.

It is important to distinguish a mental state's being conscious from our being introspectively aware of that state. Higher-order thoughts are sometimes invoked to explain introspection, which is a special case of consciousness.[5] But introspection is a more complex phenomenon than the ordinary consciousness of mental states. Intuitively, a mental state's being conscious means just that it occurs in our stream of consciousness. Introspection, by contrast, involves consciously and deliberately paying attention to our contemporaneous mental states. As Ryle remarks "introspection is an attentive operation and one which is only occasionally performed, whereas consciousness is supposed to be a constant element of all mental processes."[6]

Normally when mental states occur in one's stream of consciousness, one is unaware of having any higher-order thought about them. But when we are reflectively or introspectively aware of a mental state, we are aware not only of being in that mental state; we are also aware that we are aware of being in it. The Cartesian picture of mind and consciousness thus tacitly conflates a mental state's being conscious with our being introspectively aware of it. For on that picture the consciousness of a mental state is inseparable from that mental state. So reflective awareness, which is being aware both of a mental state and of one's awareness of that state, will be inseparable from awareness of the state which is not thus reflective. Here our common-sense intuitions diverge from the Cartesian view that consciousness is essential to mental states, since the two kinds of awareness plainly do differ.

Introspection is consciously and deliberately paying attention to mental states that are in our stream of consciousness. So, whatever else one holds about consciousness, it is natural to explain introspection as one's having a conscious higher-order thought that one is in the mental state that one is introspectively aware of. So, if these higher-order thoughts all had to be conscious, we could invoke them only to explain introspective consciousness. For only when we are introspectively aware of a mental state are we also aware of our higher-order thoughts. But higher-order thoughts are not automatically conscious, any more than other mental states are. They are conscious only when we have a yet high-order thought that we have such a thought. So there is no difficulty about using higher-order thoughts to explain not only reflective or introspective awareness, but also what it is for a mental state just to be in our stream of consciousness without our also consciously focusing on it. Introspective awareness of a particular mental state is having a thought that one is in that mental state, and also a thought that one has that thought. Having a conscious mental state without introspectively focusing on it is having the second-order thought without the third-order thought. It may seem slightly odd that each of these hierarchies of conscious mental states has a nonconscious thought at its top. But whatever air of paradox there seems to be here is dispelled by the common-sense truism that we cannot be conscious of everything at once.

One might urge against the present account that higher-order thoughts are unnecessary to explain the consciousness of mental states. Intuitively, a mental state is conscious if it is introspectible. And one might conclude from this that, to explain such consciousness, we need not posit actual higher-order thoughts, but only dispositions to have such thoughts. A mental state is conscious, on this suggestion, if one is disposed to think that one is in that state.[7] But there are several difficulties with such a dispositional account. For one thing, the consciousness of mental states is phenomenologically something occurrent. Since consciousness does not appear to be dispositional, it is ad hoc simply to posit a disposition that comes and goes as needed. We cannot, of course, save all the phenomenological appearances, but we should prefer to do so when we can. Moreover, it is unclear what explanatory work a disposition to have a higher-order thought would do, except when one actually had that thought, and the disposition would then be superfluous.

In any case, the present account readily enables us to explain the intuition that a state's being conscious means that it is introspectible. To introspect a mental state is to have a conscious thought about that state. So introspection is having a thought about some mental state one is in and, also, a yet higher-order thought that makes the first thought conscious. It is a feature of our experience that, when a mental state is conscious, we can readily come to have a conscious thought about that mental state. On the present account, we do not come to have a new thought about that mental state; we simply come to be conscious of a thought we already had, albeit nonconsciously. Higher-order thoughts are mental states we can become aware of more or less at will. A state's being conscious therefore amounts to its being introspectible. Only if being unaware of a higher-order thought meant that one simply did not have that thought would we have reason to try to make do with dispositions, rather than the actual thoughts themselves.

On the present account, conscious mental states are mental states that cause the occurrence of higher-order thoughts that one is in those mental states. And, since those higher-order thoughts are distinct from the mental states that are conscious, those thoughts can presumably occur even when the mental states that the higher-order thoughts purport to be about do not exist. But such occurrences would not constitute an objection to this account. It is reasonable to suppose that such false higher-order thoughts would be both rare and pathological. Nor would they be undetectable if they did occur. We can determine the presence of nonconscious mental states by way of their causal connections with behavior and stimuli, and with other mental states, both conscious and not. Similarly, we can detect the absence of mental states by virtue of the causal connections they would have with such other events.

By itself, the present account of consciousness does not imply a materialist or naturalist theory of mind. Indeed, the account is compatible with even a thoroughgoing Cartesian dualism of substances. But it does square nicely with materialist views. For the account holds that what makes conscious mental states conscious is their causing higher-order thoughts that one is in those mental states. And the materialist can reasonably maintain that this causal pattern is due to suitable neural connections.

Moreover, the materialist can argue that intentional and sensory properties are themselves simply special sorts of physical properties. For one thing, arguments that these mental properties are not physical properties usually rely on the unstated, and question-begging assumption that anything mental is automatically nonphysical. Independent support for this supposition is seldom attempted. Even more important, however, the characteristics that are supposed to show that intentional or sensory properties are not physical turn out, on scrutiny, to be characteristics that various indisputably physical properties also exhibit.[8] So even if no developed, satisfactory account of these properties is presently at hand, there is no reason to doubt that accurate accounts will be forthcoming that are compatible with a thoroughgoing naturalist view of mind. Together with the present explanation of the consciousness of mental states, this should make possible a reasonably comprehensive naturalist theory of mind.

It is a welcome benefit of the present account that it is hospitable to naturalist theories, but this is not its main strength. Rather, its principal advantage is just that it enables us to explain what it is for a mental state to be a conscious state. The present explanation, moreover, has precise empirical consequences that one could reasonably hope to test. For it implies not only that conscious mental states are accompanied by distinct higher-order thoughts, but also that some causal mechanism exists that connects conscious mental states to the corresponding higher-order thoughts.

Such an explanation is possible only if we adopt the non-Cartesian view that intentional and sensory character are jointly the mark of the mental. If, instead, we were to follow the Cartesian tradition in regarding consciousness itself as the key to mentality, no account of consciousness in terms of higher-order thoughts could succeed. For then one would have to deny that a mental state could occur without its being conscious. As Descartes put it, "no thought can exist in us of which we are not conscious at the very moment it exists in us."[9] But, if all mental states are conscious, and a higher-order thought exists for every conscious mental state, serious, insurmountable difficulties immediately ensue. For one thing, there would be denumerably many distinct higher-order thoughts corresponding to every conscious mental state. No mental state could be conscious without being accompanied by a higher-order thought. But that thought would itself have to be conscious, and so a yet higher-order thought would be necessary. This regress would never halt. It strains credulity to suppose that human beings can have infinitely many conscious thoughts at a particular time. And even if we could, it is hardly sensible to explain a mental state's being conscious by way of such an infinite series.

Even more damaging consequences follow for an account in terms of higher-order thoughts if all mental states are conscious states. As noted above, we are not normally aware of the higher-order thoughts that, on such an account, make mental states conscious. But, if all mental states were conscious, we would be aware of any higher-order thoughts that we have. So we could not explain why we typically seem not to have such thoughts by saying that they are simply not conscious thoughts. By requiring that all mental states be conscious states, the Cartesian conception of mentality rules out our explaining consciousness by reference to higher-order thoughts.

If consciousness were what makes a state a mental state, therefore, any account that represents that consciousness as being due to a connection that conscious mental states have with some other mental state would be radically misguided. For that other mental state would then itself have to be conscious, and we would have to invoke yet another mental state to explain its being conscious. A vicious regress would thus be unavoidable. So long as we hold that all mental states are conscious, we can prevent that regress only by maintaining that the consciousness of a mental state is not a relation that state bears to some other mental state, but rather an intrinsic property. Moreover, if consciousness is what makes mental states mental, it will be viciously circular to explain that consciousness in terms of a relation that conscious mental states bear to other mental states. An explanation in terms of other mental states would appeal to states we know to be conscious. It is plain that we cannot explain or analyze consciousness at all unless we can do so in terms of some sort of mental phenomenon. So, if consciousness is what makes a state a mental state, consciousness will not only be an intrinsic, nonrelational property of all mental states; it will be unanalyzable as well. It will, as Russell disparagingly put it, be "a pervading quality of psychical phenomena."[10] Indeed, if being mental means being conscious, we can invoke no mental phenomenon whatever to explain what it is for a state to be a conscious state. Since no nonmental phenomenon can help, it seems plain that, on the Cartesian concept of mentality, no informative explanation is possible of what it is for a mental state to be conscious.

Since consciousness is a matter of our noninferential and nonsensory knowledge of our mental states, it is tempting to describe the issue in terms of such notions as incorrigibility, infallibility, and privacy.[11] But the foregoing obstacles to explaining consciousness do not derive from any such epistemic matters. Rather, they result simply from the Cartesian idea that all mental states are conscious states.

On the Cartesian concept of mentality and consciousness, consciousness is essential to mental states. It is therefore a nonrelational property of those states that is very likely unanalyzable as well. That this conception prevents us from explaining consciousness in any useful way is the most compelling reason we can have for adopting, instead, a non-Cartesian mark of the mental. But there are other reasons as well to prefer a non-Cartesian mark. For one thing, it is impossible to conceive of a mental state, whether or not it is conscious, that lacks both intentional and sensory properties. So, even though it may not always be easy to imagine one's being in a mental state that is not conscious, intentional and sensory properties are evidently more central to our concept of a mental state than consciousness is. So, even though the characteristically mental differences among mental states are, as noted earlier, a function of their intentional and sensory properties, those properties are not only important for explaining how we distinguish among the various types of mental state. They are also necessary for explaining how mental states differ from everything nonmental.

The Cartesian might concede that we can have no notion of a mental state that has neither intentional nor phenomenal character, but go on to insist that we also can have no idea of what it would be like to be in a nonconscious mental state even if it does have intentional or sensory properties. But knowing what it would be like to be in such a state is not relevant here. Knowing what it is like to be in a state is knowing what it is like to be aware of being in that state. So, if the state in question is not a conscious mental state, there will be no such thing as what it is like to be in it, at least in the relevant sense of that idiom. This does not show, however, that intentional and phenomenal states cannot lack consciousness. Conscious states resemble and differ in respect of their intentional or phenomenal features. Accordingly, nonconscious mental states will simply be states that resemble and differ from one another in exactly these ways, but without one's being noninferentially aware of their existence and character.

Indeed, it is indisputable that inner states that resemble and differ in just these ways do occur outside our stream of consciousness. Many sorts of mental state, such as beliefs, desires, hopes, expectations, aspirations, various emotions, and arguably even some bodily feelings such as aches, often occur in us without our noticing their presence. And the only thing that makes these states the kinds of states they are is the intentional and phenomenal properties they have. So we must explain what it is for these states to be mental not by reference to consciousness, but by appeal to their having phenomenal or intentional character. As noted above, we can deny that some of these mental phenomena are properly speaking mental states at all, and instead construe them as mere dispositions for mental states to occur. But states of these sorts often have a strong effect on our actual behavior, and even influence the course and content of our stream of consciousness. These mental phenomena must presumably be nondispositional states at least on those occasions when they exercise such causal influence. So the only reason to regard them as mere dispositions would be a question-begging concern to sustain the theory that all mental states are conscious states.

Perhaps the Cartesian will counter that, even if nonconscious intentional states are unproblematic, the idea that a mental state could have sensory character and yet not be conscious is simply unintelligible. For it may seem that the very idea of a nonconscious state with sensory qualities is, in effect, a contradiction in terms. What seems to make intelligible the idea of a mental state's having phenomenal qualities at all is our immediate awareness of how such states feel, or what they are like for those who are in them. This issue will receive extended consideration in section III. For now, however, it is enough to note that, even if we understand what it is for a state to have sensory quality only because we are familiar with cases in which we are conscious of being in such states, it hardly follows that nonconscious sensory states cannot occur. That we understand a kind of phenomenon by way of a particular kind of case does not show that cases of other sorts are impossible.

14

II

On the Cartesian view, consciousness is definitive of the mental. This concept of mentality implies that consciousness cannot be a relational characteristic of mental states, and that it may well be inexplicable as well. The difficulty in explaining consciousness on that view actually results from the Cartesian strategy for dealing with mental phenomena. The main strength of the Cartesian picture is that it closely matches our presystematic, common-sense intuitions. But it achieves this close match by building those intuitions into our very concepts of mind and consciousness. And this automatically trivializes any explanation we might then give of them. We cannot very well give non-question-begging accounts of intuitions that we incorporate definitionally into our very concepts. Explanations based on the Cartesian conceptions of mind and consciousness thus rely heavily, and ineliminably, on interdefinition of such terms as 'mind', 'consciousness', 'subjectivity', and 'self'. Such interdefinition may be useful in marking out a range of interconnected phenomena, but it cannot do much to help explain the phenomena thus delineated.

On the non-Cartesian concept, by contrast, consciousness is not essential to mental states, and thus consciousness may well be an extrinsic characteristic of whatever mental states have it. The Cartesian achieves its close match with common sense at the cost of ruling out any useful explanation. No such trade-off is necessary on the non-Cartesian picture. The non-Cartesian has no trouble in giving a theoretically satisfying explanation of consciousness. And it is possible to show that this account enables us to save the phenomenological appearances at least roughly as well as the Cartesian can. Moreover, objections to an account cast in terms of higher-order thoughts can be convincingly met. In this section and the next I consider some of the most pressing of these objections, and also argue that such an account does do justice to the phenomenological data. In the present section I take up various general questions about the adequacy of the non-Cartesian account; in section III I address issues that pertain specifically to sensory qualities and to subjectivity.

One especially notable feature of our presystematic view of consciousness which the Cartesian conception seems to capture perspicuously is the close connection between being in a conscious state and being conscious of oneself. An account in terms of higher-order thoughts has no trouble here. If a mental state's being conscious consists of having a higher-order thought that one is in that mental state, being in a conscious state will imply having a thought about oneself. But being conscious of oneself is simply having a higher-order thought about oneself. So being in a conscious mental state is automatically sufficient for one to be conscious of oneself.

Any reasonable account of consciousness will presumably insist on this connection. But the Cartesian can say little that is informative about why the connection should hold. An account that appeals to higher-order thoughts has no such difficulty. Moreover, there is a well-motivated reason why the higher-order thought that the non-Cartesian invokes must be a thought about one-self. To confer consciousness of a particular mental state, the higher-order thought must be about that very mental state. And the only way for a thought to be about a particular mental state is for it to be about somebody's being in that state. Otherwise, the thought would just be about that type of mental state, and not about the particular token of it. So, in the case at hand, the higher-order thought must be a thought that one is, oneself, in that mental state.[12]

Having a thought that one is, oneself, in a particular mental state does not by itself presuppose any prior conception of the self, or of some sort of unity of consciousness. Rather, the present view allows us to explain these conceptions as themselves actually arising from our being in conscious mental states. For we can construe the second-order thoughts as each being a thought to the effect that whatever individual has this very thought is also in the target mental state. And, if a fair number of these thoughts are conscious thoughts, it is plausible to suppose that a sense of the unity of consciousness will, in time, emerge.

If one held the Cartesian view that all mental states are conscious, invoking higher-order thoughts would issue in the vicious regress noted in secion I. So, if one is tempted by both these moves, one might try to adjust things in order to avoid that outcome. The most promising way to do so would be simply to insist that the higher-order thoughts in virtue of which we are conscious of conscious mental states are actually part of those conscious states themselves. Every conscious mental state would then be, in part, about itself, and our knowledge that we are in such states would be due to that self-reference. Metaphorically, we would then conclude that a mental state's knowing itself is, in Ryle's apt metaphors, a matter of its being "self-intimating" (158) or "self-luminous" (159).[13]

This line of reasoning is particularly inviting, since it suggests that the Cartesian can, after all, give some nontrivial explanation of the consciousness of mental states. Conscious mental states are conscious, on this account, because they are about themselves. And this self-reference is intrinsic; it does not result from some connection those states have with other mental states. But anything that would support the view that conscious mental states are conscious because they know, or are in part about, themselves would provide equally good evidence that consciousness is due to an accompanying higher-order thought. Moreover, we have no nonarbitrary way to tell when one mental state is a part of another. Accordingly, there is no reason to uphold the idea that our awareness of conscious states is a part of those states other than a desire to sustain the Cartesian contention that all mental states are conscious states. Moreover, if conscious states have parts in this way, the question arises whether all the parts of such states must be conscious, or only some. If all, then the awareness of the mental state will have to be conscious. A regress would thus arise that is exactly parallel to that which arose when we construed the awareness of conscious mental states as due to a distinct higher-order thought. The only advantage of an account on which that awareness is a part of the conscious mental state is if the awareness is a nonconscious part of the conscious state. This reinforces the conclusion that there is no nonarbitrary way to distinguish this view from an account in terms of higher-order thoughts. And it undercuts the idea that the Cartesian can formulate an informative explanation of consciousness along these lines. Since the Cartesian explanation would work only if the part of each conscious state that makes it conscious were itself conscious, the regress is unvoidable.

One reason that consciousness seems intrinsic to our sensory states is that it is difficult to isolate that consciousness as a distinct component of our mental experience. When we try to focus on the consciousness of a particular sensory state, we typically end up picking out only the sensory state we are conscious of, instead. As Moore usefully put it, consciousness is "transparent," or "diaphanous."[14] Since efforts to pick out consciousness itself issue instead in the states we are conscious of, it is tempting to conclude that the consciousness is actually part of those states. But the present account gives a better explanation of the diaphanous character of consciousness. We normally focus on the sensory state and not on our consciousness of it only because that consciousness consists in our having a higher-order thought, and that thought is usually not itself a conscious thought.

There is a strong intuitive sense that the consciousness of mental states is somehow reflexive, or self-referential. But we need not invoke the idea that conscious states are conscious of themselves to explain this intuition. For a mental state to be conscious, the corresponding higher-order thought must be a thought about oneself, that is, a thought about the mental being that is in that conscious state. So, as noted above, we can construe that thought as being, in part, about itself. For it is reasonable to regard the content of the thought as being that whatever individual has this very thought is also in the specified mental state. The sense that something is reflexive about the consciousness of mental states is thus not due to the conscious state's being directed upon itself, as is often supposed. Rather, it is the higher-order thought that confers such consciousness that is actually self-directed.

The foregoing objections have all challenged whether an account based on higher-order thoughts can do justice to various ways we think about consciousness. But one might also question whether higher-order thoughts are enough to make mental states conscious. Here a difficulty seems to arise about mental states that are repressed. By hypothesis such states are not conscious. But it might seem that mental states can be repressed even if one has higher-order thoughts about them. Higher-order thoughts could not then be what makes mental states conscious. A person who has a repressed feeling may nonetheless take pleasure, albeit unconscious pleasure, from having that repressed feeling. But to take pleasure in something we must presumably think that it is so. So that person will have a higher-order thought about the repressed feeling.[15] Moreover, it appears intuitively that the feeling cannot remain unconscious unless the pleasure taken in it also does. And this suggests that, contrary to the present account, a higher-order thought can confer consciousness only if that thought is itself already conscious.

But genuine counterexamples along these lines are not all that easy to come up with. Despite the foregoing suggestion, one can take pleasure in something without having any actual thought about it. I cannot, of course, take pleasure in something I disbelieve or doubt,[16] but that does not imply the actual occurrence of a thought that it is so. Indeed, we frequently form no actual thought about the things in which we take pleasure. Sometimes, by 'thought', we mean only to speak of propositional contents, as when I talk about some thought you put forth. Taking pleasure is a propositional mental state; so taking pleasure in something does involve a thought about it, in the sense of a proposition. But it hardly follows that one also has a thought, in the sense of a particular kind of mental state. Having a thought in that sense is the holding of an assertive mental attitude, which need not occur when one takes pleasure in something. So, in the foregoing example, we have no reason to suppose that the person would actually have any higher-order thought about the repressed feeling.

The difference between taking pleasure in a mental state and having an actual thought about it is crucial for the present account of consciousness. It is natural to hold that being aware of something means having a thought about it, not taking pleasure in it. And one can take pleasure in something without knowing what it is that gives one pleasure. One may have no idea why one feels good, or be mistaken about why. One may even be unaware of feeling good at all if one is sufficiently distracted or other factors interfere. So taking pleasure in something is compatible with being unaware of that thing. Such considerations also apply to putative counterexamples based on other sorts of higher-order mental states. For example, repressed feelings are, presumably, always accompanied by higher-order desires not to be in them. But desires that something not be so do not, in general, imply any awareness that it is.

Conceiving of nonconscious mental states on the model of the repressed cases is doubly misleading. For one thing, it ignores factors that in such cases presumably block consciousness. Moreover, it conceals a tacit Cartesian premise. For it suggests that consciousness is the norm: unless exceptional pressures intervene, a mental state will automatically be conscious. Consciousness is to be presupposed unless some external factor prevents it. Thus, on this model, we can explain the forces that interfere with consciousness, but consciousness itself may very likely be inexplicable.[17]

III

Whatever one holds about intentional states, it may seem altogether unacceptable to try to explain the consciousness of sensory states by way of higher-order thoughts. Consciousness seems virtually inseparable from sensory qualities, in a way that does not seem so for intentional properties. Indeed, as noted at the end of section I, it may seem almost contradictory to speak of sensory states' lacking consciousness. This intimate tie between sensory quality and consciousness seems to hold for all sensory states, but appears strongest with somatic sensations, such as pain. Saul A. Kripke succinctly captures this intuition when he insists that "[f]or a sensation to be *felt* as pain is for it to *be* pain"[18] and, conversely, that "for [something] to exist without being *felt as pain* is for it to exist without there *being any* pain."[19] And, more generally, Kripke seems to insist that for something to be a sensation of any sort it must be felt in a particular way (*NN* p. 146).

Since consciousness seems more closely tied to sensations than to intentional states, it is tempting to consider a restricted from of the Cartesian view, on which all sensations are conscious but not all intentional states are.[20] This restricted thesis would still allow one to explain consciousness in terms of higher-order thoughts; no regress would arise, because then those thoughts could themselves be nonconscious. But the Cartesian view holds not only that all mental states are conscious, but also that consciousness is an intrinsic property of mental states. And if it is, an explanation in terms of higher-order thoughts is impossible, and all the problems about giving an informative explanation of consciousness will arise. So even if not all mental states are conscious, it is important to see whether consciousness is intrinsic to those which are.

We can, however, explain our tendency to associate consciousness and sensory qualities without having to suppose that consciousness is intrinsic to sensory states, or even that all sensory states are conscious. We are chiefly concerned to know what bodily sensations we and others have because they are highly useful indicators of bodily and general well being. People cannot tell us about their nonconscious sensations, and bodily sensations usually have negligible effect on behavior unless they are conscious. So nonconscious sensations are not much use as cues to such well being, and we thus have little, if any, interest in pains or other somatic sensations, except when they are conscious.

Things are different with other sorts of mental states, even perceptual sensations. It is often useful to know somebody's thoughts, emotions, and perceptual sensations, even when that person is unaware of them. Moreover, when mental states are not conscious, our interest in knowing about them is greatest with propositional states, less with emotions, less still with perceptual sensations, and far the least with somatic sensations. Strikingly, our sense that consciousness is intrinsic to mental states increases accordingly. The less useful it is to know about a particular kind of mental state even when the person is unaware of it, the more compelling is our intuition that that kind of mental state must be conscious. This correlation is telling evidence that, even in the case of pains and other somatic sensations, the idea that being mental entails being conscious is just a reflection of our usual interests, and not a matter of the meanings of our words or of the nature of the mental itself.

Some of our idiomatic ways of describing somatic sensations do entail consciousness. Something's hurting, for example, implies awareness of the hurt. And perhaps one cannot correctly say that somebody is in pain unless that person knows it. Phrases such as 'what a sensation is like' and 'how a sensation feels' reinforce this impression, since they refer both to a sensory quality and to our awareness of it, and seem thus to yoke the two together. But when one is in pain or when something hurts, we not only are in a sensory state, but are also aware that we are. And our idiomatic descriptions of these situations have no bearing on whether that very kind of sensory state may sometimes occur without one's being aware of it. Perhaps we would then withhold from such states the epithet 'pain'. But those states would still resemble and differ from other nonconscious states in just those ways in which conscious pains resemble and differ from other conscious sensory states. And that is what it is for a state to have sensory qualities. The intuitive simplicity of those qualities might tempt one to hold that consciousness also is simple and, hence, an intrinsic characteristic of sensory states. But it is question begging to suppose that the apparent simplicity of sensory qualities tells us anything about the nature of our consciousness of them.

Examples of sensory states that sometimes occur without consciousness are not hard to come by. When a headache lasts several hours, one is seldom aware of it for that entire time. Distractions occur, and one pays attention to other things, or just forgets for a bit. But we do not conclude that each headache literally ceases to exist when it temporarily stops being part of our stream of consciousness, and that such a person has only a sequence of discontinuous, brief headaches. Rather, when that happens, our headache is literally a nonconscious ache.[21] The same holds even more vividly for mild pains and minor bodily discomforts. So, to insist that nonconscious states are just not mental states, or that they cannot have sensory qualities, is not, as Kripke seems to urge (e.g., *NN* 152–3), the elucidation of decisive and defensible presystematic intuitions, but only the tacit expression of the Cartesian definition of mind.

Indeed, an account in terms of higher-order thoughts actually helps explain the phenomenological appearances. If a sensory state's being conscious is its being accompanied by a suitable higher-order thought, that thought will be about the very quality we are conscious of. It will be a thought that one is in a state that has that quality. So it will indeed be impossible to describe

21

that consciousness without mentioning the quality. An account in terms of higher-order thoughts actually helps explain why the qualities of our conscious experiences seem inseparable from our consciousness of them.

Moreover, we typically come to make more fine-grained discriminations as we master more subtle concepts pertaining to various distinct sensory qualities. Experiences from wine tasting to hearing music illustrate this process vividly. An account in term of higher-order thoughts explain the bearing these concepts have on our very awareness of sensory differences. If consciousness is intrinsic to sensory states, the relevance of concepts remains mysterious. The Cartesian might just deny that sensory differences exist when we are unaware of them. But it will be even more difficult to explain how learning new concepts can actually cause sensory qualities to arise that previously did not exist.

Perhaps the strongest objection to an account in terms of higher-order thoughts is that there are creatures with conscious sensations whose ability to have thoughts at all may be in doubt. Infants and most nonhuman animals presumably have a relatively rudimentary ability to think, but plainly do have conscious sensations. But one need not have much ability to think to be able to have a thought that one is in a particular sensation. Infants and nonhuman animals can discriminate among external objects, and master regularities pertaining to them. So most of these beings can presumably form thoughts about such objects, albeit primitive thoughts that are very likely not conscious. No more is needed to have thoughts about one's more salient sensory experiences. Infants and nonhuman animals doubtless lack the concepts required for drawing many distinctions among their sensory states. But, as just noted, one can be aware of sensory states and yet unaware of many of the sensory qualities in virtue of which those states differ.

The common tendency to link the ability to think with the ability to express thoughts in speech may account for the doubt we can fall into about whether infants and nonhuman animals can think at all. But the capacity for speech is hardly necessary for thinking. It is often reasonable to interpret nonlinguistic behavior, of other people and of non-language-using creatures alike, in terms of the propositional content and mental attitude we take it to express. Such behavior is convincing evidence of the occurrence of intentional states.

Forming higher-order thoughts about one's own propositional mental states takes a lot more than having such thoughts about one's sensations. For one thing, the concept of a mental state with propositional content is more complex than the concept of a sensory experience. And picking out particular mental states demands an elaborate system of concepts, whereas referring to salient sensory experiences does not. An account in terms of higher-order thoughts fits well with these points. Infants and most nonhuman species lack the ability to have the more complex higher-order thoughts needed to make intentional states conscious, though they presumably can form higher-order thoughts about their sensory states. And, though these beings plainly have conscious sensations, we have little reason to suppose that their intentional states are also conscious. Indeed, these considerations help explain why we associate consciousness so much more strongly with sensory than with intentional states. Conscious of sensory states arises far more readily, since higher-order thoughts about them are far easier to have.

Some animal species, however, lack the ability to think at all. And this may seem to support Nagel's contention that conscious "experience is present in animals lacking language and thought" (167, n. 3). But being a conscious creature does not entail being in conscious mental states. For an organism to be conscious means only that it is awake, and mentally responsive to sensory stimuli (cf. Ryle, pp. 156–7). To be mentally responsive does require that one be in mental states. And to be mentally responsive to sensory stimuli may even mean that one is in some way conscious of the objects or events that are providing such stimulation. But a creature can be in mental states without being in conscious mental states, and can be conscious of external or bodily events without also being aware of its own mental states.

Conscious experiences, as Nagel has stressed, manifest a certain subjectivity. We each experience our sensory states in a way nobody else does, and from a point of view nobody else shares. It is notoriously difficult to articulate these differences. But we understand their occurrence reasonably well, and it is far from clear that such subjectivity causes any problem for the present account.

One way differences arise in sensory experiences is from variations in sense organs, or other aspects of physical makeup. Experiences also vary from individual to individual because of such factors as background and previous experience. When these factors diverge markedly, aspects of our sensory experiences may as well. When the individuals belong to distinct species, this effect may be quite dramatic. But hard as it is to pin down precisely what these differences amount to, they do not bear specifically on the consciousness of the experiences in question. Rather, the variations are due to differences in the mental context in which the experiences occur, or, when biological endowment is at issue, they are actual differences in the sensory qualities of those experiences. Nagel holds that "the subjective character of experience" is a matter of what "it is like to *be* a particular organism – [what] it is like *for* that organism" (p. 166). But the present account can accommodate this idea. What it is like to be a particular conscious individual is a matter of the sensory qualities of that individual's conscious experiences, and the mental context in which those experiences occur. The consciousness of those experiences, by contrast, is simply that individual's being aware of having the experiences.[22]

According to Nagel, "[a]ny reductionist program has to be based on an analysis of what is to be reduced. If the analysis leaves something out, the problem will be falsely posed" (p. 167). Indeed, no account that is even "logically compatible with" the absence of consciousness could, Nagel contends, be correct (p. 166;cf. "Panpsychism," p. 189). And the present account is reductionist, since it seeks to explain conscious mental states ultimately in terms of mental states that are not conscious. But that account aims only at explaining consciousness, and not also at conceptual analysis. And satisfactory explanations do not, *pace* Nagel, require full analyses of the relevant concepts. Explanation, in science and everyday context alike, must generally proceed without benefit of complete conceptual analyses.

Nagel's language is strongly evocative of that sense we have of ourselves which can make it appear difficult to see how, as conscious selves, we could find ourselves located among the physical furniture of the universe. When we focus on ourselves in this way, there seems to be nothing more basic to our nature than consciousness itself. If nothing were more basic to us than consciousness, there would be nothing more basic in terms of which we could explain consciousness. All we could do then is try to make consciousness more comprehensible by eliciting a sense of the phenomenon in a variety of different ways. Analyzing concepts would be central to any such project, and Nagel's demand for conceptual analysis would then make sense. But consciousness could be essential to our nature only if all mental states are conscious states. If a fair number of our mental states are not conscious, we cannot define our mental natures in terms of consciousness, and there will be nonconscious mental phenomena in terms of which we can explain consciousness itself.

24

The puzzled cognitive disorientation that can result from reflecting on the gulf that seems to separate physical reality from consciousness makes any noncircular explanation of consciousness seem inadequate. How could any explanation of consciousness in terms of nonconscious phenomena help us to understand how consciousness can exist in the physical universe, or how physical beings like ourselves can have conscious states? But no other explanation can do better with these quandaries so long as an unbridgeable gulf seems to divide the conscious from the merely physical. To understand how consciousness can occur in physical things, we must dissolve the intuitive force of that gulf. And we can do so only by explaining the consciousness of mental states in terms of mental states that are not conscious. For the stark discontinuity between conscious mental states and physical reality does not also arise when we consider only nonconscious mental states. And once we have explained consciousness by reference to nonconscious mental states, we may well be able also to explain nonconscious mental states in terms of phenomena that are not mental at all.

IV

The central place consciousness has in our conception of the mental is doubtless due in large measure to the way we know about mind in general, and in particular about our own mental states. We get most of that knowledge, directly or indirectly, from introspection. And we have introspective access to mental states only when they are conscious. Since our chief source of knowledge about the mind tells us only about conscious mental states, it is natural to infer that consciousness is an important feature of mental phenomena.

But stronger claims are sometimes made about the epistemic status of introspection. Introspection may seem particularly well adapted to its subject matter, since most of our knowledge of mind derives from introspection, and all introspective knowledge is about mind. This close fit may tempt some to hold that introspection is a privileged source of knowledge that is somehow immune from error. If so, perhaps introspection reveals the essential nature of those states. And, since introspection tells us only about conscious mental states, perhaps consciousness is itself a part of that essential nature. But inviting as these Cartesian conclusions may be, they are without foundation. Introspection is simply the having of conscious thoughts that one is in particular mental states. Those thoughts can by themselves no more reveal the essences of those states than having a conscious perceptual thought that a table is in front of one can reveal the essence of the table. Nor can we infer anything from the close fit between introspection and its subject matter. Sight is an equally well adapted to knowing about colored physical objects. But there are other ways to know about those objects. And even though sight informs us only about illuminated objects, we can hardly conclude that only illuminated objects are colored.

Introspective apprehension seems to differ, however, from perceptual knowledge in a way that undermines this analogy. Perception is never entirely direct. Some causal process always mediates, even in ostensibly direct perception, between our perceptual experience and what we perceive. Introspection, by contrast, may seem wholly unmediated. And if it is, there would be no way for error or distortion to enter the introspective process. There would thus be no difference between how our mental states appear to us and how they really are. Mental states would have no nonintrospectible nature, and introspection would be an infallible and exhaustive source of knowledge about the mind. Nagel evidently endorses this view when he claims that "[t]he idea of moving from appearance to reality seems to make no sense" in the case of conscious experiences (174). Kripke too seems to hold that introspection is different from perception in this way. Thus he writes:

although we can say that we pick out [physical] heat contingently by the contingent property that it affects us in such and such a way, we cannot similarly say that we pick out pain contingently by the fact that it affects us in such and such a way ("IN" p. 161; cf. *NN* pp. 150–2).

We could not, Kripke contends, have been aware of our pains in a way different from the way we actually are.

Introspection is the reflective awareness of our mental states. So, the only way introspective apprehension of those states might be entirely unmediated would be for consciousness to be a part, or at least an intrinsic property, of such states. For nothing could then come between a mental state and our being conscious of it, nor between our being thus conscious and our also having reflective consciousness. But as noted in section II, that view is indefensible. Accordingly, consciousness must be a relational property, for example, the property of being accompanied by higher-order thoughts. And some causal process must therefore mediate between mental states and our awareness of them — in Kripke's example, between a pain and "the fact that it affects us in such and such a way." And, since mental states might have been connected causally to different high-order thoughts, we might have been aware of mental states differently from the way we are. The appearance of mental states will not, therefore, automatically coincide with their reality. Indeed, since how mental states appear is a matter of our introspective awareness of them, their appearance and reality could be the same only if our consciousness of mental states were a part or an intrinsic property of those states.

These considerations notwithstanding, we do rely heavily on introspection in picking out and describing mental states. And introspection tells us about nothing except conscious mental states. So even though consciousness is not what distinguishes mental states from everything else, it is reasonable to hold that it is by reference to a range of conscious states that we fix the extension of the term 'mental'. Similarly, even though the various kinds of mental state can all occur nonconsciously, it is also reasonable to suppose that we fix the extensions of our terms for different kinds of mental state by way of the conscious cases. As Kripke and Hilary Putnam have stressed, what fixes the extension of a general term can turn out to be distinct from what is essential to the items in that extension.[23] And, just as the way we know about things is not, in general, a reliable guide to their nature, so the way we pick out things is not, either. So, even if we fix the extensions of terms for mental states by way of the conscious instances, we could still discover that the states so determined are not all conscious, and that what is actually essential to all such states is just their sensory or intentional properties.

The idea that we fix mental extensions by way of the conscious cases plainly supports the non-Cartesian picture. But it also helps explain why consciousness seems so crucial to our mental concepts. And it even enables us to explain why we group sensory and intentional states together as mental states, despite its seeming that they have little intrinsic in common. We do so because in both cases we fix extensions by way of states to which we have noninferential and nonobservational access.

Kripke contends that what fixes the reference of terms for sensory states cannot diverge from what is essential to those states (*NN* pp. 149–54; "IN" pp. 157–61). Thus, he insists, "[i]f any phenomenon is picked out in exactly the same way that we pick out pain, then that phenomenon *is* pain."[24] But what fixes the extension of 'pain' must coicide with what is essential to

27

pains only if it is necessary that pains affect us in the way they do. And this would be necessary only if consciousness were intrinsic to them. It therefore begs the question to base the Cartesian picture on an insistence that what fixes the extensions of mental terms cannot diverge from the essences of mental states. Kripke offers no independent support for that insistence.

Relative to what we now know about other natural phenomena, we still have strikingly scant understanding of the nature of the mental. So introspection looms large as a source of information, just as sense perception was a more central source of knowledge about physical reality before the flourishing of the relevant systematic sciences. But, since not all knowledge about mind is derived from introspection, we have no more reason to suppose that mental states have no nonintrospectible nature than that the nature of physical objects is wholly perceptible. Nor, therefore, have we any reason to hold that the essences of mental states must be what fixes the extensions of mental terms. It is reasonable to conclude that whatever temptation we have to accord absolute epistemic authority to introspection derives solely from our relative ignorance about the mind. Only because we now know so little about mental processes does it make sense to suppose that, in the case of mental states, appearance and reality coincide. Accordingly, we have no reason to continue to favor that picture, or to reject an explanation of consciousness based on higher-order thoughts.[25]

NOTES

[1] 'What is it like to be a bat?', The Philosophical Review, LXXXIII, 4 (October 1974): 435–50; reprinted in Nagel's Mortal Questions (Cambridge and New York: Cambridge University Press, 1979); 165–180, p. 165. Page references to Nagel will be to Mortal Questions and, unless otherwise indicated, to that article.

[2] Richard Rorty and Paul M. Churchland, among others, have championed this view; see Rorty's Philosophy and the Mirror of Nature (Princeton: Princeton University Press, 1979), Part I, and Churchland's Scientific Realism and the Plasticity of Mind (Cambridge: Cambridge University Press, 1979).

[3] In a review of Frank Jackson's Perception [The Journal of Philosophy, LXXXII, 1 (January 1985): 28–41] I argue in detail that words for such qualities have this double use. On this point see also G. E. Moore, 'A reply to my critics', in The Philosophy of G. E. Moore ed. Paul Arthur Schilpp (LaSalle, Illinois: Open Court, 1942): 535–677, pp. 655–8; and Thomas Reid, Essays on the Intellectual Powers of Man, ed. Baruch A. Brody (Cambridge, MA: The M.I.T. Press, 1969), II, xvi, p. 244.

[4] 'Chisholm-Sellars correspondence on intentionality', in Minnesota Studies in the Philosophy of Science, II, ed. Herbert Feigl, Michael Scriven, and Grover Maxwell (Minneapolis: University of Minnesota Press, 1958): 521–39, p. 533. In 'Intentionality,'

Midwest Studies in Philosophy, X (1985), I argue that this claim is defensible if it is construed in strictly causal terms.

[5] See, e.g., David M. Armstrong, A Materialist Theory of the Mind (New York: Humanities Press, 1968), pp. 92–115 and 323–7 and 'What is consciousness?', in Armstrong's The Nature of Mind and Other Essays (Ithaca, NY: Cornell University Press, 1980): 55–67, pp. 59–61; David Lewis, 'An argument for the identity theory,' The Journal of Philosophy, LXIII, 1 (January 6, 1969): 17–25, p. 21 and "Psychophysical and Theoretical Identifications," Australasian Journal of Philosophy, 50, 3 (December 1972): 249–58, p. 258; and Wilfrid Sellars, "Empiricism and the Philosophy of Mind," in Sellars' Science, Perception and Reality (London: Routledge & Kegan Paul and New York: Humanities Press, 1963): 127–96, pp. 188–9 and 194–5.

[6] Gilbert Ryle, The Concept of Mind (London: Hutchinson and Company, 1949), p. 164.

[7] Allen Hazen has urged this line especially forcefully, in correspondence. Also, see Kant's claim that the representation 'I think' must be able to accompany all other representations (K.d.R.V., B131–2; cf. B406, though Kant insists that the representation 'I think' is a nonempirical (B132) or transcendental (B401, A343) representation.

[8] For a detailed argument to this effect, see my 'Mentality and neutrality,' The Journal of Philosophy, LXXIII, 13 (July 15, 1976): 386–415, sec. I.

[9] Fourth Replies, Oeuvres de Descartes, ed. Charles Adam and Paul Tannery (Paris: J. Vrin, 1964–75) [henceforth "AT"] VII, 246 [see The Philosophical Works of Descartes, ed. Elizabeth S. Haldane and G. R. T. Ross (Cambridge: Cambridge University Press, 1931) (henceforth "HR") II, 115]. See also the Geometrical Exposition of the Second Replies: "the word 'thought' applies to all that exists in us in such a way that we are immediately conscious of it" [AT VII, 160 (see HR II, 52)], and elsewhere, e.g.: First Replies, AT VII, 107 (HR II, 13), Letter to Mersenne, AT III, 273 [see Descartes: Philosophical Letters, ed. Anthony Kenny (Oxford: Oxford University Press, 1970), p. 90], Fourth Replies, AT VII, 232 (HR II, 105) and AT VII, 246 (HR II, 115), and Principles I, ix, AT VIII–1, 7 (HR I, 222).

[10] Bertrand Russell, The Analysis of Mind (London: George Allen & Unwin Ltd. and New York: Humanities Press, 1921), p. 9.

[11] See, e.g., Rorty's argument "that incorrigibility is the best candidate" for a satisfactory mark of the "sorts of entities [that] make up the content of the stream of consciousness" ["Incorrigibility as the Mark of the Mental," The Journal of Philosophy, LXVII, 12 (June 25, 1970): 399–424, pp. 406–7; cf. also Rorty's Philosophy and the Mirror of Nature, e.g., pp. 88–96, and Part I, passim.

[12] As Hector-Neri Castaneda and G. E. M. Anscombe have pointed out, believing something of oneself must involve the mental analogue of the indirect reflexive construction, represented here by 'oneself' [Castaneda: 'On the Logic of Attributions of Self-Knowledge to Others,' The Journal of Philosophy, LXV, 15 (August 8, 1968): 439–56 and elsewhere; Anscombe: 'The first person,' in Mind and Language, ed. Samuel Guttenplan (Oxford: Oxford University Press), 1975: 45–65]. Even when George believes that somebody who turns out to be George is F it may not be true that George believes that he, himself, is F. For example, George may truly believe somebody is F while wrongly believing that that person is not George, himself. Or he may not even believe of himself that he is George. Unlike token-reflexive constructions, these terms involve anaphora. But the clauses that contain them are grammatical transforms of sentences that do contain genuine token reflexives.

[13] See Franz Brentano, Psychology from an Empirical Standpoint, tr. Antos C. Rancurello et al. (London: Routledge & Kegan Paul and New York: Humanities Press, 1973), pp. 129–30.

[14] G. E. Moore, 'The Refutation of Idealism', in his Philosophical Studies, (London: Routledge & Kegan Paul, 1922): 1–30, pp. 20 and 25.

[15] I am grateful to Georges Rey and Eric Wefald for independently raising this point.

[16] On this point, see Robert M. Gordon's illuminating 'Emotions and knowledge,' The Journal of Philosophy, LXVI, 13 (July 3, 1969): 408–13 and 'The aboutness of emotions,' American Philosophical Quarterly, 11, 1 (January 1974): 27–36.

[17] It is noteworthy that this very attitude appear even in Freud's own writings. Freud does, indeed, "energetically den[y] the equation between what is psychical and what is conscious" ['Some elementary lessons in psycho-analysis,' in The Complete Psychological Works of Sigmund Freud, tr. and ed. James Strachey (London: The Hogarth Press, 1966–74) (henceforth "Works"), XXIII: 279–86]. And he understood that to do so one must define the mental in terms of phenomenal and intentional character; thus he insisted that "all the categories which we employ to describe conscious mental acts ... can be applied" equally well to unconscious mental states ('The Unconscious,' Works, XIV: 166–215, p. 168). Moreover, he maintained that "[t]he psychical, whatever its nature may be, is itself unconcious" ('Some elementary lessons,' p. 283), and so, "[l]ike the physical, the psychical is not necessarily in reality what it appears to us to be" ("The Unconscious," p. 171; Cf. 'Some elementary lessons', p. 282 and The Interpretation of Dreams, Works, V, p. 613). But despite all this, Freud operated with a surprisingly Cartesian concept of consciousness. Consciousness, he wrote, is a "unique, indescribable" quality of mental states ('Some elementary lessons,' p. 282), and "the fact of consciousness" "defies all explanation or description" (An Outline of Psycho-Analysis, Works, XXIII: 141–208, p. 157). In thus regarding consciousness as unanalyzable, Freud seems to have uncritically accepted the core of the Cartesian doctrine he strove to discredit. (To dissociate the present account from Freud's views I eschew here the colloquial 'unconscious mental state' in favor of the somewhat awkward term 'nonconscious'.)

[18] 'Identity and necessity,' in Identity and Individuation, ed. Milton K. Munitz (New York: New York University Press, 1971): 135–64 [henceforth "IN"], p. 163, n. 18; emphasis original here and elsewhere.

[19] Naming and Necessity (Cambridge, MA: Harvard University Press, 1980) [henceforth "NN"] p. 151. Compare Reid, Essays, II, xvi, p. 243: "When [a sensation] is not felt, it is not. There is no difference between a sensation and the feeling of it; they are one and the same thing.

[20] Even Freud does not hold that feelings can strictly speaking be unconscious, though he sees no difficulty about unconscious intentional states (The Ego and the Id, Works, XIX: 3–68, pp. 22–3; cf. An Outline, p. 197).

[21] Thus it is a parody for Wittgenstein to suppose that all we could mean by an unconscious toothache, e.g., is "a certain state of decay in a tooth, not accompanied by what we commonly call toothache" [The Blue and Brown Books (Oxford: Basil Blackwell, 1958), p. 22].

[22] On difficulties in Nagel's discussion and, especially, his notion of a point of view, see my 'Reductionism and knowledge,' in How Many Questions?, ed. Leigh S. Cauman, Isaac Levi, Charles Parsons, and Robert Schwartz (Indianapolis: Hacket Publishing Company, 1983): 276–300.

[23] NN 54–9, "IN" 156–61, and Putnam, 'The meaning of "meaning",' in Putnam's Philosophical Papers, vol. 2 (Cambridge: Cambridge University Press, 1975): 215–71, pp. 223–35.

[24] NN 153; cf. "IN" 162–3. Again, cf. Reid, Essays, II, xvi, p. 243: A "sensation can be nothing else than it is felt to be. Its very essence consists in being felt." Cf. also J. J. C. Smart: "[t]o say that a process is an ache is simply to classify it with other processes that are felt to be similar" ['Materialism,' The Journal of Philosophy, LX, 22 (October 24, 1963): 651–62, p. 655. It is striking that this Cartesian claim about our mental concepts should be shared by theorists who, in other respects, diverge as sharply and as thoroughly as do Smart and Kripke. That it is so shared suggests that this claim

may underlie much of what is, in different ways, unintuitive about each of those theories.
[25] I am greatly indebted to many friends and colleagues for comments on earlier versions of this paper, most especially to Margaret Atherton, Adam Morton, and Robert Schwartz.

The Graduate School and University Center of the City University of New York,
Ph.D. Program in Philosophy,
New York, NY 10036-8099,
U.S.A.

WHAT IS CONSCIOUSNESS?*

NORTON NELKIN[†‡]

Department of Philosophy
University of New Orleans

When philosophers and psychologists think about consciousness, they generally focus on one or more of three features: phenomenality (how experiences feel), intentionality (that experiences are "of" something, that experiences *mean* something), and introspectibility (our awareness of the phenomenality and intentionality of experience). Using examples from empirical psychology and neuroscience, I argue that consciousness is not a unitary state, that, instead, these three features characterize *different* and *dissociable* states, which often happen to occur together. Understanding these three features as dissociable from each other will resolve philosophical disputes and facilitate scientific investigation.

When philosophers and psychologists think about consciousness, they generally focus on one or more of three features: phenomenality, intentionality, and introspectibility. I argue that, rather than being three features of a single, noncomposite state, these three features characterize *different* states of human beings. While these three features can, and often do, occur together, compositely, in human experience, each of the first two can exist independently of each other and of introspectibility. Because of their frequent co-occurrence, the three are taken to be features of a single, noncomposite state. And because each feature characterizes an important way in which things like human beings differ from things like rocks, no one state has any more priority in being considered as what consciousness *really is* than the others (although the consciousness that most firmly grounds our personhood is probably introspectibility, where "personhood" is that property that grounds moral agency—see Nelkin 1987b). While one should be wary of claims that important terms are systematically ambiguous, "consciousness" really is systematically am-

*Received January 1992; revised August 1992.

[†]I would like to thank Carolyn Morillo, Alan Soble, Jim Stone, Edward Johnson, Richard Hall, Martin Davies, and A. J. Marcel for commenting on an earlier version of this paper and for providing useful suggestions for its improvement. Kent Bach provided helpful comments when a version of this paper was presented at the meetings of the Society for Philosophy and Psychology (10 June 1990). Audience comments there, at the Tulane Seminar for Current Research, and at the Louisiana State Philosophy Convention were also helpful. I would especially like to thank an anonymous reader for this journal who asked many thought-provoking questions concerning an earlier draft, resulting in an improved presentation of the ideas.

[‡]Send reprint requests to the author, Department of Philosophy, University of New Orleans, Lakefront, New Orleans, LA 70148, USA.

Philosophy of Science, 60 (1993) pp. 419–434

biguous; although, in every use mentioned here, it picks out sets of properties that distinguish beings like us from other sorts of things (see, also,
Miller 1942 and Natsoulas 1983; however, I cut the joints somewhat differently from either Miller or Natsoulas).

In a normal, everyday conscious experience, like looking at our watch
when someone asks us the time, all three features manifest themselves.
(i) Our looking at our watch just is a qualitatively different experience
for us from looking at Big Ben. Looking at the hands of our dial watch
just is qualitatively different from looking at a digital watch face, and so
on. Such experiences just "feel" different from each other. This qualitative difference is what I mean by phenomenality (many philosophers
call this property a quale [plural: qualia]—I will also use "phenomena"
and "phenomenal experiences" as synonyms of "phenomenality" and
"qualia"). Nagel (1974) alludes to this property when he says that even
if we knew a bat's neurophysiology we would not know what it is like
to be a bat, what the bat's "sonar" experience is *like*. (ii) We take our
experience to be *of* a watch and take that watch to be *out there*, independent of us. Indeed, we take it to *be* a watch itself that we are seeing.
Moreover, we could, in principle, have an experience as of an object, a
watch in this case, even if no watch were *out there*. As Descartes ([1642]
1986) forcefully brought to our attention, we have such experiences in
dreams. This sort of representation is what I am calling "intentionality".

First, note that this use of "intentionality" is related to a use of the
word "meaning" (in German, *Bedeutung*) as in a representation's *meaning* an object or state of affairs. This use is only distantly related (though
less distantly than many think) to our more common uses of "intentionality", as in intending an action, or as meaning "on purpose". Second,
note that this sort of intentionality is involved in judgements, both perceptual and otherwise (though not exclusively in judgements). It is the
sort of intentionality that I have elsewhere called first-order, linguisticlike
representation (and previously labeled as "C1"—for fuller accounts, see
Nelkin 1987b; 1989a,b). Other sorts of intentional states are not included
in this category: The intentionality involved in introspection is a *second-
order*, linguisticlike representation (which is "about" the two first-order
states); and, if imagelike representation exists (see Cooper and Shepard
1984 and Kosslyn 1980, 1987 for the arguments that it does), it is not
to be included with first-order linguisticlike intentionality. Whenever I
use "intentionality" without modifiers, I mean first-order, linguisticlike
intentionality. I will return, from time to time, to the notion of imagelike
representation. (iii) Finally, introspectibility is presented in two senses:
First, while our attention may be primarily directed toward the watch,
we at the same time believe ourselves to be seeing the watch rather than
hearing it, guessing what the watch says, or the like. If asked whether

we saw what time it was or heard the clock strike (while looking at Big Ben, say), we could certainly reply, without hesitation—in normal circumstances—that we saw the clock, even though our attention had been focused on the clock itself and not on our seeing of it.[1] Second, if we do perceive by means of representations, either linguisticlike or imagelike, and representing is an internal experience, then we are introspectively aware of our representation of the watch (even if we are not also aware *that* it is a representation that we are aware of—people can be naive realists, even if mistakenly so—for arguments that perception involves linguisticlike representation in an essential way, see Nelkin 1990, forthcoming-b).

Because our normal experience incorporates all three features that make us different from rocks (or even from roses), it is very easy to believe that these three features belong to a single, noncomposite state, and one might even think that intentionality and introspectibility require phenomenality. But this is mistaken. Both theoretical and empirical reasons suggest that intentional states can be dissociated from phenomenality and introspectibility, and even suggest that phenomenality can exist dissociated from both of the others. Of course, introspectibility cannot occur apart from both of the others, though we have reasons to think it can exist apart from each of the others.

Recently, a number of arguments have been given to show that intentionality is tied in an *essential* way to introspectibility and phenomenality (see Searle 1989, 1990; McGinn 1988, 1989; and Nagel 1986). My arguments and examples are intended, in the first place, to show that these ties cannot be essential, that we can imagine the three states as dissociated from each other. Having established such, I end the paper by indicating some reasons for preferring a theory that treats them as dissociable. Because of the attention that intentionality has recently received, a fair amount of this paper is spent on what some might conceive as flaying a dead horse; yet, all the recent attention given the issue belies the belief that the horse is dead.

The arguments presented here are condensed from lengthier versions (see Nelkin 1986; 1987a,b; 1989a,b; and 1993a). Since Searle's arguments are so well known, I focus on them. Searle puts forward four claims

---

[1]Note that the term "introspection", as used here, does not entail attentiveness. Nor does introspection involve perceptual-like mechanisms. "Introspection" is being applied only to a second-order, noninferential belief that one is having either of the first two states. In this respect, what I am calling "introspection" is similar to what Rosenthal (1986) calls, simply, "consciousness". One important difference is that Rosenthal's notion of consciousness involves self-consciousness. What I am calling "introspection" entails no such involvement. For instance, I think that infants introspect their experiences *before* they have any concept of a self. For how this difference makes a difference, see Nelkin (forthcoming-a).

that play a significant role in the discussion to follow. *First*, Searle (1980, 1983, 1989, 1990) claims that a feature that essentially characterizes intentionality is the displaying of an aspectual character (similar to what others have labeled "referential opacity": All intentional states are from a point of view). *Second*, he takes ordinary, familiar perceptual states to be intentional states (1983). *Third*, he claims that intentionality is *essentially* connected to consciousness (1989, 1990). And, *fourth*, Searle claims that unconscious states are intentional only because, and insofar as, they could become conscious (1989, 1990). I argue that there are empirical reasons to deny the last two claims, especially if we accept the first two. Let us begin with occurrent (as opposed to dispositional) intentional states.

Blindsight cases (Weiskrantz 1977, 1986) give us some reason to think that intentionality occurs without either phenomenality or introspectibility. When blindsight patients "guess" that they see an "X" or "O" in their "blind" field of view, their guesses make sense if unintrospected judgements have been made based on presentations to their blind fields. But do we have to posit intentionality in these cases? Could we not build an unthinking mechanism that would respond to "X's" and "O's" as blindsight patients do? This important question can be faced squarely only at the end of this paper. For the while, all I want to show is that it is *possible* that blindsight patients make unintrospected perceptual judgements. The fact that the patients—granted, under forced-choice conditions—"guess" an "X" or an "O" provides some evidence that these patients *see* the object under an "aspect", as an "X" or an "O", and it is the aspectual nature of an experience that Searle plausibly takes to be defining of an intentional state. Moreover, other blindsight experiments have results that make explanations involving "mechanical" responses seem less plausible. These results more clearly manifest semantic processing and an aspectual nature.

Torjussen (Weiskrantz 1986, 133–134) conducted experiments with patients who were shown a semicircle in their preserved, introspectible field and nothing in their "blind" field. Their responses were that they saw a semicircle. If shown a semicircle in their blind field and nothing in their sighted field, they denied seeing anything. But, if shown a semicircle in their preserved, introspectible field and an attached semicircle in their blind field, these patients said they saw a circle. Surely, if intentionality-characterized processes are involved in the visual processing of the preserved field, as Searle thinks, they are also involved in the visual processing of the "blind" field. Otherwise this difference would seem inexplicable since the presentation to the preserved field is the same in both cases. That is, it is reasonable to believe that the subjects see the circle, in part, because they *see* a semicircle in their blind fields.

Even more remarkable are cases of semantic priming from words shown in the "blind" field, as reported by Marcel (ibid., 142). For instance, when shown the word "river" in their blind fields and, at the same time, asked auditorily to associate the word "bank", subjects were much more likely to associate it with a body of water than with money (ibid., 139 and 149). A possible, perhaps even plausible, explanation is that semantic (i.e., intentionality-characterized) processing took place in the "blind" field.

Other blindsight experiments, as well as ones involving other hemianopias, prosopagnosia (Young and de Haan 1990), visual extinction (Volpe et al. 1979), and split-brains (Gazzaniga and LeDoux 1978) equally establish the possibility of unintrospected intentionality (see Nelkin 1993a). Searle himself recognizes that there is unintrospected intentionality. Nevertheless, Searle (1983, 1989, 1990) claims to preserve the *essential* tie between the two states by claiming that unintrospected intentional states are intentional only if they are potentially introspectible (of course, Searle uses "conscious" rather than "introspectible", even denying that he is talking about introspection).[2] A fuller discussion of this claim will be presented shortly. In the meantime, we can note that, not only do blindsight, prosopagnosic, extinction, and split-brain cases manifest the possibility of intentionality without introspectibility, there is also no reason to believe that they are the kinds of processes that could become introspectible. Blindsight, for instance, has been claimed to involve visual processing that, in any meaningful sense of "can", *cannot* become introspectible (even employing neural circuits other than those of normal perception; see Weiskrantz 1977, 1986). Our brain may be arranged so that such experiences are not introspectible. I see no a priori argument that makes this supposition impossible. Searle's claim that unconscious intentional states are intentional only if they are potentially conscious makes some sense for dispositional states like beliefs. However, we have little reason to accept such a claim for occurrent intentional states like those involved in blindsight. Given the likelihood that a "second"—midbrain—visual system is involved in blindsight, it is also likely that this system is never introspectible and never has been.

These same cases bear on the dissociability of phenomenality from either intentionality or introspection. We would presumably have one of two further possible dissociations here if we accept the possibility of intentionality's being dissociated from introspection. If phenomenality occurs only when introspected, then obviously intentionality can be dissociated from phenomenality since it would so occur in blindsight cases. In fact,

---

[2]Searle (1990) denies that he is talking about introspection, but clearly what he means by "introspection" is a perceptionlike state involving attention. He would surely agree that in seeing Big Ben we are aware of our *seeing* it, i.e., he does not deny introspection in the sense in which I use it.

blindsight patients usually report no phenomenality; and when they do report phenomenal states, the states seem wholly inappropriate (see Campion et al. 1983, who present a more or less opposing point; I think the cases support my reading, not theirs). The second option is to say that the unintrospected perception *would* also involve phenomenality. In that case, of 'course, the dissociability of phenomenality and introspection would follow. Later, I will argue that unintrospected phenomenality can occur.

There are also experiments that illustrate unintrospected intentionality but which involve intact brains. Marcel showed subjects three consecutive strings of letters. The subjects were asked whether the third string was or was not a word. The cases where the third string was a word themselves divided into "congruent" and "incongruent" cases. In each of these cases the middle string was a polysemous word. In congruent cases, the first and third strings were both words that related similarly to the polysemous middle word (HAND, PALM, WRIST). In incongruent cases, the first and third strings were words that related to the different meanings of the polysemous word (TREE, PALM, WRIST). When the polysemous word was presented normally, correct identification of the third string as a word was enhanced only in congruent cases. But if the polysemous word was shown in a visually degraded (subliminal) manner, correct identification was enhanced in *both* congruent and incongruent cases (Reingold and Merikle 1990, 24).

Consider another case from the subliminal perception literature. Dixon (1984) discusses a case where subjects were asked to select either the word "smug" or the word "cosy" to complete sentences such as "She looked . . . in her fur coat". If primed with the word "snug" such that the prime was uttered just above the audible limen, subjects invariably chose "smug" as their sentence completion. But if the prime was uttered subliminally, the subjects chose "cosy".

One could argue that these last two cases, as well as the previous semantic priming case, do not really involve intentionality. It might be claimed that the "semantic" priming is not really fully semantic, that, while based on semantic features, the processing is merely a kind of lexical, associative, merely "mechanical" process, not characterizable by intentionality. That is, what takes place in semantic priming cases may be no more *semantic* than is the "Spell-Check" feature of my word-processing program (A. J. Marcel, in conversation, presented this objection). Perhaps we could build a connectionist machine that responded in these ways, but which we would all agree was not conscious in any sense (I owe this example to an anonymous reader for this journal). These claims may be true, but at this point I am not arguing for their falsity, only for the possibility that they are false. At the end, I will more fully confront this issue. Even if the association is purely lexical, one might argue that the fact that the printed marks are taken as *words* at all already displays an aspectual nature, which gives us at least some reason to think that these states are intentional states. Second, it is reasonable to think that, if given priming in their introspectible field, the subjects' subsequent behaviors would be similar. But if introspectible perception is fully intentional—as it is on Searle's account—then, given that there are seemingly no differences in the two situations other than the lack of introspectibility in the blindsight case, there seems no good reason to think that blindsight is not characterized by intentionality if the ordinary, introspectible perception is. Granted that the responses in the semantic priming cases are "automatic", even associative, that fact is not incompatible with their *also* revealing underlying intentional states. With Marcel's subjects, it seems even more likely that the priming is semantic, and so intentional, for it is difficult to believe that there could be a very strong *automatic* correlation between "palm" and "wrist". Dixon's results show, somewhat to our surprise, that, while the liminal, introspectible association is predominantly phonological, the unintrospected, subliminal one is predominantly semantic.

The semantic priming examples not only support the claim that unintrospected intentional states are possible, but the different results in the liminal and subliminal trials suggest that the processing is different in the two conditions. This difference again belies Searle's claim that unconscious intentional states must be of an introspectible sort.

In fact, it is difficult to know just what is meant by the claim that unintrospected intentional states must at least in principle be introspectible (see f.n. 2). One might mean that by an act of attention—or perhaps by some deeper method such as psychoanalysis—one could come to introspect what one could not introspect before. But, empirically, this claim seems false for many of the relevant cases. Alternatively, one might mean that the sort of processing that takes place in the unconscious cases is the same as in conscious ones. But the blindsight and subliminal perception cases raise reasonable doubts about this claim. Surely, even if the processing were similar, the moral would be the *independence* of intentionality from introspection, not the dependence. A third possibility is that one might mean that anything displaying unintrospected intentionality is at that very time also displaying introspected intentionality of some kind or another. True or not, this claim seems irrelevant. Finally, one might mean that only beings capable of introspection can have intentional states. This claim may be true, but there would still be no reason to think that intentional states, once they become possible at all, could not then have a life of their own, independent of introspection. Indeed, the cases cited above are evidence for the independence of intentional states from introspective consciousness.

If the aspectual nature of states that are not conscious is retained even when they are not conscious, especially if they can occur without ever having been conscious, then it is difficult to comprehend what the intentionality of such states has to do with whether the switch is on or not (Searle's metaphor for consciousness), or even with whether intentional states can be causes of turning on such a switch. If Searle believes their aspectual nature is acquired or retained *out of* consciousness, why is that not an *admission* of the independence of intentionality from *both* phenomenality and introspection? When Searle gives examples of unconscious intentional states, he almost always cites *dispositional* states, such as beliefs, and not *occurrent* states, such as blindsight experience. The claim that occurrent unconscious intentional states are potentially conscious is, in many cases, much more tendentious than the claim that dispositional states are. There are reasons to think that Searle's claim that all occurrent, in-principle unconscious states *must* be mere neural states is contradicted by the empirical facts. It is at least possible to read intentionality into the relevant cases.

A final sort of case illustrating the possibility (even plausibility) of intentionality without introspectibility is that of creative thinking. Often, when working on a problem, we get stuck. We leave it for a while. Sometime later a well-formed solution comes to us. It seems as if this solution could not have been arrived at without rational processing having taken place outside of our introspective awareness. Moreover, these reasoning processes would seem to require premises thought of in the aspectual way Searle demands for intentionality. Once again, we have no compelling reason to think that these intentionality-characterized processes are of a kind that ever could become introspectible.

Nor do such unintrospected reasoning processes *feel* any way at all, that is, they have no phenomenality. One sometimes says things like, "I feel the wheels turning"; but this use of "feel" no more expresses a phenomenal feeling of belief than does "I feel he will come tomorrow". There is no phenomenal feeling that *is* one's thought that he will come tomorrow, and no feeling grounds that thought. Aristotle undoubtedly felt the wheels turning in his heart, not in his head. In these cases, the word "feel" is used because one does not know the origins of one's thoughts. This use of "feel" occurs when we are conscious of our states, but do not know how we are conscious of them.

This discussion of "feeling he will come tomorrow" underscores the truth that many of our ordinary, introspectible thoughts themselves involve no phenomenality. If one occurrently thinks that tomorrow is Tuesday, or one thinks that 1000-sided figures have more sides than 999-sided figures, no phenomenal experiences are necessary for such thoughts (the example is borrowed from Descartes [1642] 1986, 50–51, who makes a similar point). One may have phenomenal experiences that *accompany* such occurrent thoughts, but the phenomena may be different on different occasions; and if one is having phenomena at all when one expresses such a thought, the phenomena are often irrelevant to the thought being expressed (Wittgenstein 1953 argued for this same separation of the cognitive and the phenomenal). These last cases support the idea that not only do intentionality and phenomenality come apart, but introspectibility and phenomenality also come apart: We can introspect, at times, that we have an occurrent thought such as that a chiliagon has more sides than a figure with 999 sides, and this introspection itself does not *feel* any way at all. Whatever phenomenal states accompany this introspection, they are not conceptually required for introspection to take place and certainly do not constitute it. One should not confuse the subjectivity of introspective states with phenomenality. It is a mistake to identify the subjectivity of (introspective) consciousness with phenomenality (as do Searle, Nagel, and McGinn). Neither intentionality nor introspectibility would seem to require phenomenality.

Searle (1983) recognizes the force of these arguments, which demonstrate the possibility that introspection and intentionality dissociate from phenomenality (for fuller arguments concerning this point, see Nelkin 1987b, 1989a). He still does not capitulate. Instead, he argues that while no phenomenal properties are essential to thinking, there are phenomenological ones: Thoughts are "felt", even if not felt (for a similar move, but about perception, see Leon 1988). My response is much like Hume's to that of an introspectible self. If there be such properties of experience, then I fail to find them in my own experiences. I *do* find phenomenal ones. It is only phenomenological "feelings" that are somehow *other* than phenomenal ones that I deny. Nor am I denying that we have intentional states or introspective ones. I am denying only that these states "feel" like anything. Thinking that 1000-sided figures have more sides than 999-sided ones does not "feel" different from thinking that 1001-sided figures have more sides than 1000-sided ones. Nor does introspecting these two thoughts "feel" any way at all. I may have *phenomenal* states that often (but inessentially) accompany my thoughts, but I can find no phenomenological ones that do even that. This need for something "felt", if not felt, is, I believe, an unhappy relic of British empiricism.

At this point, one of my own claims may be used against me. Since I accept that there could be unintrospected phenomenality, how can I be certain that phenomenality does not always—and somehow essentially— occur along with intentionality and introspection (only nonintrospectively so)? I cannot be *certain*, but neither have I any reason to believe it. That unintrospected phenomenality is always lurking in the shadows, as it were, and *essentially* so, seems just too much to believe. It is not an impossibility, I suppose, but I would want to be given some reason to think it is *true*. As for its being phenomenological, rather than phenomenal, properties doing the lurking, I would even more want reasons, since I *never* introspect such properties. In those cases where I *will* opt for saying that unintrospected phenomenality is present, there will be, as we will see, *reasons* for saying so. The sensible conclusion to be drawn from these considerations is that intentionality and introspection can dissociate from phenomenality.

So far, the following have been shown. Intentionality can be dissociated from both introspectibility and phenomenality. Perceptual states which do have intentionality, and may or may not have phenomenality, possibly occur without being introspectible. In addition, nonperceptual intentional states, even if introspectible, most likely involve no phenomenality. Finally, the creativity cases show a reason to believe that there are nonperceptual intentional states that are neither introspectible nor phenomenal.

What remains to be shown is that phenomenality and introspectibility can also be dissociated. Actually, half of that last goal has already been accomplished: The chiliagon case shows that introspectibility can exist dissociated from phenomenality. But can phenomenality come apart from introspection? Could we feel pains or see colors, say, without being introspectively aware that we are? Indirect evidence for unintrospected phenomenal states is provided by certain blindsight cases. We know from human blindsight cases that subjects have left visual fields that are unintrospected, even though active. We also have evidence that blindsighted monkeys, whose brains are similar to our own, can recover many of their visual functions (Weiskrantz 1986). Among the functions reported to have been recovered is the ability to make color discriminations (Keating 1979). Moreover, we have (some) reason to believe that color (hue) discrimination involves phenomenal states in an essential way (for instance, see Hardin 1988, Boghossian and Velleman 1991). Putting all these facts together, we have some reason to believe that those blindsighted monkeys capable of making color discriminations experience color phenomena without being (introspectively) aware that they do. Of course, it is possible that, in recovering, the monkeys recover introspectibility of phenomenal states, but it is an empirical question as to which reading of the facts is correct. It is certainly *possible* that the dissociation occurs in the way outlined.

More direct evidence of the dissociation of phenomenality from introspection is provided by recent work on *human* blindsight subjects (Stoerig and Cowey 1989). These subjects were able to make color discriminations which tracked normal discriminations; although, when asked, they denied seeing any colors. They took themselves to be guessing. Since their responses mirror those of normal color perceivers, rather than being directly light-wave dependent, it seems reasonable to believe that they were discriminating *hues* while not being introspectively aware that they were.

Even if phenomenality comes apart from intentionality and introspection, does it also come apart from imagelike representation, and, thus, from all intentionality? If so, we may have to recognize four sorts of consciousness rather than the three I have put forward. What is important about this question for our present purposes is that it is a theoretical/empirical one, not an a priori one. In a previous paper (Nelkin 1989b), I have defended the claim that imagelike representation (I would now say only a specifiable subset of it) and phenomenality are not dissociable. Since I have defended this claim elsewhere, I will not do so again, especially as establishing it is not germane to my central task of defending the dissociability of phenomenality and intentionality from each other, and from introspectibility.

Let me say here only that phenomenality's being a property of this set of imagelike representations clarifies why there is phenomenality at all and how we find phenomenality to be instantiated when we *do* introspect it. For instance, at times when we decide to pay attention to a certain sensation, we now introspect the sensation (in these cases, attention and introspection coalesce). In these experiences we seem to discover phenomenal qualities of the sensation present all along, even when we were not introspecting them. It does not seem as if our introspecting itself *creates* the phenomenal qualities or changes their character. That it seems this way is, if I am right, the result of its being this way. Of course, one might interpret these cases differently—as one can the blindsight cases, and all the others (see Dennett 1991, for instance, for a recent and different reading of many of the cases). But all I have claimed so far is the *possibility* of my reading of the cases. I will address its correctness later.

Natsoulas (1983, 1989, 1990), who recognizes analogues of the three forms of consciousness discussed here, argues that a fourth form—reflective consciousness—is needed. He agrees that what I have called "introspective consciousness" involves no phenomenality. He also recognizes the existence of sensation states that are not introspected. However, he does not see how adding a linguisticlike state (such as we agree introspection to be) to unconscious sensations could result in the *felt* quality of a second-order awareness of sensations. To fill this gap, he enlists reflective consciousness, a second-order consciousness different from "mere" introspective consciousness.

If Natsoulas and I agree on so many things, how do we differ? The key lies in differing notions of unconscious sensations. Natsoulas thinks unconscious sensations have all the properties of sensation *other than* phenomenality (feltness). I am claiming, supported by the Keating and the Stoerig and Cowey studies, among other things, that unconscious sensations have phenomenality as well. We do not need to enlist some fourth state to find phenomenality; it was there all along. We *feel* sensations but sometimes do not know that we do. "But how can one *unconsciously feel* a sensation?" This question is itself the result of the confusion I have been trying to clarify. If by "conscious" one means what I have called "sensation consciousness" (states with phenomenality) (Nelkin 1989b), then these sensation experiences are *conscious*, not unconscious. It is only if the question means "conscious" in the sense of introspection that the states in question are unconscious. As far as the question of how there could be phenomenality that is not introspected, why should that state of affairs be any more puzzling than that there is intentionality that is not introspected? It is only if we think all three features—phenomenality, intentionality, introspectibility—must be features of a single, noncomposite state that there appears anything contradictory in the notion of unintrospected phenomenality, or unintrospected intentionality for that matter.

We have established the possibility of the dissociations of intentionality and phenomenality, and of each of these from introspectibility. It is a theoretical/empirical matter as to whether this reading of the supporting cases is the best. Its worth will not be settled by philosophical argument alone. No obvious conceptual, or other *a priori*, *essential* connection holds between any two of these states.

Yet, granting the possibility of this reading, are there any reasons to think it is the *best* reading of the supporting cases? After all, there are competing theories, including Dennett's (1991) and Natsoulas's (1983, 1989, 1990). I will not attempt in this paper to provide an adequate answer to this question (see, instead, Nelkin 1987b; 1989a,b; 1993b). I will only indicate what such an answer would have to be like.

However, let us first highlight elements of *this* paper that already move us beyond the possible to the plausible. The creative thinking cases make it likely, not merely possible, that intentionality dissociates from introspectibility. It is difficult to imagine an explanation of these cases that would not appeal to unintrospected reasoning (i.e., unintrospected thought). Equally, the discussion involving thoughts such as that 1000-sided figures have more sides than 999-sided ones goes beyond showing merely the possibility of dissociating intentionality and introspectibility on the one hand from phenomenality (or phenomenologicality) on the other. Perhaps even more interestingly, the color-discrimination blindsight cases make it plausible that phenomenality (and perhaps intentionality) dissociates from introspectibility, especially if color perception involves phenomenality in some essential way. What is striking about the color cases is that the discriminations do not track light waves per se, but track the opponent-process hue discriminations (which are at variance with light-wave—for reflectance—discriminations) of normal human perceivers. Perhaps one could build a *mere* connectionist machine (one which we could all agree is not conscious) to track light-wave or reflectance discriminations, but it is not at all obvious that we can build such a clearly nonconscious one to track ordinary color perception. Even so, the burden of proof seems to be on those who say it can be done. If it is plausible that these blindsight subjects display the relevant dissociations, it perhaps becomes more likely that the other blindsight subjects do so as well. Finally, in regard to at least one of my opponents, Natsoulas, there *is* an argument in my favor. Since Natsoulas and I agree on the dissociations proposed, my theory is preferable because of Occam's Razor: The proposed theory does not need a fourth form of consciousness, reflective consciousness, over and above the three defended here. So, all in all, some work has been done in this paper toward moving from the possible to the plausible. Still, we have a long way to go; the rest of what I say is no more than gesturing.

One virtue of reading this ambiguity into our use of "conscious", and accepting the dissociability of the various states so called, is that we can have a uniform and comprehensive theoretical treatment of all the cases. We do not have to apply one theory ad hoc to one set of cases and another theory ad hoc to another set. Undoubtedly, my reading is not the only one with this apparent virtue. Dennett's (1991) seems to have it (or to come close to having it). There are many areas of agreement between his theory and mine ("theory" means "philosophical theory", not "scientific theory"). One reason for our agreement is that Dennett comes close to accepting a theory of consciousness that stresses the importance of introspection. Even while trying to empty introspection of any substance, Dennett makes repeated use of it in his explanations (for one instance where introspection seems to be playing a bigger role in his explanations than he would acknowledge, see 1991, 168). In the end, large theoretical reasons, having to do especially with concept acquisition and possession, persuade me that my view of consciousness is closer to the truth than Dennett's or others' (see Nelkin 1993b and forthcoming-a).

As we saw, in discussing cases, many of the blindsight and other situations might be read as involving no intentionality, so no consciousness of any kind at all. Should we read them as possessing intentionality? The same answer applies: We get a kind of theoretical coherence and breadth if we do read them in this way. It is true that present-day computers lend us little reason to think of them as conscious in any way, even as possessing intentional states (though some would argue otherwise). And it is true that many of the blindsight responses might be simulated by a computer, especially by a connectionist one. On the other end of the spectrum are human beings, who, at times, display all three conscious states and perform the same discrimination tasks consciously. Unlike our attitudes toward computers, most of us accept as a given that human beings *are* conscious (what is at controversy is how that consciousness is to be analyzed). So the fact that nonconscious computers can simulate the activities of conscious discriminators does not tell us much about whether blindsight discriminations are or are not conscious, intentional-state discriminations. We know, after all, that human beings do have intentional and phenomenal states (see Chandler 1988 for a similar methodological point). Pointing this out does not establish that blindsight discriminations *are* intentional-state ones, or involve phenomenal properties, but it may remove some biases against thinking in these ways. Again, the final determination of our views will be for large-scale theoretical reasons not elaborated.

Each of intentionality, phenomenality, and introspection is a state we possess and rocks or roses do not. If consciousness is what distinguishes us from such things, then, if the proposed theory is correct, each of these states has the same priority for being called "consciousness". On the other hand, when one of these states is absent at a given moment, the subject can be said to be *un*conscious—in *that* respect. For just as there are three ways of being conscious, there are three ways of being *un*conscious. Moreover, that these are three independent, dissociable states (though they sometimes causally interact) means that we should expect to find beings who do not possess all three states, but possess only some subset of them. It is very likely that some nonhuman animals are just such creatures. So my philosophical theory begins to edge into a scientific one insofar as it has testable consequences. To the question, "But what about the unity of consciousness?" I reply that there is no unity, only the appearance of it.

REFERENCES

Boghossian, P. A. and Velleman, J. D. (1991), "Physicalist Theories of Color", *Philosophical Review 100*: 67–106.
Campion, J.; Latto, R.; and Smith, Y. M. (1983), "Is Blindsight an Effect of Scattered Light, Spared Cortex, and Near Threshold Vision?" *Behavioral and Brain Sciences 6*: 423–486.
Chandler, M. (1988), "Doubt and Developing Theories of Mind", in J. W. Astington, P. L. Harris, and D. R. Olson (eds.), *Developing Theories of Mind*. Cambridge, England: Cambridge University Press, pp. 387–413.
Cooper, L. A. and Shepard, R. N. (1984), "Turning Something Over in the Mind", *Scientific American 251*: 106–114.
Dennett, D. C. (1991), *Consciousness Explained*. Boston: Little, Brown & Company.
Descartes, R. ([1642] 1986), *Meditations on First Philosophy*, in J. Cottingham (trans. and ed.), *René Descartes: Meditations on First Philosophy, With Selections from the Objections and Replies*. Cambridge, England: Cambridge University Press.
Dixon, N. F. (1984), "Subliminal Perception", in R. L. Gregory (ed.), *The Oxford Companion to the Mind*. Oxford: Oxford University Press, pp. 752–755.

Gazzaniga, M. S. and LeDoux, J. E. (1978), *The Integrated Mind*. New York: Plenum Press.

Hardin, C. L. (1988), *Color for Philosophers: Unweaving the Rainbow*. Indianapolis: Hackett.

Keating, E. G. (1979), "Rudimentary Color Vision in the Monkey after Removal of Striate and Preoccipital Cortex", *Brain Research 179*: 379–384.

Kosslyn, S. M. (1980), *Image and Mind*. Cambridge, MA: Harvard University Press.

——. (1987), "Seeing and Imaging in the Cerebral Hemispheres: A Computational Approach", *Psychological Review 94*: 148–175.

Leon, M. (1988), "Characterising the Senses", *Mind & Language 3*: 243–270.

McGinn, C. (1988), "Consciousness and Content", *Proceedings of the British Academy 74*: 219–239.

——. (1989), "Can We Solve the Mind-Body Problem?" *Mind 98*: 349–366.

Miller, J. G. (1942), *Unconsciousness*. New York: Wiley.

Nagel, T. (1974), "What is it Like to be a Bat?" *Philosophical Review 83*: 435–450.

——. (1986), *The View from Nowhere*. Oxford: Oxford University Press.

Natsoulas, T. (1983), "Concepts of Consciousness", *Journal of Mind and Behavior 4*: 13–59.

——. (1989), "An Examination of Four Objections to Self-Intimating States of Consciousness", *Journal of Mind and Behavior 10*: 63–116.

——. (1990), "Reflective Seeing: An Exploration in the Company of Edmund Husserl and James J. Gibson", *Journal of Phenomenological Psychology 21*: 1–31.

Nelkin, N. (1986), "Pains and Pain Sensations", *Journal of Philosophy 83*: 129–148.

——. (1987a), "How Sensations Get Their Names", *Philosophical Studies 51*: 325–339.

——. (1987b), "What is it Like to be a Person?" *Mind & Language 2*: 220–241.

——. (1989a), "Propositional Attitudes and Consciousness", *Philosophy and Phenomenological Research 49*: 413–430.

——. (1989b), "Unconscious Sensations", *Philosophical Psychology 2*: 129–141.

——. (1990), "Categorising the Senses", *Mind & Language 5*: 149–165.

——. (1993a), "The Connection Between Intentionality and Consciousness", in M. Davies and G. Humphreys (eds.), *Consciousness: Psychological and Philosophical Essays*. Oxford: Blackwell, pp. 224–239.

——. (1993b), "Patterns". In preparation.

——. (forthcoming-a), "The Belief in Other Minds", in D. F. Gottlieb and S. J. Odell (eds.), *Wittgenstein and the Cognitive Sciences*.

——. (forthcoming-b), "Phenomena and Representation", *British Journal for the Philosophy of Science*.

Reingold, E. M. and Merikle, P. M. (1990), "On the Inter-Relatedness of Theory and Measurement in the Study of Unconscious Processes", *Mind & Language 5*: 9–28.

Rosenthal, D. M. (1986), "Two Concepts of Consciousness", *Philosophical Studies 49*: 329–359.

Searle, J. R. (1980), "Minds, Brains, and Programs", *Behavioral and Brain Sciences 3*: 417–457.

——. (1983), *Intentionality: An Essay in the Philosophy of Mind*. Cambridge, England: Cambridge University Press.

——. (1989), "Consciousness, Unconsciousness, and Intentionality", *Philosophical Topics 17*: 193–209.

——. (1990), "Consciousness, Explanatory Inversion, and Cognitive Science", *Behavioral and Brain Sciences 13*: 585–642.

Stoerig, P. and Cowey, A. (1989), "Wavelength Sensitivity in Blindsight", *Nature 342*: 916–918.

Volpe, B. T.; LeDoux, J. E.; and Gazzaniga, M. S. (1979), "Information Processing of Visual Stimuli in an 'Extinguished' Field", *Nature 282*: 722–724.

Weiskrantz, L. (1977), "Trying to Bridge some Neuropsychological Gaps Between Monkey and Man", *British Journal of Psychology 68*: 433–445.

————. (1986), *Blindsight: A Case Study and Implications*. Oxford: Oxford University Press.

Wittgenstein, L. (1953), *Philosophical Investigations*. London: Macmillan.

Young, A. W. and de Haan, E. H. F. (1990), "Impairments of Visual Awareness", *Mind & Language* 5: 29–48.

Philosophical Perspectives, 9, AI, Connectionism, and
Philosophical Psychology, 1995

CONSCIOUSNESS AS INTERNAL MONITORING, I

The Third *Philosophical Perspectives* Lecture

William G. Lycan
University of North Carolina, Chapel Hill

Locke put forward the theory of consciousness as "internal Sense" or "reflection"; Kant made it "[i]nner sense, by means of which the mind intuits itself or its inner state."[1] On that theory, consciousness is a perception-like second-order representing of our own psychological states and events.

The term 'consciousness' of course has many distinct uses.[2] My concern here is with that use according to which much of one's mental or psychological activity is unconscious or subconscious even when one is wide awake and well aware of other goings-on both external and internal. I shall argue that what distinguishes conscious mental activity from un- and subconscious mental activity is indeed second-order representing.

Locke's idea has been urged in our own time by philosophers such as D.M. Armstrong and by psychologists such as Bernard Baars; I have previously defended it too.[3] But some interesting criticisms have been raised against the view by a number of theorists. My most urgent task in this paper is to overcome an objection due to Georges Rey.[4]

1. Armstrong states the Inner Sense doctrine as follows.

> Introspective consciousness...is a perception-like awareness of current states and activities in our own mind. The current activities will include sense-perception: which latter is the awareness of current states and activities of our environment and our body.[5]

As I would put it, consciousness is the functioning of internal *attention mechanisms* directed upon lower-order psychological states and events. I would also add (or make more explicit) a soupçon of teleology: Attention mechanisms are devices which have the *job* of relaying and/or coordinating information about ongoing psychological events and processes.[6]

Armstrong offers a plausible Just-So Story to explain the prevalence of introspective consciousness:

> [T]he biological function of introspective consciousness...is to sophisticate our mental processes in the interests of more sophisticated action.
>
> Inner perception makes the sophistication of our mental processes possible in the following way. If we have a faculty that can make us aware of current mental states and activities, then it will be much easier to achieve *integration* of the states and activities, to get them working together in the complex and sophisticated ways necessary to achieve complex and sophisticated ends.
>
> ...[C]o-ordination [of many parallel processes] can only be achieved if the portion of the computing space made available for administering the overall plan is continuously made "aware" of the current mental state of play with respect to the lower-level operations that are running in parallel, Only with this feedback is control possible... . It is no accident that fully alert introspective consciousness characteristically arises in *problem* situations, situations that standard routines cannot carry one through.[7]

A slightly deflated version of this idea will figure in my own defense of the Inner Sense theory.

The Lockean thesis is a component of a wider project of mine: that of establishing the *hegemony of representation*. I am concerned to maintain a weak version of Brentano's doctrine that the mental and the intentional are one and the same. Weak, because I am not sure that intentionality suffices for representation; but my claim is strong enough: that the mind has no special properties that are not exhausted by its representational properties. It would follow that once representation is (eventually) understood, then not only consciousness in our present sense but subjectivity, qualia, "what it's like," and every other aspect of the mental will be explicable in terms of representation, without the positing of any other ingredient not already well understood from the naturalistic point of view.[8]

I should repeat and emphasize that my concern in this paper is solely with the notion of conscious awareness, with the distinction between conscious mental states and un-, sub- or otherwise non-conscious mental states. In particular, I am not here addressing issues of qualia or phenomenal character, which I have resolved almost entirely satisfactorily elsewhere.[9] There may be Inner Sense theorists who believe that their views solve problems of qualia; I make no such claim, for I think qualia problems and the nature of conscious awareness are mutually independent and indeed have little to do with each other.[10]

2. The Inner Sense view of consciousness has a number of advantages, the first of which is that it does distinguish awareness from mere psychology, and conscious states/events (in the sense indicated above) from mere mentation. We may plausibly suppose that many lower animals have psychologies and mentation, or at least internal representation, without awareness. Second, the view affords some *grades* of un- or subconsciousness; e.g., a state/event may be unconscious just because it is unattended, but a Freudian wish to kill one's father may have been rendered unattend*able* by some masterful Censor. —And further distinctions are available, both for animals and for human beings.

Third, the Inner Sense account affords the best solution I know to the problem of subjectivity and "knowing what it's like," raised by B.A. Farrell, Thomas Nagel and Frank Jackson; Georges Rey and I hit upon that solution independently a few years ago.[11] It involves the behavior of indexical terms in the proprietary vocabulary mobilized by the relevant attention mechanisms. But there is no time to rehearse it here.

Fourth, the Inner Sense view sorts out a longstanding issue about sensations and feeling: Consider pain. A minor pain may go unfelt, or so we sometimes say.[12] Even quite a bad pain may not be felt if attention is distracted by sufficiently pressing concerns. Yet such assertions as my last two can sound anomalous; as David Lewis once said, meaning to tautologize, "pain is a feeling." When one person's commonplace sounds to another contradictory on its face, we should suspect equivocation, and the Inner Sense model delivers: Sometimes the word 'pain' is used to mean just the first-order representation of damage or disorder, a representation which can go unnoticed. But sometimes "pain" means a conscious feeling or mode of awareness, and on that usage the phrase "unfelt pain" is simply self-contradictory; it comprehends both the first-order representation and the second-order scanning together. Thus the equivocation, which gave rise to the issue; the issue is dissolved.

3. In correspondence, Fred Dretske has asked a good pair of questions about the Inner Sense view:[13] Why is consciousness (or just representation) of certain physical states enough to make those states *themselves* "conscious"? And more specifically, what is it that is so special about physical states of that certain sort, that consciousness of them makes them—but not just any old physical state—conscious? After all, we are conscious *of* (what are in fact) physical states of our stomachs; for that matter, through ordinary perception we are conscious of physical states of our skins, such as their being freckled, but no one would distinguish between "conscious" *stomachs* and "unconscious" stomachs, or between "conscious" and "unconscious" frecklednesses.

Indeed, why does the concept work that way (assuming it does work that way)? It may have something historically to do with the fact that until the 20th century the mental/psychological was simply identified with the conscious, and so only recently have we had to adopt a taxonomic distinction between states we are aware of holding and states we are not. (I am assuming that there is such a distinction in reality, and I believe—what is not uncontroversial—that the distinction in theory applies to any ordinary mental state, not counting states described as "being consciously aware of [such-and-such].")

What is it that is so special about physical states of that certain sort, that consciousness *of* them makes them "conscious"? That they are themselves mental. Stomachs and freckled patches of skin are not mental. It seems psychological states are called "conscious" states when we are conscious of them, but nonpsychological things are not.

Given the reality of the distinction between states we are aware of being in and states we are not aware of being in, the only remaining question is that of why the *word* "conscious" is thus dragged in as an adjective to mark it. My bet is that there is a grammatical answer. Maybe it is a transferred epithet: We begin with the adverbial form, as in "consciously thought" or "consciously felt," and when we make the verb into a noun the adverb automatically becomes an adjective—as in the move from "meditatively sipped" to "took a meditative sip." That is fairly plausible; at any rate it is the best I can do for now.

In any case, it is important to see that the question pertains to the notion of conscious awareness itself; it is not a problem for or objection to the Inner Sense theory of awareness in particular.

4. An initial flaw in the version stated so far is that it makes a Cartesian assumption recently highlighted by Dan Dennett:[14] that there is some determinate stage of information-processing that constitutes the locus of conscious mental states/events. More specifically, "Cartesian materialism" is the (usually tacit) assumption that there is a *physically realized* spatial or temporal turnstile in the brain, a stage where "it all comes together" and the product of pre-processing is exhibited "to consciousness."

Dennett attacks that assumption. However natural it may be, it is gratuitous and empirically implausible: First, it is a priori unlikely that Mother Nature has furnished the human brain with any central viewing-room or single monitor to do the viewing; nor is there any positive neurophysiological sign of such an organ. Second, Dennett argues at length that the famous "temporal anomalies" of consciousness discovered by psychophysical research, such as color phi, the cutaneous rabbit and Libet's "backward referral" of sensory experiences,[15] are anomalous only so long as Cartesian materialism is being assumed; jettison the assumption, and the phenomena are readily explained. Dennett's analyses of the experimental data are not completely uncontroversial,[16] but I find them convincing on the whole, and it is hard to think how anyone might defend Cartesian materialism on purely neurophysiological grounds.

The point is not just that there is no *immaterial audience* in the brain, nor just that there is no undischargeable homunculus, but that there is no such locus at all, however physically characterized—no single Boss Unit or even CPU within the brain, to serve as chief executive of my utterings and other actions. The central nervous system is as central as it gets. There is, if you like, a "stream of consciousness": "We are more-or-less serial virtual machines implemented—inefficiently—on the parallel hardware that evolution has provided for us," "Joycean" machines that formulate synthesized reports of our own passing states,[17] though the reports are never entirely accurate.

The Inner Sense theory has it that conscious awareness is the successful operation of an internal scanner or monitor that outputs second-order representations of first-order psychological states.[18] But an "internal scanner" sounds very much as though it presupposes an internal audience seated in a Cartesian Theater, even if it and the Theater are made of physical stuff. Is the Inner Sense view not then committed to Cartesian materialism?

It is not hard to come up with a pretty damning collection of direct quotations. Armstrong spoke (above) of "*the* portion of the computing space made available for administering the *overall plan.*" And (just to save you looking) I myself wrote of an internal scanner's "delivering information about...[a first-order] psychological state to one's *executive control unit.*"[19] For shame. There may be an "executive control unit" in some functional sense, but very probably not in the sense of being: that agency, arrival at which makes information conscious.

But it should be clear that the Inner Sense view is not *per se* committed to Cartesian materialism. For even if an internal scanner resembles an internal audience in some ways, the "audience" need not be seated in a Cartesian Theater: There need be no *single*, executive scanner, and no one scanner or monitor need view the entire array of first-order mental states accessible to consciousness. Accordingly, there need be neither a "turnstile of consciousness" nor one central inner stage on which the contents of consciousness are displayed in one fixed temporal order. An internal monitor is an attention mechanism, that presumably can be directed upon representational subsystems and stages of same; no doubt internal monitors work selectively and piecemeal, and their operations depend on control windows and other elements of conative context. On this point, the Inner Sense theory has already parted with Cartesian materialism.

A qualification: We should not throw out the integration-and-control baby with the Cartesian bathwater. The operation of an internal monitor does not *eo ipso* constitute consciousness. For we can imagine a creature that has a panoply of first-order states and a rich array of monitors scanning those states, but in such a way that the monitors' output contributes nothing at all to the creature's surrounding psychology, maintenance, or welfare; the outputs might just go unheard, or they might be received only by devices that do nothing but turn patches of the creature's skin different colors. For consciousness-constituting, we must require that monitor output contribute—specifically to the integration of information in a way conducive to making the system's behavior appropriate to its input and circumstances. Though the latter formulation is terribly vague, it will do for present purposes; the requirement rules out the ineffectual monitors without falling back into the idea of a Cartesian Theater or single CPU.

(This is a good juncture at which to underscore and deepen the teleological cast I am imparting to the Inner Sense theory. I said that for an internal monitor to count in the analysis of consciousness, in the present sense of 'conscious', the monitor must have monitoring as its function, or one of its functions. But that is not all. To count in the analysis of *my* consciousness, the monitor must do its monitoring *for me*. A monitor might have been implanted in me somewhere that sends its outputs straight to Reuters and to CNN, so that the whole world may learn of my first-order psychological states as soon as humanly possible. Such a device would be teleologically a monitor, but the wire services' monitor *rather than* mine. More importantly, a monitor functioning within one of my subordinate homunculi might be doing its distinctive job for that homunculus rather than for me; e.g., it might be serving the homunculus' event memory rather than my own proprietary event memory.[20] This distinction blocks what would otherwise be obvious counterexamples to the Inner Sense view as stated so far.)

Rejection of Cartesian materialism is not only compatible with the Lockean view. In an important way, it supports the Inner Sense theory: It predicts introspective fallibility of two characteristic sorts. First, as Dennett emphasizes, the result of an introspective probe is a *judgment* made by the subject, which judgment does not (or not eo ipso) simply report a "presentation" to an inner audience. And the "temporal anomalies" alone should have made us question the reliability of introspective reports. Introspection gets small temporal details wrong. That tends to confirm rather than to impugn the Inner Sense view of consciousness. If conscious awareness is indeed a matter of introspective attention and if introspection is the operation of a monitor or self-scanner, then such anomalies were to be expected—for monitors and scanners are characteristically fallible on details, and Dennett shows admirably how such devices might corporately mix up temporal sequence in particular.

Second, if there is no single Cartesian Theater, then there should be no single optimal time of probing a first-order process. More strongly, Dennett argues that probing "changes the task," i.e., interferes with the very process it purports to be monitoring. That too is good news for the Inner Sense theory. For if introspection is the operation of a monitor or self-scanner, then revisionary effects of the present sort are again just what we should have expected; monitoring instruments (such as ammeters) typically do affect the values of the magnitudes they measure.[21]

Thus the Inner Sense theory of consciousness survives the collapse of Cartesian materialism, and is even strengthened by it.

5. On at last to Rey's objection. It is that if all it takes to make a first-order state a conscious state is that the state be monitored by a scanner that makes integrative use of the information thus gleaned, then consciousness is a lot more prevalent than we think. Any notebook computer, for example, has devices that keep track of its "psychological" states. (If it be protested that no computer has genuinely psychological states—e.g., because it has neither authentic intentional states nor sensory states—that is inessential to the point. Once we had done whatever needs to be done in order to fashion a being that does have first-order intentional and sensory states, the addition of an internal monitor or two would be virtually an afterthought, a trifling wrinkle, surely not the sort of thing that could turn a simply nonconscious being into a conscious being.) For that matter, individual subsystems of our own human psychologies doubtless involve their own internal monitors, and it is implausible to grant that those subsystems are themselves conscious.

Several replies may be made to this. First, for consciousness we should require that our monitor emit a genuine representation, not just physical "information" in the Bell-Telephone sense or a simple nomological "indication" in the Wisconsin sense. But that is of little help, since surely our subsystems do contain monitors that output genuine representations.

54

Second, it should trouble no one that s/he has proper parts that are conscious. The proper part of you that consists of you minus your left arm is conscious, as is the part consisting of you minus your skin and most of your musculature. Other (individually) expendable chunks include: your entire gastrointestinal tract, your auditory system, much of your cortex, and possibly much of a hemisphere. Each of your respective complementary proper parts is conscious, even as we speak.

But (it may be said) the second reply is of little more help than the first. For each of the large proper parts I have mentioned would qualify, mentally speaking, as being *you*, if taken on its own. Its consciousness is your consciousness; at least, there is nothing present to its consciousness that is not also present to yours. But the sort of case that worries Rey is one in which self-monitoring is performed by a *silent*, subterranean subsystem, perhaps one of "all those unconscious neurotic systems postulated in so many of us by Freud, ...[or] all those surprisingly intelligent, but still unconscious, subsystems for perception and language postulated in us by contemporary cognitive psychology" (p. 11). What troubles Rey is that he or you or I should contain subsystems that are conscious on their own though we know nothing of them, and whose conscious contents are not at all like ours.

It does sound eerie. But I am not so sure that the individuation of consciousnesses is so straightforward a business. For one thing, that the contents of one consciousness coextend with those of mine hardly entails that the first consciousness *is* (=) mine; they still may be two. For another, the commissurotomy literature has raised well-known thorny questions about the counting of consciousnesses in the first place,[22] and it is abetted in that by thought-experiments such as Dan Dennett's in his classic "Where Am I?" and a more recent one by Stephen White.[23] My own preference is to doubt there to be any fact of the matter, as to how many consciousnesses live in a single human body (or as to how many bodies can be animated by the same consciousness).

A third reply to the argument: In his own essay on Rey's objection,[24] Stephen White enforces a distinction that Rey himself acknowledged but slighted: the difference between consciousness and *self*-consciousness. Rey had argued that if we already had a nonconscious perception-belief-desire machine, the addition of a "self" concept would be trifling (just as would be that of a simple internal monitor); one need only give the machine a first-person representation whose referent was the machine itself, i.e., add the functional analogue of the pronoun 'I' to the machines language-of-thought. But White argues on the basis of an ingenious group-organism example that the matter is hardly so simple, and that the difference between consciousness and self-consciousness is far larger and more important than Rey allowed. Surprisingly, having a functional inner 'I' does not suffice for being able to think of oneself as oneself; nor does mere consciousness as opposed to self-consciousness confer personhood or any moral status. And it turns out on White's analysis that although subsystems of ours might count as conscious, they would not be self-conscious in the way we are.

That difference helps to explain and assuage our reluctance to admit them to our own country club.[25] I find White's defense of these claims quite convincing.[26]

But I do not invest much in these second and third meditations as replies to Rey's objection. I have presented them mainly for the purpose of softening you up.

6. So I turn to my fourth and (*chez* me) most important reply. It is: emphatically to deny (what John Searle has recently asserted with unsurprising boldness[27]) that consciousness is an on-off affair, that a creature is either simply Conscious or simply not conscious. (If Searle did not exist I would have to invent him, for he actually puts it that way: "Consciousness is an on/off switch; a system is either conscious or not" (p. 83).) I maintain that consciousness comes in degrees, which one might describe as degrees of richness or fullness.[28] We human beings are very richly conscious, but there might be more complex and/or more sophisticated organisms that are more fully conscious than we. "Higher animals" are perhaps less fully so; "lower" animals still less, and so forth.

In saying this (you will have noticed), I am shifting my sense of 'conscious' slightly. For there is not obviously any great spectrum of degrees of: whether something has an internal monitor scanning some of its psychological states. (Actually there probably is a *significant* spectrum, based on the extent to which monitor output contributes to integration of information and to control; as was conceded at the time, I did leave the formulation vague. But I will not rest anything on this.) The paronymy works as follows. A thing is conscious, at all, if it is conscious to any degree at all, i.e., if it has at least one internal monitor operating and contributing etc. etc.; we might call this "bare" or "mere" consciousness. The thing may be *more richly or more fully* conscious if it has more monitors, monitors more, integrates more, integrates better, integrates more efficiently for control purposes, and/or whatever.

Actually I have not yet achieved paronymy, for I have located the degrees in the modifers ('richly' and 'fully') rather than in the basic term 'conscious' itself, which so far retains its original sense. But I do still mean to shift its meaning, for I want to allow at least a very vague sense in which some "barely" conscious devices are not really conscious; I take that one to be the ordinary sense of the word. But I would insist that that sense still affords a largeish spectrum of degrees. (Granted, this needs defense, and I shall provide some shortly.)

My principal answer to Rey is, then, to deny his intuition: So long as it contributes in the way aforementioned, one little monitor does make for a little bit of consciousness. More monitors and better integration and control make for fuller and fuller consciousness.[29]

Rey conjectures (p. 24), as a diagnosis of his own chauvinist intuitions about machines, that *if* consciousness is anything, it is like an "inner light" that is on in us but could be off in or missing from other creatures that were otherwise first-order-psychologically and functionally very like us; that is why he finds it so obvious that machines are not conscious even when they have been hypothetically given a perception-belief-desire system like ours. (Naturally given his conditional assumption, he asks why we should believe that *we* are not just very complicated perception-belief-desire machines, and offers the eliminative suggestion that we are therefore not conscious either; consciousness *is not* anything.[30]) But I see no reason to grant the conditional conjecture. I have no problem saying that a device whose internal monitor is contributing integration-and-control-wise is conscious of the states reported by the monitor. There is a rhetorical difference between saying that a device is conscious *of* such-and-such and saying that it, itself, is...conscious! But, I contend, that is *only* a rhetorical difference, barring my slight paronym above. What is special about us is not our being conscious *per se*, but that we monitor so much at any given time and achieve so high a degree of integration and control.

Thus two remarks made by psychologists and quoted by Rey as "astonishing" him by their naïveté do not astonish me in the slightest:

> Perceptions, memories, anticipatory organization, a combination of these factors into learning—all imply rudimentary consciousness. (Peter H. Knapp)[31]

Depending on what Knapp meant by "anticipatory organization," this is not far wrong. If anticipatory organization implies internal monitoring that contributes, or if the "combination of...[the] factors into learning" involves such monitoring, or both, I endorse the statement.

> Consciousness is a process in which information about multiple individual modalities of sensation and perception are combined into a unified, multi-dimensional representation of the state of the system and its environment and is integrated with information about memories and the needs of the organism, generating emotional reactions and programs of behavior to adjust the organism to its environment. (E. Roy John)[32]

No quarrel there either, assuming again that the "combining" is done in part by contributory monitoring.

The main *obstacle* to agreement with my matter-of-degree thesis is that we ourselves know only one sort of consciousness from the inside, and that one is particularly rich and full. We have elaborate and remarkably non-gappy visual models of our environment; we have our other four main sense modalities, which supplement the blooming, bursting phenomenological garden already furnished by vision; we have proprioception of various sorts that orient us within our surroundings; and (most importantly) we have almost complete freedom of attention within our private worlds, i.e., we can at will attend to virtually any representational aspect of any of our sensations that we choose. (All this creates the Cartesian illusion of a complete private world of sensation and thought, a seamless movie theater. There is no such completeness even phenomenologically, what with failings like the blind spot and the rapid decay of peripheral vision, but the illusion is dramatic.) Now, since this is the only sort of consciousness we have ever known from the inside, and since the only way to *imagine* a consciousness is to imagine it from the inside, we cannot imagine a consciousness very different at all from our own, much less a greatly impoverished one. What we succeed in imagining, if we try to get inside the mind of a spider or a notebook computer, is either an implausible cartoon (with anthropomorphic talk balloons) or something that hardly seems to us to deserve the title, "consciousness." It is a predicament: we are not well placed to receive the idea that there can be very low degrees of consciousness.[33]

7. But now, finally, for a bit of argument. (1) consider the total mental states of people who are very ill, or badly injured, or suffering the effects of this or that nefarious drug. Some such people are at some times called "semiconscious." Any number of altered states are possible, many of them severely diminished mental conditions. For some of these, surely, there will be no clear Searlean "Yes" or "no" to the question, "Is the patient conscious?," but only a "To a degree" or "Sort of." (2) We could imagine thousands of hypothetical artifacts, falling along a multidimensional spectrum having at its low end ordinary hardware-store items like record-changers and air conditioners and at its high end biologic human duplicates (indistinguishable from real living human beings save by their histories).[34] Along the way(s) will be robots of many different sorts, having wildly different combinations of abilities and stupidities, oddly skewed and weighted psychologies of all kinds. Which are "conscious"? How could one possibly draw a single line separating the whole seething profusion of creatures into just two groups?

(3) For that matter, the real world provides a similar argument (for those who favor the real world over science fiction); consider the phylogenetic scale. Nature actually contains a fairly smooth continuum of organisms, ranked roughly by complexity and degree of internal monitoring, integration and efficient control. Where on this continuum would God tell us that Consciousness begins? (Appropriately enough, Searle himself declares deep ignorance regarding consciousness and the phylogenetic scale.[35]) (4) If (3) does not move you (or even if (3) does), consider *human infants* as they develop from embryo to fetus to neonate to baby to child. When in that sequence does Consciousness begin?

I do not say that any of these arguments is overwhelming. But taken together—and together with recognition of the imaginative predicament I mentioned prior to offering them—I believe they create a presumption. At the very least, they open the door to my matter-of-degree view and make it a contender. Therefore, one cannot simply assume that consciousness (if any) is an on-off switch. And Rey's argument does assume that.

Thus I do not think Rey has refuted the Inner Sense view.[36]

Notes

1.   Locke, *As Essay concerning Human Understanding*, ed. A.C. Fraser (New York:

Dover Publications, 1959), Book II, Ch. I, sec. 3, p. 123; Kant, *Critique of Pure Reason*, tr. Norman Kemp Smith (New York: St. Martin's Press, 1965), A23/B37, p. 67.

2. See my "What is 'The' Problem of Consciousness?," MS.

3. D.M. Armstrong, *A Materialist Theory of the Mind* (London: Routledge and Kegan Paul, 1968), and "What Is Consciousness?," in *The Nature of Mind and Other Essays* (Ithaca, NY: Cornell University Press, 1980). Baars, "Conscious Contents Provide the Nervous System with Coherent, Global Information," in R. Davidson, G.E. Schwartz and D. Shapiro (eds.), *Consciousness and Self-Regulation, Vol. 3* (New York: Plenum Press, 1983), pp.41-79; *A Cognitive Theory of Consciousness*. Lycan, *Consciousness* (Cambridge, MA: Bradford Books / MIT Press, 1987), Ch. 6.

4. "A Reason for Doubting the Existence of Consciousness," in Davidson, Schwartz and Shapiro, *op. cit.*, pp. 1-39.

    In a sequel to the present paper, I shall also address criticisms made by Christopher Hill, David Rosenthal, Fred Dretske, and others.

5. "What is Consciousness?," *loc. cit.*, p. 61.

6. There is an potential ambiguity in Armstrong's term, "introspective consciousness": Assuming there are attention mechanisms of the sort I have in mind, they may function automatically, on their own, or they may be deliberately mobilized by their owners. Perhaps only in the latter case should we speak of introspec*ting*. On this usage, "introspective" consciousness may or may not be a result of introspecting. Armstrong himself makes a similar distinction between "reflex" introspective awareness and "introspection proper," adding the suggestion that "the latter will normally involve not only introspective awareness of mental states but also introspective awareness of that introspective awareness" ("What is Consciousness?," *loc. cit.*, p. 63).

7. "What is Consciousness?," loc. cit., pp. 65-66. Robert Van Gulick has also written illuminatingly on the uses of consciousness, though he does not focus so specifically on introspection; see particularly "What Difference Does Consciousness Make?,". *Philosophical Topics* 17 (1989): 211-30.

8. I began this project with respect to subjectivity and qualia respectively in Chs. 7 and 8 of *Consciousness, loc. cit.* Parts of it have also been pursued by Gilbert Harman ("The Intrinsic Quality of Experience," in J.E. Tomberlin (ed.), *Philosophical Perspectives, 4, Action Theory and Philosophy of Mind, 1990* (Atascadero, CA: Ridgeview Publishing, 1990)), Michael Tye ("Qualia, Content, and the Inverted Spectrum," *Noûs*, forthcoming (1994)), and Sydney Shoemaker ("Phenomenal Character," *Noûs*, forthcoming (1994)).

9. In *Consciousness, loc. cit.*; see also my "Functionalism and Recent Spectrum Inversions," unpublished MS, and "True Colors," in preparation.

10. When I made this point emphatically after a presentation of this material at the NEH Summer Institute on "The Nature of Meaning" (Rutgers University, July, 1993), Bill Ramsey responded much as follows: "I see; once you've got the explanandum whittled all the way down, as specific and narrow as you want it, the big news you're bringing us is that what *internal monitoring* really is, at bottom, is...internal monitoring!" That characterization is not *far* wrong. Though the Inner Sense doctrine is not tautologous and faces some objections, I think it is very plausible, once it has been relieved of the extraneous theoretical burden of resolving issues that

are not directly related to the "conscious"/"nonconscious" distinction *per se*.

Incidentally, I do not offhand know of any Inner Sense proponent who does claim that the theory resolves qualia problems. Yet there is a tendency among its critics to criticize it from that quarter; I conjecture that such critics are themselves confusing issues of awareness with issues of qualitative character.

11. Rey, "Sensations in a Language of Thought," in E. Villanueva (ed.), *Consciousness (Philosophical Issues, 1, 1991)* (Atascadero, CA: Ridgeview Publishing, 1991), and "Sensational Sentences," in M Davies and G. Humphreys (eds.), *Consciousness* (Oxford: Basil Blackwell, 1992); Lycan, "What is the 'Subjectivity' of the Mental?," in J.E. Tomberlin, *op. cit.*

12. From a current trash novel: "Each step was painful, but the pain was not felt. He moved at a controlled jog down the escalators and out of the building." (John Grisham, *The Firm* (New York: Island Books, Dell Publishing, 1991), p. 443.

    David Rosenthal offers a nice defense of unfelt pain, in "The Independence of Consciousness and Sensory Quality," in Villanueva, *op. cit.* See also David Palmer's "Unfelt Pains," *American Philosophical Quarterly* 12 (1975): 289-98.

13. In "Conscious Experience" (*Mind* 102 (1993): 263-83), he has also made a couple of substantive objections to Inner Sense theory. I shall address those in the sequel to this paper.

14. *Consciousness Explained* (Boston, MA: Little, Brown & Co., 1991); D.C. Dennett and M. Kinsbourne, "Time and the Observer: The Where and When of Consciousness in the Brain," *Behavioral and Brain Sciences* 15 (1992): 183-201.

15. P. Kolers and M. von Grünau, "Shape and Color in Apparent Motion," *Vision Research* 16 (1976): 329-35; F.A. Geldard and C.E. Sherrick, "The Cutaneous 'Rabbit': A Perceptual Illusion," *Science* 178 (1972): 178-79; B. Libet, "Cortical Activation in Conscious and Unconscious Experience," *Perspectives in Biology and Medicine* 9: 77-86.

16. E.g., B.J. Baars and M. Fehling, "Consciousness is Associated with Central As Well As Distributed Processes," *Behavioral and Brain Sciences* 15 (1992): 203-04, and B. Libet, "Models of Conscious Timing and the Experimental Evidence," *ibid.*: 213-15. Dennett and Kinsbourne reply to their critics in "Authors' Response," *ibid.*: 234-43.

17. *Consciousness Explained, loc. cit.*, pp. 218, 225.

18. For convenience, I shall continue to speak of the states that get monitored as "first-order" states, but this is inaccurate, for introspective states can themselves be scanned. This will be important later on.

19. *Consciousness, loc. cit.*, p. 72.

20. On such distinctions, and for more illuminating examples, see Chs. 3 and 4 of *Consciousness, loc. cit.*

21. One might be tempted to infer (something highly congenial to Dennett himself) that introspection is *woefully* fallible, unreliable to the point of uselessness. But that inference would be unjustified. Though the "temporal anomalies" alone should have made us question the reliability of introspective reports, notice that the scope of unreliability exhibited by the anomalies is very small, tied to temporal differences within the tiny intervals involved, a small fraction of a second in each case.

22. For a survey and discussion, see C. Marks, *Commissurotomy, Consciousness and the Unity of Mind* (Montgomery, VT: Bradford Books, 1979).

23. Dennett, "Where Am I?," in *Brainstorms* (Montgomery, VT: Bradford Books, 1978),

reprinted in D.R. Hofstadter and D.C. Dennett (eds.), *The Mind's I: Fantasies and Reflections of Self and Soul* (New York: Basic Books, 1981); see also D.H. Sanford, "Where Was I?," in Hofstadter and Dennett, *op. cit.* White, "What Is It Like to Be a Homunculus?" *Pacific Philosophical Quarterly* 68 (1987): 148-74.

24. "What Is It Like to Be a Homunculus?," *loc. cit.*

25. Moreover, as he observes (p. 168), we have no access to unproblematic examples of consciousness in the absence of self-consciousness, and that fact contributes to an important predicament that I shall expound below.

26. In like wise, he maintains, no notebook computer is self-conscious even if some are conscious in a less demanding functional sense. (I believe White would accept my claim that mere consciousness is more prevalent than philosophers think; see p. 169.) But I do not see that his analysis of self-consciousness generates that result, since his main concern was to argue only that self-consciousness is restricted to the highest level of organization in a *group organism*, which result does not help deny self-consciousness to whole computers. (White has explained in conversation that his analysis alone was not intended to do that; he has other means.)

27. *The Rediscovery of the Mind* (Cambridge, MA: MIT Press, 1992).

28. I have defended this thesis before, in "Abortion and the Civil Rights of Machines," in N. Potter and M. Timmons (eds.), *Morality and Universality* (Dordrecht: D. Reidel, 1985), pp. 144-145.

    It should be noted that Searle himself goes on (*ibid.*) to qualify his "on/off" claim: "But once conscious, the system is a rheostat: there are different degrees of consciousness"; he speaks of levels of intensity and vividness. Thus, it seems, our real disagreement is over, not degrees *per se*, but the question of whether a creature or device could have a much lower degree of consciousness than is ordinarily enjoyed by human beings and still qualify as being conscious at all.

29. I should emphasize again that a monitor makes for consciousness when what it monitors is itself a psychological state or event. My suggestion that notebook computers are after all conscious is conditional upon the highly controversial assumption that such computers have psychological states such as beliefs and desires in the first place.

30. By way of further diagnosis (p. 25), Rey offers the additional conjecture that our moral concern for our living, breathing conspecifics drives us to posit some solid metaphysical difference between ourselves and mere artifacts, as a ground of that concern. He opines that we need no such ground in order to care more for human beings than for functionally similar machines, but he does not say what he thinks *would* ground that difference in care.

31. "The Mysterious 'Split': A Clinical Inquiry into Problems of Consciousness and Brain," in G. Globus, G. Maxwell and I Savodnik (eds.), *Consciousness and the Brain* (New York: Plenum Press, 1976), pp. 37-69.

32. "A Model of Consciousness," in G.E. Schwartz and D. Shapiro (eds.), *Consciousness and Self-Regulation*, Vol. 1 (New York: Plenum Press, 1976), p. 1-50.

33. Samuel Butler said, "Even the potato, rotting in its dank cellar, has a certain low cunning." But I grant the potato has no internal monitors.

34. This is the one argument I gave in "Abortion and the Civil Rights of Machines," loc. cit.

35. "I have no idea whether fleas, grasshoppers, crabs, or snails are conscious" (p. 74). He suggests that neurophysiologists might find out, by a method of apparent-

consciousness-debunking, viz., looking for evidence of "mechanical-like tropism to account for apparently goal-directed behavior in organisms that lacked consciousness" (p. 75); he pooh-poohs "mechanical-like" functional processing as being in no way mental or psychological. On this, see D.C. Dennett's review of *The Rediscovery of the Mind, Journal of Philosophy* 90 (1993): 193-205.

36. This essay was presented as the Third *Philosophical Perspectives* Lecture at California State University, Northridge, in the fall of 1994. I am grateful to James Tomberlin and his colleagues for that invigorating occasion. I am grateful to Joe Levine, Ned Block and Georges Rey for extensive comments and discussion.

The Problem of Consciousness

*It can now be approached by scientific investigation
of the visual system. The solution will require a close
collaboration among psychologists, neuroscientists and theorists*

by Francis Crick and Christof Koch

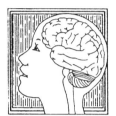

The overwhelming question in neurobiology today is the relation between the mind and the brain. Everyone agrees that what we know as mind is closely related to certain aspects of the behavior of the brain, not to the heart, as Aristotle thought. Its most mysterious aspect is consciousness or awareness, which can take many forms, from the experience of pain to self-consciousness.

In the past the mind (or soul) was often regarded, as it was by Descartes, as something immaterial, separate from the brain but interacting with it in some way. A few neuroscientists, such as Sir John Eccles, still assert that the soul is distinct from the body. But most neuroscientists now believe that all aspects of mind, including its most puzzling attribute—consciousness or awareness—are likely to be explainable in a more materialistic way as the behavior of large sets of interacting neurons. As William James, the father of American psychology, said a century ago, consciousness is not a thing but a process.

Exactly what the process is, however, has yet to be discovered. For many years after James penned *The Principles of Psychology*, consciousness was a taboo concept in American psychology because of the dominance of the behaviorist movement. With the advent of cognitive science in the mid-1950s, it became possible once more for psychologists to consider mental processes as opposed to merely observing behavior. In spite of these changes, until recently most cognitive scientists ignored consciousness, as did almost all neuroscientists. The problem was felt to be either purely "philosophical" or too elusive to study experimentally. It would not have been easy for a neuroscientist to get a grant just to study consciousness.

In our opinion, such timidity is ridiculous, so a few years ago we began to think about how best to attack the problem scientifically. How to explain mental events as being caused by the firing of large sets of neurons? Although there are those who believe such an approach is hopeless, we feel it is not productive to worry too much over aspects of the problem that cannot be solved scientifically or, more precisely, cannot be solved solely by using existing scientific ideas. Radically new concepts may indeed be needed—recall the

modifications of scientific thinking forced on us by quantum mechanics. The only sensible approach is to press the experimental attack until we are confronted with dilemmas that call for new ways of thinking.

There are many possible approaches to the problem of consciousness. Some psychologists feel that any satisfactory theory should try to explain as many aspects of consciousness as possible, including emotion, imagination, dreams, mystical experiences and so on.

Although such an all-embracing theory will be necessary in the long run, we thought it wiser to begin with the particular aspect of consciousness that is likely to yield most easily. What this aspect may be is a matter of personal judgment. We selected the mammalian visual system because humans are very visual animals and because so much experimental and theoretical work has already been done on it [see "The Visual Image in Mind and Brain," by Semir Zeki, page 68].

It is not easy to grasp exactly what we need to explain, and it will take many careful experiments before visual consciousness can be described scientifically. We did not attempt to define consciousness itself because of the dangers of premature definition. (If this seems like a cop-out, try defining the word "gene"—you will not find it easy.) Yet the experimental evidence that already exists provides enough of a glimpse of the nature of visual consciousness to guide research. In this article, we will attempt to show how this evidence opens the way to attack this profound and intriguing problem.

Visual theorists agree that the problem of visual consciousness is ill posed. The mathematical term "ill posed" means that additional constraints are needed to solve the problem. Although the main function of the visual system is to perceive objects and events in the world around us, the in-

VISUAL AWARENESS primarily involves seeing what is directly in front of you, but it can be influenced by a three-dimensional representation of the object in view retained by the brain. If you see the back of a person's head, the brain infers that there is a face on the front of it. We know this is true because we would be very startled if a mirror revealed that the front was exactly like the back, as in this painting, *Reproduction Prohibited* (1937), by René Magritte.

FRANCIS CRICK and CHRISTOF KOCH share an interest in the experimental study of consciousness. Crick is the co-discoverer, with James Watson, of the double helical structure of DNA. While at the Medical Research Council Laboratory of Molecular Biology in Cambridge, he worked on the genetic code and on developmental biology. Since 1976, he has been at the Salk Institute for Biological Studies in San Diego. His main interest lies in understanding the visual system of mammals. Koch was awarded his Ph.D. in biophysics by the University of Tübingen. After spending four years at the Massachusetts Institute of Technology, he joined the California Institute of Technology, where he is now associate professor of computation and neural systems. He is studying how single brain cells process information and the neural basis of motion perception, visual attention and awareness. He also designs analog VLSI vision chips for intelligent systems.

formation available to our eyes is not sufficient by itself to provide the brain with its unique interpretation of the visual world. The brain must use past experience (either its own or that of our distant ancestors, which is embedded in our genes) to help interpret the information coming into our eyes. An example would be the derivation of the three-dimensional representation of the world from the two-dimensional signals falling onto the retinas of our two eyes or even onto one of them.

Visual theorists also would agree that seeing is a constructive process, one in which the brain has to carry out complex activities (sometimes called computations) in order to decide which interpretation to adopt of the ambiguous visual input. "Computation" implies that the brain acts to form a symbolic representation of the visual world, with a mapping (in the mathematical sense) of certain aspects of that world onto elements in the brain.

Ray Jackendoff of Brandeis University postulates, as do most cognitive scientists, that the computations carried out by the brain are largely unconscious and that what we become aware of is the result of these computations. But while the customary view is that this awareness occurs at the highest levels of the computational system, Jackendoff has proposed an intermediate-level theory of consciousness.

What we see, Jackendoff suggests, relates to a representation of surfaces that are directly visible to us, together with their outline, orientation, color, texture and movement. (This idea has similarities to what the late David C. Marr of the Massachusetts Institute of Technology called a "2 $^1/_2$-dimensional sketch." It is more than a two-dimensional sketch because it conveys the orientation of the visible surfaces. It is less than three-dimensional because depth information is not explicitly represented.) In the next stage this sketch is processed by the brain to produce a three-dimensional representation. Jackendoff argues that we are not visually aware of this three-dimensional representation.

An example may make this process clearer. If you look at a person whose back is turned to you, you can see the back of the head but not the face. Nevertheless, your brain infers that the person has a face. We can deduce as much because if that person turned around and had no face, you would be very surprised.

The viewer-centered representation that corresponds to the visible back of the head is what you are vividly aware of. What your brain infers about the front would come from some kind of three-dimensional representation. This does not mean that information flows only from the surface representation to the three-dimensional one; it almost certainly flows in both directions. When you imagine the front of the face, what you are aware of is a surface representation generated by information from the three-dimensional model.

It is important to distinguish between an explicit and an implicit representation. An explicit representation is something that is symbolized without further processing. An implicit representation contains the same information but requires further processing to make it explicit. The pattern of colored dots on a television screen, for example, contains an implicit representation of objects (say, a person's face), but only the dots and their locations are explicit. When you see a face on the screen, there must be neurons in your brain whose firing, in some sense, symbolizes that face.

We call this pattern of firing neurons an active representation. A latent representation of a face must also be stored in the brain, probably as a special pattern of synaptic connections between neurons [see "How Neural Networks Learn from Experience," by Geoffrey E. Hinton, page 144]. For example, you probably have a representation of the Statue of Liberty in your brain, a representation that usually is inactive. If you do think about the Statue, the representation becomes active, with the relevant neurons firing away.

An object, incidentally, may be represented in more than one way—as a visual image, as a set of words and their related sounds, or even as a touch or a smell. These different representations are likely to interact with one another. The representation is likely to be distributed over many neurons, both locally, as discussed in Geoffrey E. Hinton's article, and more globally. Such a representation may not be as simple and straightforward as uncritical introspection might indicate. There is suggestive evidence, partly from studying how neurons fire in various parts of a monkey's brain and partly from examining the effects of certain types of brain damage in humans, that different aspects of a face—and of the implications of a face—may be represented in different parts of the brain.

First, there is the representation of a face as a face: two eyes, a nose, a mouth and so on. The neurons involved are usually not too fussy about the exact size or position of this face in the visual field, nor are they very sensitive to small changes in its orientation. In monkeys,

there are neurons that respond best when the face is turning in a particular direction, while others seem to be more concerned with the direction in which the eyes are gazing.

Then there are representations of the parts of a face, as separate from those for the face as a whole. Further, the implications of seeing a face, such as that person's sex, the facial expression, the familiarity or unfamiliarity of the face, and in particular whose face it is, may each be correlated with neurons firing in other places.

What we are aware of at any moment, in one sense or another, is not a simple matter. We have suggested that there may be a very transient form of fleeting awareness that represents only rather simple features and does not require an attentional mechanism. From this brief awareness the brain constructs a viewer-centered representation—what we see vividly and clearly—that does require attention. This in turn probably leads to three-dimensional object representations and thence to more cognitive ones.

Representations corresponding to vivid consciousness are likely to have special properties. William James thought that consciousness involved both attention and short-term memory. Most psychologists today would agree with this view. Jackendoff writes that consciousness is "enriched" by attention, implying that while attention may not be essential for certain limited types of consciousness, it is necessary for full consciousness.

Yet it is not clear exactly which forms of memory are involved. Is long-term memory needed? Some forms of acquired knowledge are so embedded in the machinery of neural processing that they are almost certainly used in becoming aware of something. On the other hand, there is evidence from studies of brain-damaged patients (such as H.M., described in "The Biological Basis of Learning and Individuality," by Eric R. Kandel and Robert D. Hawkins, page 78) that the ability to lay down new long-term episodic memories is not essential for consciousness.

It is difficult to imagine that anyone could be conscious if he or she had no memory whatsoever of what had just happened, even an extremely short one. Visual psychologists talk of iconic memory, which lasts for a fraction of a second, and working memory (such as that used to remember a new telephone number) that lasts for only a few seconds unless it is rehearsed. It is not clear whether both of these are essential for consciousness. In any case, the division of short-term memory into these two categories may be too crude.

If these complex processes of visual awareness are localized in parts of the brain, which processes are likely to be where? Many regions of the brain may be involved, but it is almost certain that the cerebral neocortex plays a dominant role. Visual information from the retina reaches the neocortex mainly by way of a part of the thalamus (the lateral geniculate nucleus); another significant visual pathway from the retina is to the superior colliculus, at the top of the brain stem.

The cortex in humans consists of two intricately folded sheets of nerve tissue, one on each side of the head. These sheets are connected by a large tract of about half a billion axons called the corpus callosum. It is well known that if the corpus callosum is cut, as is done for certain cases of intractable epilepsy, one side of the brain is not aware of what the other side is seeing.

In particular, the left side of the brain (in a right-handed person) appears not to be aware of visual information received exclusively by the right side. This shows that none of the information required for visual awareness can reach the other side of the brain by traveling down to the brain stem and, from there, back up. In a normal person, such information can get to the other side only by using the axons in the corpus callosum.

A different part of the brain—the hippocampal system—is involved in one-shot, or episodic, memories that, over weeks and months, it passes on to the neocortex, as described in the article by Eric R. Kandel and Robert D. Hawkins. This system is so placed that it receives inputs from, and projects to, many parts of the brain.

Thus, one might suspect that the hippocampal system is the essential seat of consciousness. This is not the case: evidence from studies of patients with damaged brains shows that this system is not essential for visual awareness, although naturally a patient lacking one, such as H.M., is severely handicapped in everyday life because he cannot remember anything that took place more than a minute or so in the past.

In broad terms, the neocortex of alert animals probably acts in two ways. By building on crude and somewhat redundant wiring, produced by our genes and

by embryonic processes [see "The Developing Brain," by Carla J. Shatz, page 60], the neocortex draws on visual and other experience to slowly "rewire" itself to create categories (or "features") it can respond to. A new category is not fully created in the neocortex after exposure to only one example of it, although some small modifications of the neural connections may be made.

The second function of the neocortex (at least of the visual part of it) is to respond extremely rapidly to incoming signals. To do so, it uses the categories it has learned and tries to find the combinations of active neurons that, on the

WILLIAM JAMES, the father of American psychology, observed that consciousness is not a thing but a process.

basis of its past experience, are most likely to represent the relevant objects and events in the visual world at that moment. The formation of such coalitions of active neurons may also be influenced by biases coming from other parts of the brain: for example, signals telling it what best to attend to or high-level expectations about the nature of the stimulus.

Consciousness, as James noted, is always changing. These rapidly formed coalitions occur at different levels and interact to form even broader coalitions. They are transient, lasting usually for only a fraction of a second. Because coalitions in the visual system are the basis of what we see, evolution

has seen to it that they form as fast as possible; otherwise, no animal could survive. The brain is handicapped in forming neuronal coalitions rapidly because, by computer standards, neurons act very slowly. The brain compensates for this relative slowness partly by using very many neurons, simultaneously and in parallel, and partly by arranging the system in a roughly hierarchical manner.

If visual awareness at any moment corresponds to sets of neurons firing, then the obvious question is: Where are these neurons located in the brain, and in what way are they firing? Visual awareness is highly unlikely to occupy all the neurons in the neocortex that happen to be firing above their background rate at a particular moment. We would expect that, theoretically, at least some of these neurons would be involved in doing computations—trying to arrive at the best coalitions—while others would express the results of these computations, in other words, what we see.

Fortunately, some experimental evidence can be found to back up this theoretical conclusion. A phenomenon called binocular rivalry may help identify the neurons whose firing symbolizes awareness. This phenomenon can be seen in dramatic form in an exhibit prepared by Sally Duensing and Bob Miller at the Exploratorium in San Francisco.

Binocular rivalry occurs when each eye has a different visual input relating to the same part of the visual field. The early visual system on the left side of the brain receives an input from both eyes but sees only the part of the visual field to the right of the fixation point. The converse is true for the right side. If these two conflicting inputs are rivalrous, one sees not the two inputs superimposed but first one input, then the other, and so on in alternation.

In the exhibit, called "The Cheshire Cat," viewers put their heads in a fixed place and are told to keep the gaze fixed. By means of a suitably placed mirror [see box on next two pages], one of the eyes can look at another person's face, directly in front, while the other eye sees a blank white screen to the side. If the viewer waves a hand in front of this plain screen at the same location

65

in his or her visual field occupied by the face, the face is wiped out. The movement of the hand, being visually very salient, has captured the brain's attention. Without attention the face cannot be seen. If the viewer moves the eyes, the face reappears.

In some cases, only part of the face disappears. Sometimes, for example, one eye, or both eyes, will remain. If the viewer looks at the smile on the person's face, the face may disappear, leaving only the smile. For this reason, the effect has been called the Cheshire Cat effect, after the cat in Lewis Carroll's *Alice's Adventures in Wonderland.*

Although it is very difficult to record activity in individual neurons in a human brain, such studies can be done in monkeys. A simple example of binocular rivalry has been studied in a monkey by Nikos K. Logothetis and Jeffrey D. Schall, both then at M.I.T. They trained a macaque to keep its eyes still and to signal whether it is seeing upward or downward movement of a horizontal grating. To produce rivalry, upward movement is projected into one of the monkey's eyes and downward movement into the other, so that the two images overlap in the visual field. The monkey signals that it sees up and down movements alternatively, just as humans would. Even though the motion stimulus coming into the monkey's eyes is always the same, the monkey's percept changes every second or so.

Cortical area MT (which Semir Zeki calls in his article V5) is an area mainly concerned with movement. What do the neurons in MT do when the monkey's percept is sometimes up and sometimes down? (The researchers studied only the monkey's first response.) The simplified answer—the actual data are rather more messy—is that whereas the firing of some of the neurons correlates with the changes in the percept, for others the average firing rate is relatively unchanged and independent of which direction of movement the monkey is seeing at that moment. Thus, it is unlikely that the firing of all the neurons in the visual neocortex at one particular moment corresponds to the monkey's visual awareness. Exactly which neurons do correspond remains to be discovered.

We have postulated that when we clearly see something, there must be neurons actively firing that stand for what we see. This might be called the activity principle. Here, too, there is some experimental evidence. One example is the firing of neurons in cortical area V2 in response to illusory contours, as described by Zeki. Another and perhaps more striking case is the filling in of the blind spot. The blind spot in each eye is caused by the lack of photoreceptors in the area of the retina where the optic nerve leaves the retina and projects to the brain. Its location is about 15 degrees from the fovea (the visual center of the eye). Yet if you close one eye, you do not see a hole in your visual field.

Philosopher Daniel C. Dennett of Tufts University is unusual among philosophers in that he is interested both in psychology and in the brain. This interest is much to be welcomed. In a recent book, *Consciousness Explained,* he has argued that it is wrong to talk about filling in. He concludes, correctly, that "an absence of information is not the same as information about an absence." From this general principle he argues that the brain does not fill in the blind spot but rather ignores it.

Dennett's argument by itself, however, does not establish that filling in does not occur; it only suggests that it might not. Dennett also states that "your brain has no machinery for [filling in] at this location." This statement is incorrect. The primary visual cortex (V1) lacks a direct input from one eye, but normal "machinery" is there to deal with the input from the other eye.

Ricardo Gattass and his colleagues at the Federal University of Rio de Janeiro have shown that in the macaque some of the neurons in the blind-spot area of V1 do respond to input from both eyes, probably assisted by inputs from other parts of the cortex. Moreover, in the case of simple filling in, some of the neurons in that region respond as if they were actively filling in.

Thus, Dennett's claim about blind spots is incorrect. In addition, psychological experiments by Vilayanur S. Ramachandran [see "Blind Spots," SCIENTIFIC AMERICAN, May] have shown that what is filled in can be quite complex depending on the overall context of the visual scene. How, he argues, can your brain be ignoring something that is in fact commanding attention?

Filling in, therefore, is not to be dismissed as nonexistent or unusual. It probably represents a basic interpolation process that can occur at many levels in the neocortex. It is, incidentally, a good example of what is meant by a constructive process.

How can we discover the neurons whose firing symbolizes a particular percept? William T. Newsome and his colleagues at Stanford University have done a series of brilliant experiments on neurons in cortical area MT of the macaque's brain. By studying a neuron

The Cheshire Cat Experiment

This simple experiment with a mirror illustrates one aspect of visual awareness. It relies on a phenomenon called binocular rivalry, which occurs when each eye has a different input from the same part of the visual field. Motion in the field of one eye can cause either the entire image or parts of the image to be erased. The movement captures the brain's attention.

in area MT, we may discover that it responds best to very specific visual features having to do with motion. A neuron, for instance, might fire strongly in response to the movement of a bar in a particular place in the visual field, but only when the bar is oriented at a certain angle, moving in one of the two directions perpendicular to its length within a certain range of speed.

It is technically difficult to excite just a single neuron, but it is known that neurons that respond to roughly the same position, orientation and direction of movement of a bar tend to be located near one another in the cortical sheet. The experimenters taught the monkey a simple task in movement discrimination using a mixture of dots, some moving randomly, the rest all in one direction. They showed that electrical stimulation of a small region in the right place in cortical area MT would bias the monkey's motion discrimination, almost always in the expected direction.

Thus, the stimulation of these neurons can influence the monkey's behavior and probably its visual percept. Such experiments do not, however, show decisively that the firing of such neurons is the exact neural correlate of the percept. The correlate could be only a subset of the neurons being activated. Or perhaps the real correlate is the firing of neurons in another part of the visual hierarchy that are strongly influenced by the neurons activated in area MT.

These same reservations apply also to cases of binocular rivalry. Clearly, the problem of finding the neurons whose firing symbolizes a particular percept is not going to be easy. It will take many careful experiments to track them down even for one kind of percept.

It seems obvious that the purpose of vivid visual awareness is to feed into the cortical areas concerned with the implications of what we see; from there the information shuttles on the one hand to the hippocampal system, to be encoded (temporarily) into long-term episodic memory, and on the other to the planning levels of the motor system. But is it possible to go from a visual input to a behavioral output without any relevant visual awareness?

That such a process can happen is demonstrated by the remarkable class of patients with "blindsight." These patients, all of whom have suffered damage to their visual cortex, can point with fair accuracy at visual targets or track them with their eyes while vigorously denying seeing anything. In fact, these patients are as surprised as their doctors by their abilities. The amount of information that "gets through," however, is limited: blindsight patients have some ability to respond to wavelength, orientation and motion, yet they cannot distinguish a triangle from a square.

It is naturally of great interest to know which neural pathways are being used in these patients. Investigators originally suspected that the pathway ran through the superior colliculus. Recent experiments suggest that a direct albeit weak connection may be involved between the lateral geniculate nucleus and other cortical areas, such as V4. It is unclear whether an intact V1 region is essential for immediate visual awareness. Conceivably the visual signal in blindsight is so weak that the neural activity cannot produce awareness, although it remains strong enough to get through to the motor system.

Normal-seeing people regularly respond to visual signals without being fully aware of them. In automatic actions, such as swimming or driving a car, complex but stereotypical actions occur with little, if any, associated visual awareness. In other cases, the information conveyed is either very limited or very attenuated. Thus, while we can function without visual awareness, our behavior without it is rather restricted.

Clearly, it takes a certain amount of time to experience a conscious percept. It is difficult to determine just how much time is needed for an episode of visual awareness, but one aspect of the problem that can be demonstrated experimentally is that signals received close together in time are treated by the brain as simultaneous.

A disk of red light is flashed for, say, 20 milliseconds, followed immediately by a 20-millisecond flash of green light in the same place. The subject reports that he did not see a red light followed by a green light. Instead he saw a yellow light, just as he would have if the red and the green light had been flashed simultaneously. Yet the subject could not have experienced yellow until after the information from the green flash had been processed and integrated with the preceding red one.

Experiments of this type led psychologist Robert Efron, now at the University of California at Davis, to conclude that the processing period for perception is about 60 to 70 milliseconds. Similar periods are found in experiments with tones in the auditory system. It is always possible, however, that the processing times may be different in higher parts of the visual hierarchy and in other parts of the brain. Processing is also more rapid in trained, compared with naive, observers.

Because it appears to be involved in some forms of visual awareness, it

To observe the effect, a viewer divides the field of vision with a mirror placed between the eyes (*a*). One eye sees the cat; the other eye a reflection in the mirror of a white wall or background. The viewer then waves the hand that corresponds to the eye looking at the mirror so that the hand passes through the area in which the image of the cat appears in the other eye (*b*). The result is that the cat may disappear. Or if the viewer was attentive to a specific feature before the hand was waved, those parts—the eyes or even a mocking smile—may remain (*c*).

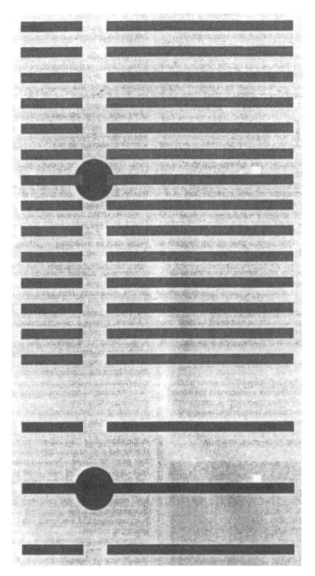

OPTICAL ILLUSION devised by Vilayanur S. Ramachandran illustrates the brain's ability to fill in, or construct, visual information that is missing because it falls on the blind spot of the eye. When you look at the patterns of broken green bars, the visual system produces two illusory contours defining a vertical strip. Now shut your right eye and focus on the white square in the green series of bars. Move the page toward your eye until the blue dot disappears (roughly six inches in front of your nose). Most observers report seeing the vertical strip completed across the blind spot, not the broken line. Try the same experiment with the series of just three red bars. The illusory vertical contours are less well defined, and the visual system tends to fill in the horizontal bar across the blind spot. Thus, the brain fills in differently depending on the overall context of the image.

would help if we could discover the neural basis of attention. Eye movement is a form of attention, since the area of the visual field in which we see with high resolution is remarkably small, roughly the area of the thumbnail at arm's length. Thus, we move our eyes to gaze directly at an object in order to see it more clearly. Our eyes usually move three or four times a second. Psychologists have shown, however, that there appears to be a faster form of attention that moves around, in some sense, when our eyes are stationary.

The exact psychological nature of this faster attentional mechanism is at present controversial. Several neuroscientists, however, including Robert Desimone and his colleagues at the National Institute of Mental Health, have shown that the rate of firing of certain neurons in the macaque's visual system depends on what the monkey is attending to in the visual field. Thus, attention is not solely a psychological concept; it also has neural correlates that can be observed. Several researchers have found that the pulvinar, a region of the thalamus, appears to be involved in visual attention. We would like to believe that the thalamus deserves to be called "the organ of attention," but this status has yet to be established.

The major problem is to find what activity in the brain corresponds directly to visual awareness. It has been speculated that each cortical area produces awareness of only those visual features that are "columnar," or arranged in the stack or column of neurons perpendicular to the cortical surface. Thus, area V1 could code for orientation and area MT for motion. So far, as Zeki has explained, experimentalists have not found one particular region in the brain where all the information needed for visual awareness appears to come together. Dennett has dubbed such a hypothetical place "The Cartesian Theater." He argues on theoretical grounds that it does not exist.

Awareness seems to be distributed not just on a local scale, as in some of the neural nets described by Hinton, but more widely over the neocortex. Vivid visual awareness is unlikely to be distributed over every cortical area because some areas show no response to visual signals. Awareness might, for example, be associated with only those areas that connect back directly to V1 or alternatively with those areas that project into each other's layer 4. (The latter areas are always at the same level in the visual hierarchy.)

The key issue, then, is how the brain forms its global representations from visual signals. If attention is indeed crucial for visual awareness, the brain could form representations by attending to just one object at a time, rapidly moving from one object to the next. For example, the neurons representing all the different aspects of the attended object could all fire together very rapidly for a short period, possibly in rapid bursts.

This fast, simultaneous firing might not only excite those neurons that symbolized the implications of that object but also temporarily strengthen the relevant synapses so that this particular pattern of firing could be quickly recalled—a form of short-term memory. (If only one representation needs to be held in short-term memory, as in remembering a single task, the neurons involved may continue to fire for a period, as described by Patricia S. Goldman-Rakic in "Working Memory and the Mind," page 110.)

A problem arises if it is necessary to be aware of more than one object at exactly the same time. If all the attributes of two or more objects were represented by neurons firing rapidly, their attributes might be confused. The color of one might become attached to the shape of another. This happens sometimes in very brief presentations.

Some time ago Christoph von der Malsburg, now at the Ruhr-Universität Bochum, suggested that this difficulty would be circumvented if the neurons associated with any one object all fired in synchrony (that is, if their times of firing were correlated) but out of synchrony with those representing other objects. More recently, two groups in Germany reported that there does appear to be correlated firing between neurons in the visual cortex of the cat, often in a rhythmic manner, with a frequency in the 35- to 75-hertz range, sometimes called 40-hertz, or γ, oscillation.

Von der Malsburg's proposal prompted us to suggest that this rhythmic and synchronized firing might be the neural correlate of awareness and that it might serve to bind together activity in different cortical areas concerning the same object. The matter is still undecided, but at present the fragmentary experimental evidence does rather little to support such an idea. Another possibility is that the 40-hertz oscillations may help distinguish figure from ground [see "The Legacy of Gestalt Psychology," by Irvin Rock and Stephen Palmer; SCIENTIFIC AMERICAN, December 1990] or assist the mechanism of attention.

Are there some particular types of neurons, distributed over the visual neocortex, whose firing directly symbolizes the content of visual awareness? One

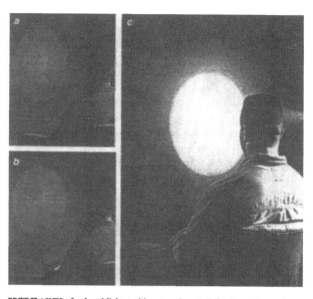

BRIEF FLASHES of colored light enable researchers to infer the minimum time required for visual awareness. A disk of red light is projected for 20 milliseconds (*a*), followed immediately by a 20-millisecond flash of green light (*b*). But the observer reports seeing a single flash of yellow (*c*), the color that would be apparent if red and green were projected simultaneously. The subject does not become aware of red followed by green until the length of the flashes is extended to 60 to 70 milliseconds.

very simplistic hypothesis is that the activities in the upper layers of the cortex are largely unconscious ones, whereas the activities in the lower layers (layers 5 and 6) mostly correlate with consciousness. We have wondered whether the pyramidal neurons in layer 5 of the neocortex, especially the larger ones, might play this latter role.

These are the only cortical neurons that project right out of the cortical system (that is, not to the neocortex, the thalamus or the claustrum). If visual awareness represents the results of neural computations in the cortex, one might expect that what the cortex sends elsewhere would symbolize those results. Moreover, the neurons in layer 5 show a rather unusual propensity to fire in bursts. The idea that the layer 5 neurons may directly symbolize visual awareness is attractive, but it still is too early to tell whether there is anything in it.

Visual awareness is clearly a difficult problem. More work is needed on the psychological and neural basis of both attention and very short term memory. Studying the neurons when a percept changes, even though the visual input is constant, should be a powerful experimental paradigm. We need to construct neurobiological theories of visual awareness and test them using a combination of molecular, neurobiological and clinical imaging studies.

We believe that once we have mastered the secret of this simple form of awareness, we may be close to understanding a central mystery of human life: how the physical events occurring in our brains while we think and act in the world relate to our subjective sensations—that is, how the brain relates to the mind.

FURTHER READING

PERCEPTION. Irvin Rock. Scientific American Library, 1984.
CONSCIOUSNESS AND THE COMPUTATIONAL MIND. Ray Jackendoff. The MIT Press/Bradford Books, 1987.
COLD SPRING HARBOR SYMPOSIA ON QUANTITATIVE BIOLOGY, Vol. LV: THE BRAIN. Cold Spring Harbor Laboratory Press, 1990.
TOWARDS A NEUROBIOLOGICAL THEORY OF CONSCIOUSNESS. Francis Crick and Christof Koch in *Seminars in the Neurosciences*, Vol. 2, pages 263-275; 1990.
THE COMPUTATIONAL BRAIN. Patricia S. Churchland and Terrence J. Sejnowski. The MIT Press/Bradford Books, 1992.

Images and Subjectivity: Neurobiological Trials and Tribulations

Antonio R. Damasio and Hanna Damasio

This chapter is not for or against anything the Churchlands have ever said or written but rather about something that we, the Churchlands and their critics, are passionately interested in: consciousness. More specifically, it is about the possibility of investigating a neurobiology of consciousness.

We shall begin by considering the results of some experiments, in both humans and animals. The first is quite close to the Churchlands. Figure 7.1 shows a section taken through a 3-dimensional reconstruction of Pat Churchland's own and very-much-living brain. The image is based on a fine-resolution magnetic resonance scan, manipulated by Brainvox, a collection of computational techniques that allow the fine neuroanatomical analysis of the brain (H. Damasio and Frank, 1992). The stippled area marks the part of that section which contains Pat's "early" visual cortices. (The early visual cortices are a set of heavily interlocked cortical regions whose cellular architecture and connectivity are distinctive; one cortex within the set, the primary visual cortex, also known as V_1 or area 17, is the main recipient of visual input from the lateral geniculate nucleus.) The white circle marked within the early visual cortices signifies a peak of neural activation detected during a positron emission tomography (PET) experiment.

The activation peak shown on Pat's visual cortex was obtained as follows: In the course of an experiment aimed at understanding how the brain processes stimuli with emotional value, Pat and four other normal persons (not all philosophers, we might add, but all smart, normal people) were asked to visualize, in as much detail as they could muster, a series of familiar places and faces while their eyes were not only closed but covered. That is, they were asked to form mental images when neither their visually-related subcortical structures, such as the lateral geniculate nucleus or the superior colliculus, nor their primary visual cortices could receive any stimuli from the exterior. We will omit the details of the other experimental specifications and of the long and laborious process of data analysis that finally yielded the image you have in the figure (see H. Damasio et al., 1993a; H. Damasio et al., 1993b). We will simply say that the activation peak in the picture is an average for the activation peaks of the five experimental subjects, Pat included, plotted onto Pat's brain.

This striking image is an appropriate excuse for us to ask the following question: What is the neurophysiological meaning of this focal activation, within a specific set of brain structures, during the experience of a

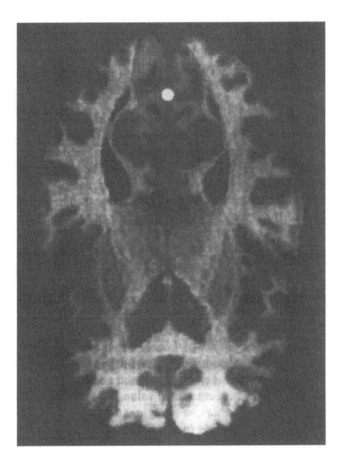

Figure 7.1   Pat Churchland's brain

particular set of images? Does the figure, along with the voluminous numerical data on the basis of which it was generated, mean that the images in the subjects' minds were being generated at this particular brain site? And if the answer is affirmative, were the subjects' "experiences of seeing" also generated at the same site? Or might neither be the case? Might it be that mental images as such cannot be correlated with any specific set of structures and that the experience of an image is even less correlatable with neural structures?

To make a long answer short we will say this. First, we *suspect* that a considerable part of the process of generating visual images was in fact taking place in the early visual cortices, in and about the area where the activity peaks were found. More about this later. Second, we *know* that the experience of those images did not depend on those early visual cortices *only*. With this short answer we encapsulate many of our assumptions, prejudices, hunches, and hypotheses concerning the neurobiology of consciousness. Rightly or wrongly, we believe consciousness can be approached neurobiologically. The effort has already begun, theoretically and practically, in the work of several colleagues. Rightly or wrongly, we believe that neurobiology must deal with two critical issues if it is to make progress in the understanding of consciousness. The first issue is the process whereby our brains create images in our minds. The second issue is the process that makes those images our own, in other words the process that lets images acquire the property of subjectivity. We do not suggest for a moment that the neurobiological solutions to both problems, were they to come, would explain comprehensively the neural basis of consciousness, but they would help elucidate the basis for two of its indispensable aspects. We do not think it is reasonable to discuss consciousness without invoking images *and* subjectivity (although we can conceive of images being formed in a brain that is pathologically deprived of subjectivity and thus deprived of consciousness).

Making Images

As we talk about images, some qualifications are important. First, when we use the term "images" we do not refer to visual images only but rather to images based on any sensory modality, e.g., sound images and images of movement in space.

Second, images convey both nonverbal and verbal entities. Wordforms or sign denotations, in auditory or visual modes, are images.

Third, there are images to describe both the world external to an organism as well as the world within the organism – that is, images of visceral states, images of musculoskeletal structure, body movement, and so forth.

Fourth, images may or may not be attended. When they are not attended we may still discover indirectly that they were formed and displayed (as happens in a priming experiment or when we turn attention to an unattended image and it suddenly is "on," from one moment to the next).

Fifth, images may or may not be held in working memory. When they are, they can be displayed for long periods – many seconds or minutes; this really means that the pattern of neural activity that constitutes them is being reiterated.

Sixth, images are usually not a luxury. In general, they help us optimize our responses. They probably evolved and endured because they allowed organisms to perfect movements and plan even more perfect movements. Humans use images for numerous purposes, from motor behavior to the long-term planning of actions and ideas.

Finally, as was probably made clear by our choice of illustration above, when we talk about images we refer both to images generated in perception and to images generated from memory, during recall.

Let us return to the experiment and discuss the comments we made on its interpretation. Why is it that we believe that a substantial part of image making depends on early cortices, such as those of vision, in the case of the experiment? The reasons are manifold. We cannot discuss all of them, but the main ones are as follows. We know that the partial destruction of early visual cortices precludes both the perception and the recall of certain aspects of vision. After damage in cortical areas V_2 and V_4, for instance, color is neither perceivable nor recallable. The patients so affected have no experience of color, looking *out* or thinking *in*. No consciousness of color is possible even if other aspects of vision can be appreciated and even if you are aware of the absence of the color experience. The fact that perception and recall are compromised by damage at the same site and that no other site of damage encountered so far produces such a defect is one among several findings to suggest that early sensory cortices are the critical base for processes of image making in the brain. Other relevant findings we must mention include the fact that damage to higher-order association cortices, which are located outside the early sensory region, does not preclude the making of images (see Damasio and Damasio, 1993; 1994).

Based on lesion studies in humans, neurophysiological studies in nonhuman primates, and reflection on the patterns of neuroanatomical connectivity currently known, we have hypothesized that the early sensory cortices of each modality construct, with the help of structures such as the thalamus and the colliculi, neural representations that are the basis for images. As far as the cortex is concerned, the process seems to require the concerted activity of several early cortical regions that are massively interconnected among themselves. The precise mechanisms behind this process are not known, but several current proposals suggest the problem is treatable (see Churchland, Sejnowski, and Ramachandran, 1994; Crick, 1994; Tononi, Sporns, and Edelman, 1992; Zeki, 1993). Whatever the mechanisms are, the temporally-coordinated activity of those varied early cortices and of the subcortical stations they are interconnected with yields a particular type of representation we call an "image." An important characteristic of such image representations is that they have spatially and temporally organized patterns. In the case of visual, somatosensory, and auditory images, those patterns are topographically organized (according, we believe, to a space grounded on body structure and body movement). The correspondence between the structure of the neural activity pattern in early sensory cortices and the structure of the stimulus that evoked the pattern can be quite striking, as shown by Roger Tootell and his colleagues (1988).

Topographic representations can be committed to memory in the form of dispositional representations and can be stored in dormant form, in cortical regions or subcortical nuclei. Their subsequent reactivation followed by signaling from their storage sites back to early sensory cortices can regenerate topographically-organized representations (the process uses the rich connectional patterns of feedforward and feedback that characterize the architecture of cortical regions and subcortical nuclei). In short, nontopographically organized representations (which we generally refer to as dispositional representations) can trigger topographically-organized ones; they can also trigger movements or other dispositional representations. Topographic representations arise as a result of signals external to the brain, in the perceptual process; or they arise, in the process of recall, in signals inside the brain, coming from memory records held in dispositional representation form.

Given the fact that most of our experiences are based on images of several sensory modalities occurring within the same window of time, and that the early sensory cortices for each sensory modality are not contiguous or directly interconnected, it follows that our polymodal experiences must result from concurrent activity in several separate brain regions rather than in a single one. The making of images is a spatially parcellated process. But since our experiences are "integrated" rather than parcellated – at least that is how they appear to our mind – we should consider how the integration occurs. Our idea is that timing – that is, a fair amount of synchronization of separate activities – is probably an essential condition for integration. However, synchronization does not happen by magic. We suspect that the neural mechanism behind it is based on signaling from both cortical and subcortical neuron ensembles, capable of simultaneous firing towards *many* separate neuron populations. Such ensembles are a neuroanatomical reality, and we have chosen to call them convergence zones (Damasio, 1989 a,b). They are the receivers of convergent signals and the originators of divergent signals toward the sites in which convergent signals originate. Convergence zones are located throughout the association cortices and subcortical nuclei, including the thalamus, a structure we believe to be critical both for making images and for generating the process of subjectivity, and they contain a storehouse of knowledge in the form of dispositional representations, ready to be activated. Knowledge and timing properties help convergence zones play a critical role in the coherent integration of perceptual images. Naturally, they play an indispensable role in generating images in recall and in making perceived and recalled images cohere.

Subjectivity

We turn now to the issue of subjectivity, one of the basic properties of consciousness. Inevitably we must also take on the issue of self. Without subjectivity we do not know what consciousness means, nor, for that matter, do we know what awareness is. But if our brain/mind is capable of generating states of subjectivity, it is not unreasonable to presume that a crucial agent in the process is something we may call "self," whatever the neurobiological nature of self may turn out to be. We know that the word "self" is ambiguous, even treacherous, but perhaps the word will be less of a problem if we can define what it means for *us*.

Jerome Kagan has keenly described the self as "the universal emergence in the second year of an awareness that one can have an effect upon people and objects, together with a consciousness of one's feelings and competences" (Kagan, 1989). What we have in mind when we talk about self and what we would like to understand in neurobiological terms is *the antecedent and foundation* for the process Kagan describes. It is the neural structure and neurobiological states that help engender the consistent perspective automatically conferred upon images as the brain processes them; it is what allows us to know unequivocally, without the help of inferences based on language, that the images we perceive are our own rather than somebody else's. This is our general idea of self – the core structures and operations necessary but not sufficient for subjectivity to emerge – and it does not include concepts such as "self-esteem" and "social self," although we see it as likely that the processes to which those terms refer develop from the same core.

Postulating that there is a cognitive sense of self and neurobiological structures and operations that support such a sense does not mean, we believe, that all images and image manipulations that take place in a brain are controlled by a single and central "knower" and "master," and even less that such a knower/master would be in a single brain place sitting in judgment as the audience in a Cartesian theater. Our view is thus compatible with the criticisms leveled at single-brain-region views of self or consciousness (Damasio, 1989 a,b; Dennett, 1991; Churchland and Sejnowski, 1992; P. M. Churchland, 1984; P. S. Churchland, 1986). We simply mean that many of the images that constitute our mind are perceived from a *consistent perspective*, which we identify as that of our individual persons, and that experiences have a *consistent point of view*, which is only diminished or suspended in pathological brain states (which include extreme forms of anosognosia, some types of seizure, multiple personality disorder, and schizophrenia).

Our idea of self does not require the pronouns "I" or "me" as anchors. Many nonlinguistic creatures must have a self, in the sense described above, or something like it. Most of the animals commonly used in neurobiological experimentation do, in our view. No single experiment of which we are aware can contradict this statement, although several experiments suggest, predictably, that the complexity of the self is maximal in higher primates and leaps to the highest level in humans. Needless to say, language does enrich the human self even if it does not serve as its base. In fact, it is hard to imagine how language-making devices would have been selected and would have evolved if animals had not possessed prelinguistic "selves."

The appropriately-maligned solution to the problem of the self, that is, the solution to the problem of ownership and inspection of images, is the *homunculus*. It consists of postulating a spatially-defined creature to whom images are referred within the brain and assuming that the creature is equipped with the knowledge necessary to interpret the images. The solution would have been fine were it not for the problem that the thinking homunculus would need to have its own brain and its own knowledge so that, in turn, its images could be interpreted. This "spatial homunculus" solution is no solution at all, since it simply removes us from the problem by one step, whereupon the problem starts again *ad infinitum*. In recent years, recognition of the homunculus pitfall has given way to homunculus phobia, presumably an adaptive mental condition. But because the self has traditionally been conceptualized in homuncular terms, the attempt to avoid homuncular thinking has entailed a denial of anything that sounds like the self and, by extension, subjectivity. This is simply not reasonable. Rejecting the idea of a homunculus in our brains does not alter the fact that most images in our minds are processed from a consistent perspective. To evade the problem by saying that our brains just form images and that we are aware of those images is not a satisfactory solution. The nature of the neural entity that is aware of those images remains unclear.

The task at hand then is to propose a plausible and preferably testable hypothesis for a neural structure of the self-related processes, such that the problems of the spatial homunculus may be avoided. The solution we propose includes the following features and components.

We conceptualize the self as a collection of images about the most invariant aspects of our organism and its interactions – certain aspects of body structure; certain aspects of body operation, including the repertoire of motions possible with the whole body and its varied parts; and identity-defining traits (kinships to certain persons, activities, places; typical motor and sensory patterns of response). These images have a high probability of being evoked repeatedly and continuously by direct signaling (as happens in body states) or by signals arising from stored dispositional representations (as happens with records concerning identity and typical response patterns).

In our framework, the cognitive/neural self is the cognitive/neural instantiation of a *concept*, no different in its essence from the concept of a particular object whose representation relies on the segregated mapping of its properties (such as shape, size, color, texture, characteristic motion, etc.) in varied neural systems from which they can be conjointly retrieved, momentarily, as the concept is activated. But if the essence of the concept of self and of, say, orange, need not be different in basic cognitive and neural specifications, they are different in one important respect. Objects come and go from the vicinity of the organism, and thus they come and go from the sensory sheaths that can signal their presence. Yet the body, its parts, and some of its operations, as well as the stable aspects of our autobiography, remain with us, the former signaling incessantly to the brain, the latter indelibly represented and ready to be reinstantiated.

We propose that the core components of the concept of self concern the body structure (the viscera and musculoskeletal frame) and fundamentals of one's identity (one's usual activities, preferences, physical and human relationships, etc.). These core components change substantially during childhood and adolescence and then change far less, and far more gradually, throughout the remaining life span. The anchor lies with visceral states, which change almost not at all, and with the neural mechanisms that represent and regulate basic biological processes, whose modifiability is minimal. The former are continuously signaled to the complex of somatosensory cortices in the insula, parietal operculum, and post-rolandic parietal cortices (the signaling is bilateral but there is a right hemisphere dominance effect in humans). The biological regulating machinery is represented in the brain core (hypothalamus and brain stem).

Skeptics may counter that we are usually unaware of our body states and that body signaling is thus an odd choice to anchor subjectivity. As argued elsewhere, however, the objection is weak (A. Damasio, 1994). Firstly, although our attention is more often than not centered in nonbodily signals, its focus may shift rapidly, especially in conditions such as pain or emotional upheaval. Secondly, the argument we are making is especially concerned with the historical development of the sense of self, in evolutionary and individual development terms, rather than with the situation of an adult. Thirdly, since we all agree that the mechanisms behind the emergence of subjectivity are hidden, there is no reason why the body states we propose as their scaffolding should be easily revealed in consciousness. The important issue to decide is whether the mechanisms we propose are a plausible base, not whether we are or should be aware of them.

As images corresponding to a newly perceived entity (e.g., a face) are formed in early sensory cortices, *the brain reacts to those images*. In our framework this happens

1   because signals arising in those images (rather than the images themselves) are relayed to several subcortical nuclei (for instance the amygdala and the thalamus) and multiple cortical regions (in temporal, parietal, and frontal sectors);
2   because those nuclei and cortical regions contain dispositions for response to certain classes of signal.

The end result is that dispositional representations in nuclei and cortical regions are activated and, as a consequence, induce some collection of changes in the state of the organism. In turn, those changes alter the body image momentarily, and thus perturb the *current* instantiation of the concept of self. In other words, the multifarious process of recognizing an object generates sets of responses – autonomic, hormonal, motor, imagetic – and those responses change the state of the organism for a certain time interval. We suspect that the essence of the neural mechanisms of consciousness may well reside with the perturbation of self images by newly occurring images.

Although the responding process we outline above implies knowledge (there is indeed abundant knowledge recorded in innate as well as experience-driven dispositional representations throughout the brain), it certainly does not imply, *per se*, that any brain component "knows" that responses are being generated to the presence of an entity. In other words, when an organism's brain generates a set of responses to an entity, the existence of a representation of self does not make that self *know* that its corresponding organism is responding. The self as described above cannot *know*. And here we arrive at the critical question in this discussion. How can the current image of an entity, on the one hand, and a set of images of the organism's state, on the other, both of which exist as momentary activations of topographically-organized representations, generate subjectivity? Our current answer to this question is in the form of a hypothesis. It consists of

1   having the brain create some kind of *description of the perturbation of the state of the organism* that resulted from the brain's responses to the presence of an image;
2   having the description *generate an image of the process of perturbation*;
3   displaying the image of the *self perturbed* together or in rapid interpolation with the image that triggered the perturbation.

Does the brain have the structures required for this proposed mechanism to operate? Indeed it does. Firstly, the brain possesses neural structures that support the image of an object and neural structures that support the images of the self, but it also has neural structures that support neither yet are reciprocally interconnected with both. In other words, the brain has available the kind of third-party neuron ensemble, which we have called a convergence zone, and which we have invoked as the neural substrate for building dispositional representations all over the brain, in cortical regions as well as in subcortical nuclei.

Secondly, such a third-party ensemble receives signals from both the representation of an object and the representations of the self, *as the latter are perturbed by the reaction to the object*. In other words, the third-party ensemble can build a *dispositional representation of the self in the process of changing, while the organism responds to an object*. This dispositional representation would be of precisely the same kind as the one the brain continuously holds, makes, and remodels. The information necessary to build such a dispositional representation is readily available (shortly after we see an object and hold a representation of it in early visual cortices, we also hold many representations of the organism reacting to the object in varied somatosensory regions).

As is the case with all dispositions, the dispositional construction we envision has the potential, once formed, to reactivate an image in any early sensory cortex to which it is connected. The basic image in the description would be that of the organism's body in the process of responding to a particular object, i.e., a somatosensory image.

We propose that all the ingredients described above – an object that is being represented, an organism responding to the object of representation, and a description of the organism in the process of changing in response to the object – are held simultaneously in working memory and are placed, side by side or in rapid interpolation, in early sensory cortices. Subjectivity would emerge during this latter step, when the brain is simultaneously producing not just images of an entity, of self, and of the organism's responses, but also another kind of image: *that of an organism in the act of perceiving and responding to an entity.* The latter kind of image (one might call it the metaself) would be the source of subjectivity incarnate (see A. Damasio, 1994, for details).

The description we have in mind is neither created nor perceived by a homunculus, and it does not require language. The third-party disposition provides a schematic view of the main protagonists from a perspective external to both, a nonverbal narrative document of what is happening to those protagonists, accomplished with the elementary representational tools of the sensory and motor systems.

We cannot think of any reason why the brains of birds and mammals would not be able to make such narratives. In effect, subjectivity would emerge in any brain equipped with some representation of self, with the capacity to form images and respond to them, and with the capacity to generate some kind of dispositional description in a third-party neuron ensemble.

The second-order narrative capacities provided by language would allow humans to engender verbal narratives out of nonverbal ones, and the refined form of subjectivity that is ours would emerge from this process. The virtual serial machine mechanism proposed by Dennett would operate at that high level rather than at the basic level we postulate here. It would not be the source of subjectivity, although it might contribute to important aspects of thinking and reasoning. In short, as we have noted elsewhere, language would not be the source of the self but it certainly would be the source of the "I."

Would a machine equipped with image-making devices, the ability to represent its physical structure and physical states imagetically, and dispositional knowledge about its past be capable of generating subjectivity if it were to construct images of itself perturbed, as described above? Probably it would not, unless the machine's body were a living body, with properties derived from its precarious homeostatic balance, from its need for survival, and from its inherent sense that what promotes survival is valuable. In the end, the neural device we propose for generating subjectivity serves to connect images with the process of life, and that may well be what consciousness is most about.

References

Churchland, P. M. (1984). *Matter and Consciousness*. MIT Press. Cambridge.

Churchland, P. S. (1986). *Neurophilosophy: Toward a Unified Science of the Mind-Brain*. MIT Press: Cambridge.

Churchland, P. S. and Sejnowski, T. J. (1992). *The Computational Brain: Models and Methods on the Frontiers of Computational Neuroscience*. MIT Press: Cambridge.

Churchland, P. S., Sejnowski, T. J., and Ramachandran, V. S. (1994). The critique of pure vision. In *Large-Scale Neuronal Theories of the Brain*, C. Koch (ed.). MIT Press: Cambridge.

Crick, F. (1994). *The Astonishing Hypothesis: The Scientific Search for the Soul*. Charles Scribners: New York.

Damasio, A. R. (1989a). Time-locked multiregional retroactivation: a systems level proposal for the neural substrates of recall and recognition. *Cognition*, **33**: 25–62.

Damasio, A. R. (1989b). The brain binds entities and events by multiregional activation from convergence zones. *Neural Computation*, **1**: 123–32.

Damasio, A. R. (1994). *Descartes' Error: Emotion and Reason in the Human Brain*. Putnam: New York.

Damasio, A. R. and Damasio, H. (1993). Cortical systems underlying knowledge retrieval: evidence from human lesion studies. In *Exploring Brain Functions: Models in Neuroscience*, P. Poggio and D. Glaser (eds). 233–48. Wiley: New York.

Damasio, A. R. and Damasio, H. (1994). Cortical systems for retrieval of concrete knowledge: the convergence zone framework. In *Large-Scale Neuronal Theories of the Brain*. C. Koch (ed.). MIT Press: Cambridge.

Damasio, H. and Frank, R. (1992). Three-dimensional *in vivo* mapping of brain lesions in humans. *Archives of Neurology*, **49**: 137–43.

Damasio, H., Grabowski, T. J., Damasio, A. R., Tranel, D., Boles-Ponto, L. L., Watkins, G. L., and Hichwa, R. D. (1993a). Visual recall with eyes closed and covered activates early visual cortices. *Society for Neuroscience*, **19**: 1604.

Damasio, H., Grabowski, T. J., Frank, R. J., Knosp, B., Hichwa, R. D., Watkins, G. L., and Boles-Ponto, L. L. (1993b). PET-Brainvox, a technique for neuroanatomical analysis of positron emission tomography images. *Quantification of Brain Function: Tracer Kinetics and Image Analysis in Brain PET. Proceedings of PET '93 Akita*, K. Uemura (ed.). 465–73. Elsevier: Amsterdam.

Dennett, D. C. (1991). *Consciousness Explained*. Little, Brown: Boston.

Kagan, J. (1989). *Unstable Ideas: Temperament, Cognition, and Self*. Harvard University Press: Cambridge.

Tononi, G., Sporns, O., and Edelman, G. (1992). Reentry and the problem of integrating multiple cortical areas: simulation of dynamic integration in the visual system. *Cerebral Cortex*, **2**: 310–35.

Tootell, R. B. H., Switkes, E., Silverman, M. S., and Hamilton, S. L. (1988). Functional anatomy of macaque striate cortex. II. Retinotopic organization. *The Journal of Neuroscience*, **8**: 1531–68.

Zeki, S. (1993). *A Vision of the Brain*. Blackwell Scientific Publications: Oxford.

0028-3932(95)00051-8

CONSCIOUSNESS AND THE NATURAL METHOD

OWEN FLANAGAN

Department of Philosophy, Duke University, Durham, NC 27708, U.S.A.

(*Received* 19 *September* 1994; *accepted* 1 *March* 1995)

Abstract—'Consciousness' is a superordinate term for a heterogeneous array of mental state types. The types share the property of 'being experienced' or 'being experiences'—'of there being *something that it is like for the subject* to be in one of these states.' I propose that we can only build a theory of consciousness by deploying 'the natural method' of coordinating all relevant informational resources at once, especially phenomenology, cognitive science, neuroscience and evolutionary biology. I'll provide two examples of the natural method in action in mental domains where an adaptationist evolutionary account seems plausible: (i) visual awareness and (ii) conscious event memory. Then I will discuss a case, (iii), dreaming, where I think no adaptationist evolutionary account exists. Beyond whatever interest the particular cases have, the examination will show why I think that a theory of mind, and the role conscious mentation plays in it, will need to be built domain-by-domain with no *a priori* expectation that there will be a unified account of the causal role or evolutionary history of different domains and competences.

Key Words: consciousness; dreams; evolution; memory; naturalism; vision.

CONSTRUCTIVE NATURALISM AND THE NATURAL METHOD

'Consciousness' is a superordinate term for a heterogeneous array of mental state types. The types share the property of 'being experienced' or 'being experiences'—'of there being *something that it is like for the subject* to be in one of these states.' Conscious mental states necessarily have phenomenal properties; they seem or feel or are experienced in certain ways. There are mental states and processes that involve information-processing but that do not involve consciousness. For example, most of the processes involved in typing these very words involve complex information-processing, but there is *nothing* it seems like for me, this organism, to be in these information-processing states.

The multifarious kinds of conscious mental states are obvious when we think about what different kinds of conscious states *seem* like—when we focus on their phenomenal features. Consciousness in the sensory modalities differs across modalities. Sensory experience is heterogeneous in kind yet coordinated—as when I smell, taste, see and eat supper—a supper I ordered by reading words on a menu and making articulated, memorable judgments to a waiter. So there is sensation, perception, propositional thought, and there are moods, emotions, dreams, and introspection, and much else besides. That these differ in phenomenal kind is obvious; how exactly they differ is however hard to articulate—we are inclined to say they *seem* different from one another.

From the first personal point of view consciousness only has *phenomenal-aspects*. The story of the brain side of conscious states will need to be provided by the neuroscientists, and the functional–causal role(s) of different kinds of consciousness (now taking both the

85

phenomenal and brain sides together) will need to be nested in a general psychological and evolutionary (both natural and cultural) account.

The idea is to deploy *the natural method* [7]. Start by treating different types of analysis with equal respect. Give *phenomenology* its due. Listen carefully to what individuals have to say about how things seem. Also, let the psychologists and cognitive scientists have their say. Listen carefully to their descriptions about how mental life works, and what jobs if any consciousness has in its overall economy. Third, listen carefully to what the neuroscientists say about how conscious mental events of different sorts are realized, and examine the fit between their stories and the phenomenological and psychological stories. But phenomenology, psychology, and neuroscience are not enough. Evolutionary biology and cultural and psychological anthropology will also be crucial players as the case of dreams will make especially clear. Embedding the story of consciousness into theories of evolution (biological and cultural), thinking about different forms of consciousness in terms of their ecological niche, and in terms of the mechanisms of drift, adaptive selection, and free-riding will be an important part of understanding what consciousness is, how it works, and what, if anything, it is good for—again, taking as axiomatic that there will be no unified theory for all the kinds of conscious mental state types.

Now *the natural method*, the tactic of 'collapsing flanks' will, I claim, yield success in understanding consciousness if anything will. The expectation that success is in store using this method is what makes my kind of naturalism *constructive* rather than *anti-constructive*, as is the naturalism of philosophers like Colin McGinn [19] who thinks that although consciousness is a natural phenomena, we will *never* be able to understand it.

In this paper I'll provide two examples of the natural method in action in mental domains where an adaptationist evolutionary account seems plausible: for (i) visual awareness and (ii) conscious event memory. Then I will discuss a case, (iii) dreaming, where I think no adaptationist evolutionary account exists. Beyond whatever interest the particular cases have, the examination will show why I think that a theory of mind, and the role conscious mentation plays in it, will need to be built domain-by-domain with no *a priori* expectation that there will be a unified account of the causal role or evolutionary history of different domains and competences.

TWO EXAMPLES: NEURAL CORRELATES OF SUBJECTIVE VISUAL ATTENTION AND CONSCIOUS EVENT MEMORY

Neural correlates of subjective visual attention

One way to study conscious experience is to correlate particular qualitative types of experience with particular kinds of activity. One fascinating set of studies linking particular types of awareness with particular types of neural activity has recently been done on rhesus macaques and exploits the well-known phenomenon of rivalry. The stimulus pattern stays the same, but perception flip-flops. Binocular rivalry is a particular type of rivalry that exploits the fact that the visual system tries to come up with a single percept even though the eyes often receive different visual information. In experiments involving binocular rivalry the visual input presented to two eyes is incompatible. For example, a line moving upward is presented on the left, and a line moving downward is presented on the right. "Because such stimuli cannot be fused by the cyclopean visual system, the perception alternates between the right eye alone or the left eye alone" [18, p.

761]. Humans report such alternations in perception. But how, one might ask, does one find out a monkey's phenomenology? How does one get at how things seem to the monkey. The answer is that one trains the monkey prior to the experiment to be a reliable reporter of whether it perceives a line moving up or a line moving down. This can be done by training monkeys to give bar-press reports or, more surprisingly, by training the monkeys to execute a saccade (a quick eye movement) to a spot on the left if a downward movement is perceived and to a spot on the right if an upward movement is perceived. A monkey's report of how things appear at any moment (its phenomenology) provides physical data about the rate of perceptual shifting and raises interesting questions about why there is shifting perception as opposed to a winner-take-all lock on one of the rival perceptual interpretations. But the phenomenological data 'The line is now moving upwards' and the physical data about the time between perceptual switches yield no information about what is going on in the brain, and in particular they yield no information about what neuronal events are involved in the shifting perceptions. This is where looking at the brain helps.

It is well known that monkeys have as many as 20–25 visual areas in the neocortex. The bulk of retinal output projects through the lateral geniculate nucleus of the thalamus to the primary visual cortex at the back of the brain that computes edges. Other areas are interested in color and shape, position relative to the organism itself, facial features, and motion and depth. Motion detection occurs primarily in the middle temporal areas in the superior temporal sulcus (STS). So the activity of neurons in the STS were monitored as the monkeys reported upward or downward motion. The study of the effect of the rival stimuli on 66 single neurons indicated that the "activity of many neurons was dictated by the retinal stimulation. Other neurons, however, reflected the monkeys' reported perception of motion direction" [18, p. 761]. The principle of supervenience says that every difference at the level of mentality must be subserved by a difference at some lower level (but perhaps not conversely). This experiment indicates how the robust phenomenological difference between an upward or downward moving image might be subserved (although undoubtedly not exclusively) by small but detectable changes in activity in the cortical areas subserving motion detection.

The experiment is an excellent example of how subjective awareness can be studied by drawing together information gathered at different levels of analysis and by distinctive techniques. First, there is the assumption that *there is something it is like for a monkey to have visual experience.* Second, good old-fashioned psychological techniques of operant conditioning are used to train the monkeys to provide reports about what they see. Finally, these reports are linked with detailed observations of the activity of 66 distinct neurons to yield information about the distinct brain processes subserving perceptual experiences of upward and downward moving lines. The natural method works.

Conscious event memory

Many years ago a famous neurological patient H.M. had medial-temporal-lobe surgery to cure his epilepsy. H.M.'s ability to remember new events and facts died with the excision. Still, H.M. remembered how to perform standard motor tasks and how to use language. Indeed, to this day H.M. is something of a crossword puzzle aficionado. His intact semantic memory is tapped by questions directly before his eyes, and his place in the puzzle is visually available to him. When H.M. is away from a puzzle he is working on, he cannot remember how far along on it he is or even that he is working on a puzzle.

H.M's good general intelligence and his semantic memory can sustain him during a short period of newspaper reading. But minutes after he has read the newspaper, he is blank about what he read. It is tempting to say that H.M. has an intact short-term memory but no long-term memory. But this is too simplistic for a variety of reasons. First, he retains information over the long term about how to do the certain puzzles, e.g. the Tower of Hanoi puzzle. The information is just not recoverable consciously. Second, when he does a crossword puzzle or reads a newspaper, he depends on something longer than short-term memory, which on standard views is very short. As long as H.M. is paying attention to a task, like a crossword puzzle or a newspaper article, he can retain information about it. When his attention is removed from the overall task his memory goes blank.

H.M. can give us a phenomenology, a set of reports about what he remembers and what he does not. He also reveals in his behavior that he remembers certain things that he can't consciously access. Other human amnesiacs provide further phenomenological and behavioral data. Putting together the phenomenological and behavioral data with the knowledge that there is damage or excision to the medial temporal lobe leads to an initial hypothesis that this area plays some important role in fixation and retrieval of memories of conscious events.

The medial temporal lobe is a large region that includes the hippocampus and associated areas, as well as the amygdala and related areas. Magnetic resonance imaging (MRI) has allowed for very precise specification of the damaged areas in living human amnesiacs and in monkeys. This research reveals that the hippocampus is the crucial component. When there is serious damage or removal of the hippocampal formation, the entorhinal cortex, and the adjacent anatomically related structures, the perirhinal and parahippocampal cortices, the ability to consciously remember novel facts or events is lost. Removal or serious lesions of the amygdala profoundly affect emotions, but not memory. It is not as if memories were created and set down in the hippocampal formation. The hippocampal formation is necessary to lay down memories, but it is not remotely sufficient for the conscious memory of facts and events. For habit and skill learning, it is not even necessary.

The complex network involved in fixing such memories has now been mapped in some detail by using monkey models and by drawing inferences about human memory function based on comparative deficit and normal data. Larry Squire and Stuart Zola-Morgan [21] report an elegant experiment in which monkeys were given a simple recall task. Each monkey was presented with a single object. Then after a delay (15 sec, 60 sec, 10 min) it was presented with a pair consisting of the original object and a new one. The monkeys were trained to pick the novel object to get a reward. The task is trivial for monkeys with intact hippocampal formations, but there is severe task impairment (increasing with latency) in monkeys with destruction to the hippocampal formation. The procedure is a reliable test of perceptual memory.

A skeptic could admit this but deny that there is any evidence that monkeys *consciously* remember what they saw. To be sure, the information about the original stimulus is processed and stored. The monkeys' behavior shows that these memories are laid down. But for all we know, these memories might be completely unconscious, like H.M.'s memories for the Tower of Hanoi puzzle. They might be. However, on the basis of anatomical similarities between monkey and human brains, the similarity of memory function, and evolutionary considerations, it is credible that the monkeys' selection of the appropriate stimulus indicates what they consciously remember, and thus that it can be read as providing us with a phenomenology, with a set of reports of how things seem to them.

The next step is to put the phenomenological and behavioral data from humans and monkeys together with the data about specific types of lesions suffered and join both to our best overall theories of the neural functions subserving memory and perception. Coordination of these theories and data yield the following general hypothesis. When a stimulus is observed, the neocortical areas known to be sensitive to different aspects of a visual stimulus are active. For example, in the simple matching task, the areas responsible for shape and color detection are active. In a task where the object is moving, the areas involved in motion detection become involved. Activity of the relevant cortical areas is sufficient for perception and immediate memory. "Coordinated and distributed activity in neocortex is thought to underline perception and immediate (short-term) memory. These capacities are unaffected by medial temporal lobe damage. . . . As long as a percept is in view or in mind, its representation remains coherent in short-term memory by virtue of mechanism intrinsic to neocortex" [21, p. 1384]. A memory is set down only if the information at the distributed cortical sites is passed to three different areas close to the hippocampus and then into the hippocampus itself. The hippocampus then passes the message back through the medial temporal lobe out to the various originating sites in the neocortex that processed the relevant aspects of the stimulus in the first place. The memory is laid down in a distributed fashion. It is activated when the connections between the hippocampus and the various areas it projects to are strengthened by, for example, the request for recall. Once the memory is laid down and especially after it is strengthened, "Proust's principle" comes into play. The memory can be reactivated by the activation of any important node, e.g. the one subserving object shape or color, even smell or sound, without having to be turned on by the hippocampus directly. In this way the hippocampal formation passes the "burden of long-term (permanent) memory storage" to the neocortex and is freed to make new memories [21, p. 1385]. This theory is credible and powerful. It explains how damage to the hippocampal formation can destroy the capacity to form new memories, and it explains why old memories are retained despite hippocampal destruction—their storage has been assumed by the neocortex.

Here again is an elegant example of the natural method. There are the phenomenological and behavioral data coming from H.M. and other human patients. There are the behavioral data coming from the monkeys, which I claim ought to be read as informing us about how things *seem* to the monkeys. There is the prior work linking visual processing with certain brain processes. And there is the eventual theory that explains how conscious memories are fixed, how they are recalled, what they are doing when they are not active (they are dispositions laid down in neural nets), and so on.

DREAMS: A DOUBLE ASPECT MODEL

Vision, especially the aspect that involves phenomenal seeing, is more comprehensible when we use the natural method. V1 is crucial, apparently necessary, to conscious sight, although it is *not* necessary to pick-up visual information. Furthermore, evidence from blindsighted patients who lack phenomenal vision but show that they process visual information provides some support for the idea that visual-*experience* serves a function. Memory retention, storage, and consolidation are also easy to give adaptationist accounts for. Why exactly certain components of memory involve consciousness is less clear. We know or have good reason to believe that robotic visual or memory devices can be built, systems that lack phenomenal or qualitative experience altogether. So conscious experience is not in any obvious sense metaphysically necessary to get these jobs done. The fact remains that for humans and other animals we have evolved with visual and memory systems that have phenomenal aspects and we perform less well when the phenomenal aspects drop out. My own view is that there is a credible case for the adaptive emergence, selection, and maintenance for certain, possibly many or most, kinds of conscious experience. But because 'consciousness' names a superordinate type some phenomenal state types may well not be biological adaptations. Dreams provide a credible example. I now turn to them, my third and last example:

Phenomenal-dreaming

Phenomenal-dreaming (henceforth *p-dreaming*) is a good example of one of the heterogeneous kinds of conscious experience, and it is at the same time, given neuroscientific evidence and evolutionary considerations, a likely candidate for being given epiphenomenalist status from an evolutionary point of view. *P-dreaming* is an interesting side-effect of what the brain is doing, the function(s) it is performing during sleep.

To put it in slightly different terms: *p-dreams*, despite being experiences, have no interesting biological function—no evolutionary proper function. The claim is that *p-dreams* (and possibly even rapid eye movements after development of the visual system is secured) are likely candidates of epiphenomena. Since I think that all mental phenomena supervene on neural events I don't mean that *p-dreams* are non-physical side-effects of certain brain processes, I mean in the first instance that *p-dreaming* was probably not selected for, that *p-dreaming* is neither functional nor dysfunctional in and of itself, and thus that whether *p-dreaming* has a function depends not on Mother Nature's work as does, for example, the phenomenal side of sensation and perception. It depends entirely on what we as a matter of cultural inventiveness—*memetic selection* [5], one might say—do with *p-dreams* and *p-dream reports*. We can, in effect, create or invent functions for dreams. Indeed, we have done this. But as temporally significant aspects of conscious mental life, they are a good example, the flip side say of awake perceptual consciousness which is neither an evolutionary adaptation nor ontogenetically functional or dysfunctional until we do certain things with 'our dreams'—for example use them as sources of information about 'what's on our mind', utilize dream mentation in artistic expression, and the like.

Despite being epiphenomena from an evolutionary perspective the way the brain operates during sleep guarantees that the noise of *p-dreams* is revealing and potentially useful in the project of self understanding. Thus many things stay the same on the view I am staking out. But there is a paradox: *p-dreams* are evolutionary epiphenomena, noise the system creates while it is doing what it was designed to do, but because the cerebral cortex is designed to make sense out of stimuli it tries half successfully to put dreams into narrative structures already in place, structures which involve modes of self-representation, present concerns, and so on. But the cortex isn't designed to do this for sleep stimuli, it is designed to do it for stimuli *period* and it is ever vigilant. The idea is that it did us a lot of good to develop a cortex that makes sense out of experience while awake, and the design is such that there are no costs to this sense-maker always being ready to do its job.

So it works during the chaotic neuronal cascades of part of the sleep-cycle that activate certain sensations and thoughts. So *p-dreams* despite their bizarreness, and epiphenomenal status, are meaningful and interpretable up-to-a-point.

WHAT IS A DREAM?

Thus far I've been using *p-dreaming* to refer to any mentation that occurs during sleep. But the term *p-dreaming* despite being useful for present purposes ultimately won't carve things in a sufficiently fine-grained way. Research has shown that perseverative anxious rehearsal—thinking about tomorrow's agenda—is most likely to occur during NREM sleep, the sleep standardly divided into four stages which occupies about 75% of the night. Night terrors, a common affliction of young children are very puzzling since the child seems totally awake, eyes wide open, running about speaking alternately sense and nonsense, but almost impossible to comfort and wake up entirely and, on most every view, suffering terrifying hallucinations (which even if the child is finally awakened are remembered much less well than hallucinatory REM dreams). The anomaly is that the terrorized child is almost certainly in stage III or IV NREM sleep. Sleep-walking, sleep-talking, and tooth-grinding are also NREM phenomena—and no one knows for certain whether we should say persons walking and talking in sleep are *phenomenally-conscious* or not. I'm inclined to say they are, and thus that the answer to William James's question: are we ever wholly unconscious?—is 'no'—at least for sleep.

So the first point is that mentation occurs during NREM sleep as well as during REM sleep, and we report mentation occurring in both states as 'dreams' [10]. Now since the discovery of REM sleep, and its close association with reports of vivid fantastic dreaming, many have simply identified dreaming with REM-ing or with mentation occurring during REM sleep [1, 6, 8, 14]. But this goes against the grain of our folk psychological usages of the term 'dream' where the term refers to any mentation that takes place while we are asleep. Furthermore, some NREM states, like night terrors, probably involve hallucinations and bizarre mentation so they have the right phenomenological character to be aligned with what one might be tempted to call 'real dreams', namely, those associated only with REM sleep.

Having recognized the complexities of the concept of 'dream', it will do no harm for present purposes if I continue to use *p-dreams* to refer to any mentation occurring during sleep recognizing full well that since mentation occurs in all stages of NREM and REM sleep, *p-dreaming* isn't precise enough ultimately to type mentation during sleep from either a phenomenological or neuroscientific point of view.

In REM-sleep pulsing signals originate in the brainstem and reach the lateral geniculate body of the thalamus. When awake this area (G) is a relay between the retina—on certain views part of the brain itself—and visual processing areas. Other pulses go to the occipital cortex—the main visual processing area of the brain. So PGO waves are the prime movers of REM-ing. This much accounts for the saliency of visual imagery in the dreams of sighted people. But the PGO noise is going to lots of different places and reverberating every which-way. This is why people who work at remembering dreams will report loads of auditory, olfactory, tactile kinesthetic, and motor imagery as well as visual imagery. There is nice convergence of neuroscientific and phenomenological data here. Recent studies have shown that the parts of the brain that reveal robust activity on PETs MRIs or magneto-encephalographs indicate that 'mentation during dreaming operates on the same anatomical substrate as does perception during the waking state' [17]. The main point is that PGO waves are dominant during REM sleep and quiescent during NREM sleep and this explains by inference to the best available explanation a good deal about why the mentation of REM sleep involves vivid, bizarre, and multimodal imagery, and why NREM-dreams standardly lack these features [17].

DREAMS AND FUNCTIONAL EXPLANATION

So far I've said something about the phenomenology of dreams, about the differences between the phenomenology of REM-mentation and NREM-mentation, and about how PGO waves account for most of the phenomenological differences.

The question remains: what is the function of sleep, sleep-cycling, and of dreams? Actually these are really three distinct questions. There is the function of sleep, generally, of the different parts of the sleep-cycle, and of the mentation that accompanies the different parts of sleep. Here is how things look to me regarding the function question. The fact that NREM is the oldest form of sleep and is hypometabolic suggests the following hypothesis: It was selected to serve for restorative and/or energy conservation and/or body building functions. Some people find this hypothesis empty—akin to saying sleep is for rest which although true is thought to be uninformative. But things are not so gloomy if we can specify some of the actual restorative/conservatory/building mechanisms and processes in detail. And we can. The endocrine system re-adjusts all its levels during sleep. For example, testosterone levels in males are depleted while awake regardless of whether any sexual or aggressive behavior has occurred, and are restored during sleep, indeed levels peak at dawn. Pituitary growth hormone does its work in NREM sleep. Growth hormone promotes protein synthesis throughout the body—new cell growth helps with tissue repair—for example, cell repair of the skin is well-studied and known to be much greater while sleeping than awake. Protein synthesis in the cerebral cortex, and the retina follow the same pattern of having a faster rate in sleep than while awake. And, of course, the amount of food needed for survival is lowered insofar as metabolic rate is. To be sure, much more needs to be said, and can be found in medical textbooks about the restorative/ conservatory/building processes that are fitness enhancing and associated with NREM sleep [16].

Regarding REM-sleep, two functions suggest themselves. First, the much larger percentage of REM-ing in development across mammals suggests that it is important in helping build and strengthen brain connections, particularly ones in the visual system, that are not finished being built in utero. Second, there is a credible case that REM-sleep, but not necessarily the associated mentation, supports neurochemical stockpiling necessary for attending, learning, and remembering during the day.

I focus only on the second function here. It is known that one of the most significant differences between waking, NREM sleep, and REM sleep has to do with the ratios of different types of neurochemicals, modulators, and transmitters in the soup [11, 12]. In particular, the ratios of cholinergic and aminergic neurochemicals flip-flop. Neurons known to release serotonin and norepinephrine shut off in the brainstem during REM and neurons secreting acetylcholine are on. What good could this do? Here's one possible answer. The best theory of attention, namely Posner's and Peterson's [20] says that norepinephrine is crucial in getting the frontal and posterior cortical subsystems to do a good job of attending. Furthermore, both norepinephrine and serotonin are implicated in thermoregulation as well as in learning, memory, and attention; and dopamine has been shown to play an essential role in learning at least in sea slugs. Now what happens in REM sleep that is distinctive in addition to the dream-mentation part is that there is a complete shift in the ratios of certain classes of neurochemicals. In particular, in waking serotonin is working hard as are dopamine and norepinephrine. The aminergic neurons that release these neurochemicals quiet down in NREM sleep and turn off during REM sleep—this helps explain why memory for dreams is degraded.

Meanwhile, cholinergic neurons, for example, those releasing acetylcholine turn on. Here is a credible hypothesis for why this might be: By a massive reduction in firing during REM sleep the neurons releasing the neurochemicals most directly involved in attention, memory, and learning get a rest. While resting they can synthesize new neurotransmitters. The evidence points to a major function of REM sleep as involving 'stockpiling' the neurotransmitters that the brain will need in the morning for the day's work.

Another hypothesized function of sleep and of REM sleep in particular, that I have not mentioned, is that something like disk maintenance, compression, trash disposal, and memory consolidation take place [4, 13]. These seem like good things for the system to do. But it's pie in the sky hypothesizing until some mechanism is postulated that could do the job. How could such memory consolidation or junkyard functions work? What sort of mechanism could govern such a process or processes. One idea is this: for memories to be retained they must be converted from storage in the halfway house of distributed electrical patterns into stable protein structures within neurons, in particular at the synapses. The idea is that memory reactivation involves the reactivation of the neural networks whose synaptic strengths have been altered. What happens during REM sleep is that the cholinergic neurons that are on and releasing acetylcholine interact with the temporary but connected electrical synaptic hot spots constituting a memory from the day, and change those hot spots to a more stable form—to some sort of protein structure.

NATURAL FUNCTIONS

The hypothesis can be formulated somewhat more precisely given what has been said so far. It is that sleep and the phases of the sleep cycle—NREM and REM sleep—were selected for *and* are maintained by selective pressures. They are *adaptations* in the biological sense [3, 24]. However, the phenomenal aspects of sleep, the thoughts that occur during NREM sleep, as well as the dreams, and lucid dreams (dreams which contain the awareness that one is dreaming) that occur during REM sleep are probably epiphenomena in the sense that they are serendipitous accompaniments of what sleep is for.

Now some things that were originally selected for to serve a certain function, end up being able—with some engineering modifications—to serve another function. Selection pressures then work as it were to select and maintain the adaptation because it serves both purposes, or to put it another way both the original phenotypic feature and the extended one, serve to increase fitness. For example, feathers were almost certainly selected for thermoregulation, but now selective pressures work to maintain feathered wings because they enable flight.

It is standard in evolutionary biology to say of some 'automatic sequelae', pleiotropic or secondary characteristic, that it is a *non-adaptation* only if it is a concomitant of a trait that was selected for and if in addition no concurrent positive selection or independent modification operate on the trait. So the capacity to fly may have been a sequelae of selection pressures to design efficient thermoregulation, but feathered wings are an adaptation because despite being a secondary characteristic they were (and are) subject to positive selection and modification pressures. But the color of blood and the human chin are common examples of sequelae that are non-adaptations.

The biological notion of an *adaptation* and even a *non-adaptation* needs to be marked off from the concept of *adaptiveness* or *functionality*. The biological notion is tied to selection pressures which contribute to reproductive success in a particular environment or set of environments [2]. But we also say of mechanical devices, intentional human acts or act-types, and of cultural institutions that they are *adaptive* or *functional*. Here we mean that the device, act, act-type, institution *does what it is designed to do* [9, 15].

We need to draw one further distinction within the nest of meanings of the terms 'function' and 'functional': this between (1) a causal contribution sense of function and (2) a functional vs dysfunctional sense. So to use Kitcher's example, mutant DNA causing tumor growth is functioning as it is supposed to from the 'point of view' of certain cell lineages; it is making the causal contribution we expect, but it is dysfunctional—bad biologically and psychologically for the organism in which the tumor is growing.

Now my argument is this: sleep and sleep-cycling is an adaptation for reasons given above—it restores, conserves, and builds and we can specify some of the specific things it does and the mechanisms these are done by. There is some reason to wonder whether REMing and NREMing, that is the moving or non-moving of eyes is an adaptation. And there is very good reason to be positively dubious about the adaptive significance of the phenomenal experiences that supervene on REM and NREM-sleep. Dreaming, broadly construed, is pleiotropic, an automatic sequelae, a spandrel. It is doubtful that dream-consciousness once in play as a sequelae of causal processes originating in the brain stem which tickle the visual areas producing REMs was subjected to positive selection pressures and modification. Actually I should put it this way: for reasons discussed earlier the brain-stem is designed to activate the visual system to finish building it during the first year of life. Once the system is built the continuation of the activation of the visual system serves no obvious further developmental function. Furthermore, whereas the PGO waves of REM-sleep are implicated in the processes of stockpiling neurochemicals for the next day's work, for making what is learned more stable so that it can be remembered, and possibly for trash disposal, there is no reason to believe that these jobs require mentation of any sort. Rats that have learned certain spatial tasks have the relevant neural circuits worked on during sleep. Why the rat would need to be thinking the 'right' spatial thoughts in addition to having the 'right' circuits worked on is utterly obscure [25, 26]; a tempting inference, but totally unwarranted as best I can see.

Assuming, tentatively, that the stabilizing idea is right, there is no phenomenological evidence that as electrical patterns are transformed into protein structures that the associated mentation involves the activation of the thoughts worth remembering. People remember nonsense syllables better after sleep than if tested right after learning but before sleep. But, to the best of my knowledge, people never report *p-dreaming* about nonsense syllables. Nor do students of mathematics work through the proofs of the previous day in dreams. It may well be that the proof of the Pythagorean theorem would go in one ear and out of the other if we didn't sleep in between. But I would place large bets that one will have trouble getting any phenomenological reports of sophomore geometry students dreaming through the steps of the theorem in REM sleep. The point is that PGO waves are causally implicated in the neurochemical stockpiling of amines (serotonin, norepinephrine, etc.) and in setting acetylcholine and its friends to the task of bringing stability to what has been learned. But there is no reason, so far as I can see, to think that the mentation caused by the PGO waves is causally relevant to these processes. The right circuits need to be worked on, but no mentation about the information that those circuits contain is needed—and typically such mentation does not occur. The visual, auditory, propositional, and sensory-motor mentation that occurs is mostly noise. One might be drawn to a different conclusion if the mentation was, as it were, about exactly those things one needs to stabilize for memory storage, but phenomenologically that seems not to be the case. It can't be the actual thoughts that occur during the bizarre mentation associated with REMing which the system is trying to stabilize, remember or store—most of that is weird. Sure some is not weird, and of course the so-called day's residue makes occasional appearances in dreams. It would be surprising if it didn't—it's on your mind. The incorporation of external stimuli is also easily explained—the system is designed to be relatively insensitive to outside noise, but it would be a pathetic survival design if it was completely oblivious to outside noise. So dripping faucets, cars passing on the street outside, are being noticed but in a degraded way, they won't wake you, but a growling predator at your campsite will.

P-dreams are a special but by no means unique case where the epiphenomenalist suspicion has a basis. *P-dreaming* is to be contrasted with cases where phenomenal awareness was almost certainly selected for. Take normal vision, for example. It is, I think, a biological adaptation. Blindsighted persons who have damage to area V1 in the visual cortex get visual information but report no phenomenal awareness of what is in the blindfield. They behave in degraded ways towards what's there if asked to guess what is there, or reach for it, which is why we say they are getting some information. But the evidence suggests that the damage to V1 which is essentially implicated in phenomenal visual awareness explains why the performance is degraded. And this suggests that the phenomenal side of vision is to be given an adaptationist account along with, and as part of, an adaptationist account of visual processing generally. This is not so with *p-dreaming*.

INVENTED FUNCTIONS

The phenomenal aspects associated with sleeping are nonadaptations in the biological sense. The question remains does *p-dreaming* serve a function. If it does it is a derivative psychological function constructed via mechanisms of cultural imagination, and utilization of the fact that despite not serving a direct biological or psychological function, the content of dreams is not totally meaningless (this has to do with what the system is trying to do during sleep and sleep-cycling) and thus dreams can be used to shed light on mental life, on well-being, and on identity. What I mean by the last remark is this: *the cortex's job is to make sense out of experience and it doesn't turn off during sleep.* The logically perseverative thoughts that occur during NREM sleep are easy for the cortex to handle since they involve real, but possibly 'unrealistic' ideation about hopes, worries and so on. Indeed, from both a phenomenological and neuroscientific perspective awake mentation and NREM sleep mentation differ more in degree than in kind: worrying and wondering and problem solving while awake are less likely than their NREM kin to get caught in perseverative ruts or involve running around in circles. The point remains: we express ourselves, what's on our minds when we wonder and worry and plan and rehearse. This is something we *learn* to do. Thinking comes in handy while awake, and apparently the costs of doing so while asleep don't exceed the benefits, unless of course perseveration is keeping us from sleep in which case we do need to learn how to prevent or control the thinking.

REM mentation is a different story. It differs in kind from ordinary thinking. Phenomenologically and brain-wise it is a radically different state—closer to psychosis than to any other mental state types. Still the cortex takes what it gets during REM sleep and tries to fit it into the narrative, script-like structures it has in place about how my life goes and how situations go, e.g. restaurant scenes, visits to amusement parks, to the beach, how sex and romance go, and so on. The basic ideas that dreams have narrative structure and that they utilize self-models, scripts, and narrative structures already in place has been corroborated by recent work by Hobson's group [22, 23].

Your cortex (but not alone) is expressing what's on your mind, how you see things. Your dreams are expressions of the way you uniquely cast noise that someone else would cast differently. So things remain the same. *Phenomenal-dreams* make a difference to your life. They may get you thinking in a certain way upon waking. You may find yourself in a hard to shrug-off mood despite learning that the imagined losses of loved-ones causing that mood were dreamed and didn't really happen. You may be inspired to write a poem or a mystical text, and you may work at the project of interpretation. This is not silly. What you think while awake or while asleep is identity-expressive. The project of self-knowledge is important enough to us that we have learned to use the serendipitous mentation produced by a cortex working with the noise the system produces to further the project of self-knowledge and identity location. This is resourceful of us.

CONCLUSION

'Consciousness' is the misleading name, misleading because it seems to name a faculty. But, as I have insisted, it is just the superordinate term that names a vast and heterogeneous array of mental-state types that share the property of being experienced, or experiences, or states such that there is something it is like to be in these states for the subject of them. I have discussed three examples: visual awareness, conscious event memory, and dreaming. The examples exemplify first the fact that conscious mental state types differ in how they seem, in their phenomenology. So visual experience has a different phenomenology from consciously remembering the tune to a song, and both differ phenomenologically from a perseverative anxious NREM-dream, which differs in turn from a bizarre multimodal REM-dream. Conscious mental state types are phenomenologically diverse.

But the phenomenal features of conscious states are only part of the story. Again take visual processing. We know that the phenomenal aspects of vision are only a small part of what the system is doing as it 'sees'. For each conscious mental-state type we will need to build information-processing and neuroscientific models that explain what the system is doing, what jobs are getting done, as well as how at the neural level it is doing the sum total of cognitive-processing that sometimes yields, as a proper part of itself, conscious experience. In addition to questions of phenomenal type and causal role, there are also questions about the evolutionary history and function, both natural and cultural, of each conscious mental-state type. Deploying the natural method, blending and coordinating the resources of phenomenology, psychology, cognitive science, neuroscience, evolutionary biology and cultural anthropology on each mental-state type while attending to what, if any role, conscious states play in the overall functioning of the system in question is the only way, as far as I can see, to build a theory of consciousness. One consequence of this approach, of course, is that we will be building a science of the mind and finding the place of consciousness in it; not the other way around.

REFERENCES

1. Aserinsky, E. and Kleitman, N. Two types of ocular motility occurring in sleep. *J. appl. Physiol.* **8**, 1–10, 1955.
2. Brandon, R. *Adaptation and Environment.* Princeton University Press, New Jersey, 1990.
3. Burian, Richard. Adaptation: Historical Perspectives. *Keywords in Evolutionary Biology,* Evelyn Fox Keller and Elisabeth Lloyd (Editors), pp. 7–12. Harvard University Press, New York, 1992.
4. Crick, F. and Mitchison, G. The function of dream sleep. *Nature* **304**, 111–114, 1983.
5. Dawkins, R. *The Selfish Gene.* Oxford University Press, New York, 1976.
6. Dement, W. The occurrence of low voltage, fast, electroencephalogram patterns during behavioral sleep in the cat. *Electroenceph. clin. Neurophysiol.* **10**, 291–296, 1958.
7. Flanagan, O., *Consciousness Reconsidered.* MIT Press, Cambridge, Mass, 1992.
8. Foulkes, D. *Dreaming: A Cognitive-psychological Analysis.* Lawrence Erlbaum, Hillsdale, New Jersey, 1985.
9. Godfrey-Smith, Peter. A modern history theory of functions. *Nous* **XXVVII**, 344–362, 1994.
10. Herman, J. H., Ellman, S. J. and Roffwarg, H. P. *The Problem of NREM Dream Recall Re-examined.* A. Arkin, J. Antrobus and S. Ellman (Editors). *The Mind in Sleep,* pp. 59–92, Lawrence Erlbaum, Hillsdale, New Jersey, 1978.
11. Hobson, J. A. *Sleep.* Scientific American Library, New York, 1989.
12. Hobson, J. A. *The Dreaming Brain.* Basic Books, New York, 1988.
13. Hopfield, J. J., Feinstein, D. I. and Palmer R. G. Unlearning has a stabilizing effect in collective memories. *Nature* **304**, 158–159, 1983.
14. Jouvet, M. Récherches sur les structures nerveuses et les mécanismes résponsables des différentes phases du sommeil physiologique. *Arch. Ital. Biol.* **100**, 125–206, 1962.
15. Kitcher, Philip. Function and design. (Manuscript to be submitted.)
16. Kryger, M. H., Roth, T., and Dement, W. *Principles and Practice of Sleep Medicine,* 2nd Edn. W. B. Saunders, London, 1994.
17. Llinás, R. and Paré, D. Of dreaming and wakefulness. *Neuroscience* **44**, pp. 521–535, 1991; *Proc. Natl. Acad. Sci.* **90**, 2078–2081, 1993.
18. Logothetis, N. and Schall, J. D. Neuronal correlates of subjective visual perception. *Science* **245**, 761–763, 1989.
19. McGinn, C. Can we solve the mind–body problem? *Mind* **98**, 349–366, 1989; reprinted in McGinn, *The Problem of Consciousness.* Blackwell, Oxford, 1991.
20. Posner, M. I. and Peterson, S. E. The attention system of the human brain. *Ann. Rev. Neurosci.* **13**, 25–42, 1990.
21. Squire, L., and S. Zola-Morgan. The medial temporal lobe memory system. *Science* **253**, 1380–1386, 1991.
22. Stickgold, R., Rittenhouse, C. D. and Hobson, J. A. Constraint on the transformation of characters, objects, and settings in dream reports. *Conscious. Cognit.* **3**, 100–113, 1994.
23. Stickgold, R., Rittenhouse, C. D. and Hobson, J. A. Dream splicing: a new technique for assessing thematic coherence in subjective reports of mental activity. *Conscious. Cognit.* **3**, 114–128, 1994.
24. West Eberhard, Mary Jane. Adaptation: Current usages. *Keywords in Evolutionary Biology,* Evelyn Fox Keller and Elisabeth Lloyd (Editors), pp. 13–18. Harvard University Press, Mass., 1992.
25. Wilson, M. A. and McNaughton, B. L. Dynamics of the hippocampal ensemble code for space. *Science* **261**, 1055–1058, 1993.
26. Wilson, M. A. and McNaughton, B. L. Reactivation of hippocampal ensemble memories during sleep. *Science* 676–679, 1994.

The Evolution of Consciousness

DANIEL C. DENNETT

How is human consciousness situated in the natural world? This is a great, ancient problem of philosophy, but also one of the most perplexing unanswered questions of science. There is no longer any serious doubt that the seat of consciousness is the brain, and nothing but the brain. In the last few decades, the explosive growth of our knowledge of the brain—and of some of the mechanisms, processes, and principles of cognition—has brought us tantalizingly within range of the big question. It is now possible, for the first time, to formulate testable hypotheses about how activities in the brain can *add up to* the phenomena of consciousness.

Some philosophers are offended by the suggestion that science could solve *our* problem, the philosophical mystery of the mind, but this proprietary attitude, which never made good sense, is particularly out of place now. Today the philosophical problem is everybody's problem. Scientists have made great progress on the mechanisms of vision and the other senses, motor control, memory, and attention; nevertheless, as they close in on the remaining big question of how all these mechanisms contribute to consciousness, they find themselves beset by deep conceptual (or philosophical) puzzles. For instance, we now understand how very complex and even apparently intelligent phenomena, such as genetic coding, the immune system, and low-level visual processing, can be accomplished without a trace of consciousness. But this seems to uncover an enormous puzzle of just what, if anything, consciousness is for. Can a conscious entity do anything for itself that an unconscious (but cleverly wired up) simulation of that entity couldn't do for itself? This is just one of the philosophical illusions that are now starting to bedevil the scientists.

It is true that most of the progress has been made by scientists resolutely ignoring the philosophical issues and concentrating on questions that they know how to answer, but this backlog of postponed puzzles is beginning to get in the way of further progress. As Darwin observed in 1838, "Experience shows that the problem of the mind cannot be solved by attacking the citadel itself."[1] However enough indirect progress has been made since Darwin for us to prepare realistically for a final assault on the citadel.

Darwin is, in fact, one of the two heroes of this essay; the other is another brilliant Englishman, Alan Turing, the mathematician who could justly be called the inventor of the computer. The subject here is the philosophical offspring created by the union of their great ideas: an answer to the question with which I began.

How is human consciousness situated in the natural world? Human consciousness is a "virtual machine" (something rather like a computer program) implemented by the brain; it is the product of three distinct but interacting processes of evolution by natural selection.

Comparisons between brains and computers, and between minds and software, have become hackneyed. The media called the first computers, those heroic monstrosities of Turing's day, "giant electronic brains"; ever since then the comparisons have been irresistible, for reasons good and bad. I will try to show that if we invert some of the standard points of comparison, and refresh our sense of the biological processes that have created human minds, we can rediscover the analogy; it is deeper, and more fruitful, than the threadbare slogans.

Three processes of evolution conspire to create consciousness:

1. *The evolution of species with genetic transmission.* Standard neo-Darwinian evolution by natural selection
2. *The evolution of neural patterns.* Patterns of activity within individual brains, in processes of learning and training
3. *The evolution of "memes"** Ideas available in the cultural environment, culturally transmitted and "implanted" in individual brains by social interactions

In each process the mechanism is classically Darwinian. The process consists of

---

* A term coined by Richard Dawkins.

1. A wellspring of variation
2. Heritability or transmission of traits
3. Differential replication due to blind, mechanical factors that are nevertheless often noncoincidentally linked (probabilistically) to *reasons*

Thus, in all three processes, there is not just natural selection, but natural selection *for* features that have some discernible raison d' être.

The three distinct processes are listed in order of their antiquity. First came species evolution; one of its discoveries was the plasticity that provided the medium for neural evolution. That process, in turn, yielded a "technology" of learning and transmission that provided the medium for cultural evolution; the latter eventually created the virtual machine of human consciousness, which, when it was reflected back onto all three processes, rendered them all swifter and more efficient.

Let us start with the first process, and remind ourselves that nervous systems are for controlling the activities in time and space of organisms that have taken the strategic path of locomotion. (The juvenile sea squirt wanders through the sea searching for a suitable rock or hunk of coral to cling to and make its home for life. It has a rudimentary nervous system. When it finds its spot and takes root, it doesn't need its brain anymore—so it eats it! This puts one in mind of a similar effect among tenured professors.*

The key to control is the ability to track, or even anticipate, the important features of the environment; all brains are, in essence, "anticipation machines." The clam's shell is fine armor, but it cannot always be in place; the hard-wired reflex that snaps the shell shut is a crude but effective harm anticipator-avoider. Brains that have some plasticity are better than fixed reflex links; they are capable not just of stereotypic anticipation, but also of adjusting to trends. Even the lowly toad has some small degree of freedom in how it responds to novelty, gradually altering its patterns of activity to track—with a considerable time lag—those features of its environment that matter

* The analogy between the sea squirt and the associate professor was first pointed out, I think, by Rodolfo Llinas. Patricia Churchland has helped to spread this most infectious meme through the academisphere.

most to its well-being. [2] But for more high-powered control, what you want is an anticipation machine that will think ahead, avoid ruts in its own activity, solve problems before encountering them, recognize novel harbingers of good and ill, and in general stay more than just one step ahead of the environment. For all our foolishness, we human beings are vastly better equipped for that task than any other self-controllers, and there can be little doubt that it is our enormous brains that make this possible. But how?

Our brains differ dramatically from those of our nearest ancestors among the hominids and other primates, but primarily in size, rather than in structure. Just why our ancestors' brains should have grown so large so fast (in the evolutionary time scale, their development was more an explosion than a blossoming) is a matter of some controversy, but little controversy surrounds the nature of the product: an enormously complex brain of unrivaled plasticity. That is to say, while its gross architecture is genetically fixed, its microstructure of trillions of connections is remarkably unfixed. Some of this variability is required simply to provide a medium for the moment-to-moment transient patterns of brain activity that somehow register, or represent, or at any rate track, the important, variable features of the environment—the bird that flies by, the drop in air temperature—and of the organism's own states—the drop in blood sugar, the increase of carbon dioxide in the lungs. Thus, some of the brain's variability is used to hold on to the fleeting data about current events that should matter to any self-controller. But much of it is available for relatively permanent fixing by events and features of that environment, and this is the job of our second evolutionary process.

While the first process of evolution "hard-wires" the nervous systems of lower organisms, in our case (more than in any other organism), it provides incompletely designed nervous systems, which postpones the completion of the design until after birth. Then specific, highly variable features of the environment can be counted on to provoke further appropriate fixings of microconnections, turning the basic brain into, say, an English speaker's brain rather than a Japanese or Swahili speaker's, to take the best-known example of postnatal design.

The processes of postnatal design fixing appear to be various, giving rise to incessant and largely unnecessary controversy: How much is innate and how much is learned? Does one really *learn* one's mother tongue, or is this process best seen as one of differential switch setting that is continuous with embryological development? (Birds don't *learn* to grow feathers; do they really *learn* to fly? Babies don't *learn* to shed their baby teeth; do they really *learn* to walk or to speak?) One can blandly rise above the fray and make the following point: Some things come quite effortlessly to us all (such as acquiring a mother tongue, or learning that we cannot walk on water) and others require focused practice or rehearsal (such as playing the violin, doing long division, or mastering a second language in adulthood). A good theory must explain all the varieties of postnatal design fixing; while some variations are apparently due to processes that are simply continuous with the gene-directed construction of the phenotype, others can be fruitfully viewed as arising from a distinct evolutionary process, on a much faster time scale (measured in seconds, not generations), and with suitable surrogates for offspring and the selective effect of the grim reaper. In short, as many have argued, the brain is itself a "Darwin Machine," (to borrow William H. Calvin's term) in which novel structures evolve and are revised, extinguished, propagated, co-opted for novel tasks, and so on.[3]

I will not say any more here about the presumed details of those evolutionary processes. Instead, I will turn to the relationship between that second process and the third: cultural evolution. Plasticity, as we have seen, is a good thing, for it makes learning possible. It is all the better if somewhere in the environment there is something to learn that is itself the product of a prior design process, so that each of us does not have to reinvent the wheel. We human beings have used our plasticity not just to learn, but to learn how to learn better, and then we've learned better how to learn better how to learn better, and so forth. We have also learned how to make the fruits of this learning available to novices. We somehow install an already invented and largely "debugged" *system* of activity habits in the partly unstructured brain.

This particularly sophisticated process of postnatal design can be best understood if we contrast it with a similar process, invented by Alan Turing: the process of creating virtual machines on computers. A computer has a basic fixed, or hard-wired, architecture. However, it also has huge amounts of plasticity thanks to its memory, which can store both programs (software) and data (the merely transient patterns that are made to track whatever is to be represented). Computers, too, are thus incompletely designed "at birth," with flexibility that can be used as a medium to create more specifically disciplined architectures, special-purpose machines, each with a strikingly individual way of responding (via the CRT screen or other output devices) to the environment's stimulation (via the keyboard or other input devices).

These temporary structures are made of rules rather than wires, you might say, and computer scientists call them virtual machines. A virtual machine is what results when a particular pattern of rules (or dispositions, or transition regularities) is imposed on all that plasticity. Anyone who is familiar with a word processor is acquainted with at least one virtual machine. If you have used more than one word-processing program, or used a spreadsheet or played a game on the very same computer you use for word processing, you are acquainted with several virtual machines, which take turns existing on a particular real machine.

Everybody knows that different programs endow computers with different powers, but not everybody knows how a computer works. A few of the details are important to our story, so I must beg your indulgence and provide a brief, elementary account of the marvelous process invented by Alan Turing.

Turing was not trying to invent the word processor or the video game when he made his beautiful discoveries; he was thinking, self-consciously and introspectively, about just how he, a mathematician, went about solving mathematical problems (or performing computations), and he took the important step of trying to break down the sequence of his mental acts into a sequence of primitive components. He was an extraordinarily well-organized thinker, but his stream of consciousness, like yours or mine or James Joyce's, was a variegated jumble of images, decisions, hunches, reminders, and so forth, from which he tried to distill the mathematical essence: the bare-bones, minimal sequence of operations that could accomplish the goals he accomplished in the florid and meandering activities of his conscious mind. The result was the specification of what we now call a Turing machine, a brilliant idealization and simplification of a hyperrational, hyperintellectual phenomenon: a mathematician performing a rigorous computation. The basic idea had five components: (1) a *serial process*, occurring in (2) a severely restricted *work space*, to which (3) *data* and *instructions*, are brought from (4) an inert but superreliable *memory*, to be operated on by (5) a finite set of *primitive operations*.

The set of primitives (the acts "atomic to introspection," if you like) was deliberately impoverished, so that there could be no question of their mechanical realizability. In other words, it was essential to Turing's mathematical purposes that each step in the processes he was studying be, without doubt, so simple, so stupid, that it could be performed by a simpleton—by someone who could be replaced by a machine.

Turing saw, of course, that his ideal specification could serve, indirectly, as the blueprint for an actual computing machine. Others saw this as well, in particular, the mathematician John von Neumann, who modified Turing's basic ideas to create the abstract architecture for the first realistically realizable digital computer. That architecture, the von Neumann machine, is illustrated in the figure below.

Briefly, the machine operates as follows. The memory is just a very large array of identical pigeonholes, or registers, each with an address, and each capable of having a number as its "contents." The CPU, or

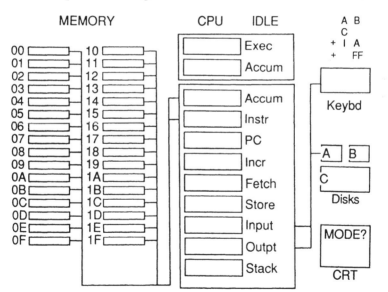

A von Neumann machine

105

central processing unit, is where all the action takes place. Here instructions (coded in numbers) are brought, one at a time, to the instruction register (Instr), where they are decoded and executed; the results appear in the accumulator (Accum). The instruction register and accumulator together are the famous "von Neumann bottleneck"; only one instruction at a time can be executed, and only one result at a time can appear in the accumulator. For instance, one such instruction would be to subtract the contents of register 12 from the number already in the accumulator (leaving the result in the accumulator). The next instruction could be to see if this result was greater than zero and, depending on the outcome, to fetch as the next instruction the contents of either register 25 or register 48.

All digital computers are direct descendants of the von Neumann design; and while they carry many modifications and improvements, they share (like all vertebrates) a fundamental underlying architecture. It is a considerable historical irony that the popular press misdescribed this architecture from the moment it was created. These fascinating new von Neumann machines were called, as I remarked earlier, "giant electronic brains," but they were, in fact, giant electronic *minds*, electronic imitations of the Joycean stream of consciousness: serial processors in which a single operation occurred at any one time. The architecture of the brain, in contrast, is massively parallel, with millions of simultaneously active channels of operation.

Is this difference theoretically important? In one sense, no. Turing had proven—and this is probably his greatest contribution—that one of his machines (called today a universal Turing machine) could compute any function that any computer, with any architecture, could compute. In effect, the universal Turing machine is the perfect mathematical chameleon, capable of imitating any other computing machine and doing, during that period of imitation, exactly what that machine does. You have only to feed the universal Turing machine a suitable description of another machine, and it produces a perfect imitation based on that description; it becomes, virtually, the other machine. A computer program can thus be seen either as a list of primitive instructions to be followed or as a description of a machine to be imitated.

In fact, once you have a von Neumann machine on which to build, you can nest virtual machines like Chinese boxes. For instance, you can first turn your von Neumann machine into a Unix machine (the Unix operating system) and then implement a Lisp machine in the Unix machine—along with WordStar, Lotus 1-2-3, and a host of other virtual machines. Each virtual machine is recognizable by its user interface—the way it appears on the screen of the CRT and the way it responds to input. This self-presentation is often called the user illusion, since a user cannot tell—and does not care—how the particular virtual machine he is using is implemented in the hardware. It doesn't matter to him whether the virtual machine is one, two, three, or ten layers away from the hardware.[4] (For instance, WordStar users can recognize, and interact with, the WordStar virtual machine wherever they find it, no matter what variation exists in the underlying hardware.)

Now, since any computing machine at all can be imitated by a virtual machine on a von Neumann machine, it follows that if the brain is a massive parallel processing machine, it can be imitated by a von Neumann machine. It is easy enough to see how this is done: by a process analogous to knitting. Suppose the parallel processor being simulated is ten channels wide, as in the figure below. The von Neumann machine is instructed to perform the operations handled by the first node of the first channel, saving the output in a buffer memory, and then the operations of the second node, and so forth, until the outputs of all the first row of nodes have been calculated and saved; these then become the inputs to the second row of nodes, whose operations are calculated one at a time, "knitting back" and forth, trading off time against space. A ten-channel machine will take at least ten times as long to simulate as a one-channel machine, and a

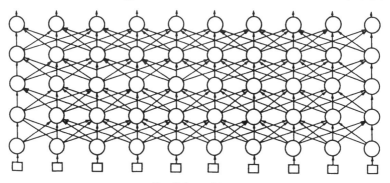

Parallel machine

million-channel machine (like the brain) will take at least a million
times as long to simulate. Turing's proof says nothing about the speed
with which the imitation will be accomplished, and for some architectures, even the blinding speed of modern digital computers is over-
whelmed by the task. That is why artificial intelligence researchers
interested in exploring the powers of parallel architectures are today
turning to *real* parallel machines—artifacts that really perform thousands or millions of operations at the same time, and thus might with
more justice be called "giant electronic brains"—on which to com-
pose their simulations.

The trouble with merely virtual parallel machines running on real
von Neumann machines is that they are terribly inefficient and waste-
ful. Von Neumann machines were not designed to perform this sort of
work. Of course, von Neumann machines were also not meant to be
word processors! Nothing in the design of their central processing
units and memories (as opposed to their peripheral organs—the
keyboard, for instance) was designed with an eye to making word
processing possible; but now, probably the most common activity of
these architectures is word processing. And here we can see a familiar
evolutionary principle applied to human technology: An organ or tool
that was initially designed for one purpose or set of purposes comes to
be valued more for a new purpose, and is appropriated, inefficiencies
and all, for that new purpose.[5]

▭

The stage is now set for the central inversion of metaphors that
promises to illuminate our investigation of human consciousness.
Having just noted the inefficiencies of virtual parallel machines implemented on serial von Neumann machines, let us consider the
hypothesis that human consciousness is nothing but the activity of a
virtual serial machine (which we might call the Joycean machine,
since it is the stream of consciousness we are modeling) implemented
in the parallel architecture of the human brain.

Quite clearly, our brains (which do not differ much in basic structure from those of other mammals) weren't designed—except for
some relatively recent peripheral organs—for word processing. Now,
however, a large portion, perhaps even the lion's share, of the activity
that takes place in adult human brains is involved in a sort of word

processing, namely, speech production and comprehension, and the serial rehearsal and rearrangement of linguistic items or, better, their neural surrogates.

Language, then, is one of the most important innovations in the environment that provoked the development of this virtual machine. But it is not only words that we process in our streams of consciousness: There are images, diagrams, and other wordless mental items that defy both novelists' and phenomenologists' efforts at direct description. We might do better, in fact, to leave the citadel of consciousness, turn our backs temporarily on the beguiling and ineffable items that swim by in the stream, and see if we can discover something useful by the indirect method of trying to answer the standard questions one would raise about such an evolutionary novelty:

1. How did this virtual machine, to Joycean machine of consciousness, arise?
2. What alternatives are there in the "gene pool" of virtual machines?
3. How is it transmitted?
4. How is it implemented in the brain?
5. What—if anything—is it good for?

If I could answer any of these questions definitively at this time, it would be a sure sign that I hadn't come up with a very fruitful idea. I present the idea as a guide to future empirical research, not as an a priori truth deduced from first principles, so I certainly do not want to be understood as claiming to have proved anything about this virtual machine. I will offer some speculations and suggestions, however, which I hope will eventually lead to confirmed answers to all these questions.

1. How did this virtual machine arise? Since, *ex hypothesi*, the human beings who discovered this technology were not already conscious (in the way we are), we shall have to suppose that it had its beginnings in unconscious invention. This is no paradox; all the inventions and discoveries of species evolution have been unconscious, and all but the most recent innovations in the behavior of individual phenotypes (the novel actions of animals, human and otherwise) must have arisen unconsciously. When Julian Jaynes speculates that the date of this unconscious invention was less than ten thousand years ago, he might be off by an order of magnitude (though I suspect he is close to the mark). In any case, there undoubtedly was a time before which our ancestors lacked our sort of consciousness and after which they had it. The intervening processes must have been unconscious, and calling them inventions is just honoring them as appreciated innovations. But what sort of processes were they? I have speculated that they were various modes of relatively undirected self-exploration (stimulating oneself over and over and seeing what happened). Because of the brain's plasticity, coupled with the no doubt innate restlessness and curiosity that motivates us to explore every nook and cranny of our environment (of which our own bodies are such an important and ubiquitous element), it is not surprising, in retrospect, that we hit upon strategies of self-stimulation, or self-manipulation, that led to the inculcation of habits and dispositions that radically altered the internal communicative structure of our brains.[6] For instance, if some subsystem of the brain had information that would be valuable to another, but genetic evolution had provided no hard-wired communication links between the subsystems, it would be immensely useful to develop a behavior (such as talking out loud to oneself) that "broadcast" the relevant information, making it available to any system that had access to auditory input. Once this became habitual, it would create a "virtual wire" linking the subsystems informationally. That is just one instance of the sort of role I suppose virtual machine elements to play in restructuring the functional architecture of the brain.

2. *What alternatives are there in the "gene pool" of virtual machines?* First of all, many varieties of self-stimulation and self-manipulation that presumably have no clearly valuable effect on cognition or control, and hence do not lead to the creation of new virtual machine elements, fail to be extinguished, for standard Darwinian reasons, and may even drift to fixation in certain subpopulations. Likely candidates are painting yourself blue, beating yourself with birch boughs, cutting patterns in your skin, starving yourself, saying a "magical" formula to yourself over and over again, and staring at your navel. In our own culture we should not overlook such apparently "noisy" but possibly quite functional activities as scratching one's head, biting one's fingernails, twisting one's hair, and doo-

dling. One man's mere fidget may be another's brilliant mnemonic crutch.

More dramatically, we can note the current population explosion of a clear rival to the standard Joycean operating system: multiple personality disorder (MPD). In 1944 approximately one hundred cases of MPD were reported (that is, in books and periodicals of some general accessibility in research libraries). By 1980, about a thousand cases were reported in the literature—again, the clinical literature, but also in the popular press. (Over the years, MPD cases have been the subjects of such popular books, and films, as *The Three Faces of Eve*, *Sybil*, and *The Minds of Billy Milligan*.) Four years later, more than two thousand cases were reportedly in professional therapy in the United States; a back-of-the-envelope calculation of one knowledgeable observer suggests that in 1988 more than four thousand cases were in therapy and perhaps another twelve thousand existed in the United States alone. (The disorder is almost unknown in other contemporary cultures.)

If we think of MPD as (whatever else it is) an alternative operating system—a different way of organizing all the "files" and "subroutines" and "applications programs"—then we can begin to trace the production and reproduction of this software, and its elaboration and transmission through the culture.[7] (We might well compare it to the software "viruses" of which we have heard so much.) As usual, looking at pathology is a good way of getting hints about the normal case.

3. How is the normal, Joycean virtual machine transmitted? This machine is not, of course, transmitted in sequences of zeroes and ones (like von Neumann software), on diskettes, or over networks. It is important to acknowledge this fundamental *disanalogy* between the virtual machines of computer science and the virtual machines of psychobiology. A von Neumann machine has a random access memory (RAM) that is extremely demanding about the form of whatever it stores (chunks of eight, sixteen, or thirty-two bits, for instance), and a central processing unit (CPU), that is even more rigid in the way it responds to those bit-strings, treated as instructions in its entirely proprietary and fixed machine language. These characteristics are definitive of the stored-program digital computer, and a human brain is no such thing. For one thing, the plasticity that somehow subserves its memory is not isolated as a passive storehouse; and, for another, there is nothing that could plausibly serve as the machine language of the brain—if by that we mean some uniform, generative, systematically compositional way of building larger processing dispositions out of smaller ones.

The methods of transmission that guarantee a fairly uniform virtual machine operating throughout the culture must be social, highly context-sensitive, and to some degree self-organizing and self-correcting. Getting different computers—say, a MacIntosh and an IBM-PC—to "talk to each other" is a matter of intricate, fussy engineering that depends on precise information about the internal machinery of the two systems. Insofar as human beings can "share software" without anyone's having such knowledge, it must be thanks to a high degree of system lability and format tolerance. A variety of methods are plausible: learning by imitation, learning as a by-product of subtle reinforcement (encouragement, disapproval, threat) in the course of communicative encounters, and even learning as the result of explicit instruction in a natural language that has already been learned via the first two methods.[8]

4. How is this virtual machine implemented in the brain? I suspect that some of the most puzzling features of human consciousness can be explained only if we go back and consider how this virtual machine may exploit features of the human nervous system designed long ago with different purposes in mind.

One of the yawning chasms in current cognitive theories of the conscious mind is the absence of any clear role for what is somewhat lamely called "affect" or "emotion," or even just "the way things feel" or (in philosophers' jargon) "qualia." Concentrating as we do these days on cognition, we have tended to incorporate its all but forgotten twin, conation, the "faculty of volition and desire," into the colorless ranks of information-bearing states; We thus tend to ignore as unacknowledged implementational noise all the zest, aversion, joy, and anguish that so typically accompany the information-bearing states of our cognitive apparatus. "What could their role be?" we ask ourselves, and, drawing a blank, we divert our attention from them. But perhaps we can make sense of their presence as a vestige of an earlier internal economy.[9]

Back in the early days of nervous systems, before the development of advanced locomotion and its concomitant need for long-range, or distal, perception, stimuli more or less wore their meanings on their sleeves: bad things triggered alarms by directly contacting—perhaps even injuring—the receptors or transducers designed to signal their presence, while the harbingers of good were stimulated by the very presence, in or adjacent to the organism, of the things they informed about. There was no such thing as a value-free, purely objective, informing "report" from a sense organ to an inner executive. Rather, every report had imperative connotations: *Grab me!* or *Flee me!* or some variation on those themes—*Look out!* or *Yummy!* or *Yuck!* As behavior became more devious, though, nervous systems came to need more dispassionate, detached sorts of signals to guide the control of behavior. But perhaps all these new armies of "reporters" were recruited from the ranks of "warners" and "applauders," and hence carried with them, now for no particular reason, a leftover trace of the positive or negative. That is, the price the organism paid for being informed about this or that was being irritated somewhat, in a particularly informing way, or, on other occasions, soothed or encouraged by the very process of being informed. All this need mean, non-metaphorically, is that some information-providing states would happen to cause the creature to strive to get out of them (intrinsically annoying or irritating states, one might call them, but it would all be just a matter of how they were wired in), while others would happen to cause the creature to strive to stay in them (intrinsically pleasant states—*other things being equal*). Suppose, however, other things weren't equal; suppose the value of staying informed about these various "irritating" matters was greater than the original disvalue that had underwritten the avoidance link, and suppose that the disvalue of lingering in the "intrinsically pleasant" information-providing states on occasion outweighed the value that had made them pleasing in the first place. In any system that began to track these higher-order values and disvalues, a host of tensions would arise, played off against each other, and keeping the whole system in a more or less incessant struggle of countervailing reactions to its own internal states. In lower creatures, this all might play itself out with no spectators and no partisans, but once the loops of self-monitoring began to feed back recognitions of these patterns, a logical space would be created for the first time in which enjoyment and disgust and their paler kin could find some room to play. (This does begin to sound more like home, doesn't it?)[10]

One shouldn't suppose that this imposition of the new cognitive economy on top of the older conative economy was simply a matter of tolerating the anachronistic pressures and vibrations and whatnot thereby created. It could very well be that the new virtual machinery opportunistically harnessed these ancient features. This could explain a great deal. As Jacob Bronowski once said, "The most powerful drive in the ascent of man is his pleasure in his own skill. He loves to do what he does well and, having done it well, he loves to do it better."[11] The meme that unlocked this storehouse of creative, exploratory energy was bound to pay for itself in both the short and the long term.

5. What—if anything—is this virtual machine good for? The upshot of my account is that all this self-stimulation creates a much more powerful virtual architecture, one that disposes the brain to ever more effective self-stimulation. It creates an architecture that is incessantly reorganizing itself, trying out novel combinations, sometimes idly, sometimes with great purpose casting about through the huge search space for a good—pleasing—new combination. And what is all this good for? It seems to be good for the sorts of self-monitoring that can protect a flawed system from being victimized by its own failures. Nicholas Humphrey has also seen it as providing a means for exploiting what might be called social simulations—using introspection to guide one's hunches about what others are thinking and feeling.[12] Julian Jaynes has argued, persuasively, that its capacities for self-exhortation and self-reminding are a prerequisite for the sorts of elaborated and long-term bouts of self-control without which agriculture, building projects, and other civilized and civilizing activities could not be organized.

These are all very plausible claims, and I am sure other good and essential uses for the architecture of consciousness can also be found.[13] As good Darwinians, however, we should remember that an innovation such as this does not *have* to be good for anything except self-replication.[14] The Joycean machine *might* just be a parasitical meme, an irresistible fad that happens to thrive in the medium provided by our huge and plastic brains, while not paying its keep with any great advantage to the creatures it infests.[15]

▭

Still, you may object, all this has little or nothing to do with consciousness. After all, a von Neumann machine is entirely unconscious. Why should implementing it, or something like it (a Joycean machine), make the brain any more conscious? I do have an answer: the von Neumann machine, by being wired from the outset as a serial machine with maximally efficient informational links, didn't have to become the object of its own elaborate perceptual systems. The workings of the Joycean machine, on the other hand, are just as "visible" and "audible" to it as any of the things in the external world that it is designed to perceive—for the simple reason that they have much of the same perceptual machinery focused on them.[16]

Now, this appears to be a trick with mirrors, I know. And it certainly is counterintuitive, hard to swallow, initially outrageous— just what one would expect of an idea that could break through centuries of mystery, controversy, and confusion. If the key to the mystery of consciousness were obvious, we would have settled on it long ago.

Let me try to make the idea a little clearer, at least, and perhaps a little more plausible. A little earlier, I mentioned the user illusion by which different machines are rendered perceptible. I am suggesting that human consciousness is a virtual machine in the brain—but *who is the user?* We seem to need selves, or egos, or subjects, to be the "victims" (or beneficiaries) of this user illusion. If this is so, we are right back where we started: at the mystery of the conscious mind and how it is situated in the natural world.

But this is not so. The user illusion is defined relative to a user, to be sure, but not necessarily to a *conscious* user. All we need posit in order to make sense of the concept of a user illusion is an information-gathering agent of some sort (it might be a robot, or even just another sedentary computer), with a limited access to features of the entity that presents the user illusion. Suppose, for instance, we make a robot—as unconscious as you please, but equipped with a televisual "eye" that provides it with "visual" information—and train it to type

115

letters for us, using WordStar. What it will "know" (in its unconscious way) about WordStar is only what it "sees," only those aspects of WordStar that are presented to it on the CRT or that it can manipulate from the keyboard. The rest of WordStar is all backstage and invisible—and irrelevant. (That, remember, is the chief beauty of virtual machines.) Such a user is external to the system in question— adopting a third-person point of view toward it—and there are indeed similarly placed agents relative to each of us: namely, the people with whom we interact socially. For whom is *my* Joycean machine a user illusion? Among others, for you.

The potency of this third-person role is revealed with particular clarity when one is confronted with a living human body that runs a different operating system: the MPD user illusion. I quote a few passages from the user's manual:[17]

> Treat all alters with equal respect and address the patient as he or she wishes to be addressed. . . .
>
> Make it clear that the staff is not expected to recognize each alter. Alters must identify themselves to staff members if they find such acknowledgment important.
>
> Explain ward rules personally, having requested all alters to listen, and insist on reasonable compliance.

In short, according to the clinical literature on MPD, adopting the view that more than one subject inhabits a single body is a strategy that works with these patients when no other strategy does. And we who are normal can all attest to the strategic wisdom of others' treating us as if we were each a single subject inhabiting a body, with only limited capacity to report on what is going on inside us. Thus other people are certainly users for whom our Joycean user illusions are valuable.

Notice also that each human brain, posed the multifaceted task of controlling a highly active and complex body through a complicated world, needs information about certain of its own states and activities—but can readily make use of only certain edited sorts of information. It needs a user illusion of itself. And thanks to the indoctrination it receives during its apprenticeship, it gets one. We call it a mind. As Douglas Hofstadter has said, "Mind is a pattern perceived by a mind."[18]

Where does this mind come from? Is it real? What is it made of? It is a virtual machine, running in the brain, a product of three sorts of evolution. And it is *its own* user illusion.

The ideas—memes—presented here have been undergoing population explosion in the academisphere in the last few years, and the task of determining their lines of ancestry—or, alternatively, their convergent evolution—is beyond me. Among the major contemporary phenotypes are those found in the following works, in addition to the works cited in the notes.

Richard Dawkins. *The Selfish Gene.* Oxford: Oxford University Press, 1976.

Gilles Fauconnier. *Mental Spaces.* Paris: Les Editions de Minuit, 1984.

Michael Gazzaniga. *The Social Brain.* New York: Basic Books, 1985.

Geoffrey Hinton (Touretzky and Hinton). "Symbols Among the Neurons: Details of a Connectionist Inference Architecture." In *Proceedings of the Ninth Joint Conference on Artificial Intelligence*, Los Angeles, California, August 1985, pp. 238–243.

Douglas Hofstadter. *Godel Escher Bach.* New York: Basic Books, 1979.

———. *Metamagical Themas.* New York, Basic Books, 1985.

Ray Jackendoff. *Consciousness and the Computational Mind.* Cambridge, Massachusetts: Massachusetts Institute of Technology Press, 1987.

Julian Jaynes. *The Origins of Consciousness in the Breakdown of the Bicameral Mind.* Boston: Houghton Mifflin, 1976.

Stephen Kosslyn. Paper presented to the Society for Philosophy and Psychology, 1984.

———. *Science* 240: 1621–1626, June 17, 1988.

Marvin Minsky. *The Society of Minds.* Cambridge, Massachusetts: Massachusetts Institute of Technology Press, 1986.

Allen Newell. "William James Lectures, 1987." In *Science* 241: 296–298, July 15, 1988; and *Science* 241: 27–29, July 1, 1988.

Gilbert Ryle. *On Thinking.* K. Kolenda, ed. New Jersey: Rowan and Littlefield, 1979.

NOTES

1 Quoted in William H. Calvin, *The River that Flows Uphill* (New York: Macmillan, 1986), p. 159.

2 Ewert, "The neuroethology of prey-capture in toads," *Behavioral and Brain Sciences* 10:3, (November 1987) and commentary. See in particular my commentary, "Eliminate the Middletoad!" *Behavioral and Brain Sciences* 10:3 (September 1987): 372–374.

3 Different versions of this story can be found in D. C. Dennett, *Content and Consciousness* (London: Routledge and Kegan Paul, 1969), Ch. 3, "Evolution in the Brain," and "Why the Law of Effect Will Not Go Away," *Journal of the Theory of Social Behavior*, 1975 (reprinted in *Brainstorms* [Cambridge, Massachusetts: Massachusetts Institute of Technology Press, 1978]); G. Edelman, *Neural Darwinism* (New York: Basic Books, 1987); J. P. Changeux and S. Dehaene, "Neuronal Models of Cognitive Functions" (forthcoming in *Cognition*); William H. Calvin, "The Brain as a Darwin Machine," *Nature* (1987), pp. 33–34.

4 Or it might not be a *virtual* machine at all. It might be a made-to-order hard-wired special-purpose real machine, such as a Lisp machine, which is a descendant of Lisp *virtual* machines, and which is designed right down to its silicon chips to run the programming language Lisp.

5 See, for example, Stephen Jay Gould, *The Panda's Thumb* (New York: Norton, 1980).

6 In *Elbow Room* (Cambridge, Massachusetts: Massachusetts Institute of Technology Press, 1984), pp. 38–43, I tell a "Just So Story" about the birth of this variety of consciousness, involving processes of talking to oneself and drawing diagrams for oneself, for instance. Many other tactics of self-manipulation are also plausible.

7 Sometimes the route of transmission is itself documented: On the American television program *Geraldo* (a morning talk show) the husband of a multiple (as one calls them in the trade) recalled his bafflement about his wife's behavior until he saw a report on MPD on *60 Minutes* (another American television program) and guessed, instantly, what was wrong with his wife. He got in touch with the therapist interviewed on *60 Minutes*, and soon this therapist had diagnosed and begun treating his wife. On need not suppose, skeptically, that the therapist *created* the condition of MPD in this woman; the patient probably already exhibited quite a robust and well-defined multiplicity. But there seems to be very

good reason to suppose that therapeutic encounters are often responsible for the typical explosion that creates the following striking statistic: The mean number of personalities in a well-studied sufferer from MPD is thirteen. See also Nicholas Humphrey and Daniel Dennett, "Speaking for Our Selves," *Raritan*, 1989; and F. W. Putnam, "The Clinical Phenomenology of Multiple Personality Disorder: Review of 100 Recent Cases," *Journal of Clinical Psychiatry* 47 (1986): 253–293.

[8] Just as the wheel is a fine bit of technology that is really quite dependent for its utility on rails or paved roads or other artificially planed surfaces, so the virtual machine that I am talking about can exist only in an environment that has not merely language and social interactions, but probably writing and diagramming as well—since the demands on memory and pattern recognition for its implementation require it to "off-load" some of its memories into buffers in the environment. (See David Rumelhart, *Parallel Distributed Processing* (Cambridge, Massachusetts: Massachusetts Institute of Technology Press, 1986), Ch. 14, for observations on this topic.)

[9] The following discussion owes a great deal to discussions with Nicholas Humphrey.

[10] Nietzsche offered a similar speculation a century ago: "thus it was that man first developed what was later called his 'soul.' The entire inner world, originally as thin as if it were stretched between two membranes, expanded and extended itself, acquired depth, breadth, and height, in the same measure as outward discharge was *inhibited*." Nietzsche, *Genealogy of Morals*, Essay 2, section 16, trans. Kaufman (Princeton: Princeton University Press, 1950), p. 84.

[11] Quoted in Calvin, p. 147.

[12] Nicholas Humphrey, *Consciousness Regained* (Oxford: Oxford University Press, 1984) and *The Inner I* (London: Oxford University Press, 1986).

[13] It is important to note that this innovation by three processes of natural selection of good software for the incompletely designed brain does not just enhance the individual's chances of coping, but increases the speed and efficiency of the first process of natural selection as well. This is known as the Baldwin effect: Thanks to the plasticity of design, species can be said to pretest the efficacy of particular designs by phenotypic (individual) exploration of the nearby possibilities. If a particularly winning setting is thereby discovered, organisms that are "closer" to it in the adaptive landscape will have a clear advantage over those more distant (and hence less likely to design themselves to the winning position during their lifetimes).

I mention the Baldwin effect because it has a clear counterpart in computer science in the principle that software design is faster and cheaper than hardware design. Thanks to their programmability, general purpose computers have permitted the cheap, swift exploration of varieties of computational architectures that would be extremely costly to build as hardware and then test. Once a design has proven its prowess as a software virtual machine, it sometimes pays to use it as the blueprint for a hard-wired machine. Lisp machines are a good case in point. And DEC computers are perhaps the beneficiaries of a true Baldwin effect—since they were selected for their capacity to turn themselves swiftly and effortlessly into Lisp virtual machines.

14 We might note, for instance, that the value we place on self-consciousness, and on being "reflective," while it might be independently justified, is certainly the result of advertisement by a self-serving or selfish meme. "The unexamined life is not worth living," Socrates proclaimed, thereby suggesting that one should acquire and foster any habits of self-examination that one encounters.

15 It is clearly arguable—but by no means certain—that its rival, MPD, does more harm than good to those who implement it; in any event, the proliferation of MPD through the culture does not in any direct way depend on its being of value, or even on its subjects' (mis)perceiving it as valuable.

16 David Rosenthal, in "Thinking That One Thinks" (paper presented at the Tufts Colloquium, 1988), develops the case for the initially counterintuitive thesis that conscious mental states are all those and only those that are the *objects of* second-order *unconscious* mental states. That is, to be conscious, one must be capable of thinking that one thinks, noticing that one is happy, noticing that one is daydreaming, and so forth, but these second-order mental states do not themselves need be conscious. Only their "perceptual" objects are thereby moved into the class of conscious mental states and events.

17 Directions to ward staff, from R. Kluft: *Directions in Psychiatry* vol. 4, lesson 24, "The Treatment of MPD: Current Concepts": 1986.

18 Douglas Hofstadter, *The Mind's I* (New York: Basic Books, 1981), p. 200.

THE JOURNAL OF PHILOSOPHY

VOLUME XCIII, NO. 5, MAY 1996

THE REDISCOVERY OF LIGHT

There is a family of seven arguments advanced by John Searle urging the ontologically distinct and physically irreducible nature of conscious phenomena. These are joined by three arguments from Frank Jackson and David Chalmers which tend to the same conclusion. My aim in what follows is to construct systematic and unitary analogs of all ten arguments, analogs that support a parallel family of antireductive conclusions about the nature of light. Since those analogous conclusions are already known to be false in the case of light (its physicalist reduction is one of the many triumphs of electromagnetic theory), it becomes problematic whether the integrity of the original family of antireductionist arguments is any greater than the purely specious integrity of their deliberately constructed analogs.

I. A SEARLE-LIKE FAMILY OF ARGUMENTS CONCERNING THE NATURE OF LIGHT

(A) A fundamental distinction:

original (intrinsic) visibilit· versus derivative (secondary) visibility

Only light itself has original visibility; for light alone is visible, when directed into the eyes, without the causal intervention of any mediating agent. By contrast, any physical object, physical configuration, or physical event is visible only when and only because light is somehow reflected from or emitted by that object, configuration, or event. Such physical items have at most *derivative* visibility, because they are utterly and forever *in*visible, save as they interact appropriately with the one thing that has *original* visibility, namely, light itself.

0022-362X/96/9305/211–28

121

These conclusions reflect the obvious fact that, if the universe contained no light at all, then absolutely nothing would be visible, neither intrinsically nor derivatively.[1]

(B) The original visibility of light marks it off as belonging to a *unique ontological category*, distinct in its essential nature from the essential nature of any physical phenomenon, which must always *lack* original visibility. In other words, for any physical object, configuration, or event, it is always a contingent matter whether or not it happens to be visible on this occasion (it is a matter of whether or not it happens somehow to be illuminated). By contrast, light itself is always and essentially visible. The ontology of light is an ontology of things and features that are uniquely accessible from the visual point of view.

This means that the phenomenon of light must be *irreducible* to any complex of purely physical or not-essentially-visible phenomena. You simply cannot get *original* visibility from things that have, at most, derivative visibility.[2]

(C) The consequence just reached is denied by a celebrated research program called *Strong EM*. This program claims not only that light can be "instructively simulated" by the behavior of interacting electric and magnetic fields (to which all may agree); it makes the stronger claim that light is actually *identical with* electromagnetic (EM) waves. The folly of Strong EM can be seen in the following obviously sound argument.

(1) Electricity and magnetism are physical forces.
(2) The essential nature of light is original visibility.
(3) Physical forces, no matter how they are deployed, are neither identical with, nor sufficient for, original visibility.

Therefore,

(4) Electricity and magnetism are neither identical with, nor sufficient for, light.

Premises (1) and (2) are obvious. That premise (3) is obvious can be seen by the following thought experiment. According to EM theory, an oscillating magnet or charged particle will generate an expanding sphere of oscillating EM fields: an EM wavefront. And by

---

[1] Cf. Searle, "Intrinsic Intentionality: Reply to Criticisms of Minds, Brains, and Programs," *Behavioral and Brain Sciences*, III (1980): 450-56, and *The Rediscovery of the Mind* (Cambridge: MIT, 1992), pp. 78-82.

[2] Cf. Searle, "Is the Brain's Mind a Computer Program?" *Scientific American*, CCLXII, 1 (January 1990): 26-31, here pp. 26-27; and *The Rediscovery of Mind*, pp. 93-95.

Figure 1. The "Luminous Room" thought-experiment. The room remains completely dark, despite the oscillating magnet. (Thanks to *Scientific American* for permission to reprint this figure.)

the same theory, this is strictly sufficient for the existence of light. But imagine a man in a pitch-black room who begins to pump a bar magnet back and forth (figure 1). Clearly, it will do nothing to illuminate the room. The room will remain wholly devoid of light.[3]

(D) The ontologically distinct nature of light is further reflected in the fact that the distinction between (visual) appearance and real-

[3] Cf. Searle, "Intrinsic Intentionality," pp. 417-57, and "Is the Brain's Mind a Computer Program?" pp. 26-31. This analogy was earlier deployed in Paul Churchland and Patricia Churchland, "Could a Machine Think?" in *Scientific American*, CCLXII, 1 (January 1990): 32-37. We did not then appreciate that the analogy was a member of a systematic and much larger family.

ity, which holds for any broadly physical phenomenon, cannot be drawn in the case of light itself. It there disappears. For while light is an agent that typically *represents* the physical objects, configurations, or events from which it has been differentially reflected or emitted, light does not represent *itself*. It is neither reflected nor emitted from itself. It thus cannot possibly *mis*represent itself, as it may occasionally misrepresent things other than itself from which it has been reflected or emitted. Accordingly, where the reality at issue is light itself (as opposed to any and all physical phenomena), the appearance just *is* the reality.[4]

(E) The irreducibility here claimed can be further seen as follows. Suppose we tried to say that the redness or blueness of light was *nothing but* a specific wavelength of EM waves. Well, if we tried such an ontological reduction, the essential features of the light would be left out. No description of the extrinsic wavelengths of EM waves could possibly convey the intrinsic character of (objective) visible redness and visible blueness, for the simple reason that the *visible* properties of light are distinct from the *physical* properties of EM waves. This argument is ludicrously simple and quite decisive.[5]

(F) Light is always and necessarily visible: there can be no such thing as *invisible* light. Granted, not all light is visible at any given time or place: light can be "shallowly" invisible to me simply because its path does not lead into my eyes. But if light exists at all, then there is some perspective from which it will be directly visible. Let us call this *the connection principle*, since it unites (*i*) being light and (*ii*) being accessible-from-the-visual-point-of-view.[6]

(G) Considerations (A)–(F) indicate that light is a phenomenon that is ontologically distinct from and irreducible to any purely physical phenomena. And yet, while nonphysical in itself, light is plainly *caused* by certain special physical phenomena, such as very high temperatures or the electrical stimulation of gases. Let us call our position here *nonreductive physical naturalism*: it holds that light is a natural (but irreducible) phenomenon caused to occur within certain special kinds of physical systems—specifically, within *self-luminous* objects, such as the sun, fires, and incandescent filaments. The aim of a scientific account of light should be to explain how such a

---

[4] Cf. Searle, *The Rediscovery of Mind*, p. 122, and "The Mystery of Consciousness: Part II," *The New York Review of Books*, XLII, 18 (November 16, 1995): 54-61, here p. 58.

[5] Cf. *The Rediscovery of Mind*, pp. 117-18.

[6] Cf. *ibid.*, pp. 132, 151-56.

nonphysical phenomenon is *caused* to occur within such highly special physical systems as stars and light bulbs.[7]

II. THREE JACKSON/CHALMERS-LIKE ARGUMENTS CONCERNING THE NATURE OF LIGHT

(H) In the study of the nature of light, there is a distinction to be drawn between the "easy" problems and "the hard problem." The first class concerns such problems as the emission, propagation, and absorption of light, its reflection and refraction, its velocity, its carrying energy, its self-interference, and so forth. These are all causal, relational, functional, and in general *extrinsic* features of light, features variously accessible by a wide variety of physical instruments and techniques; and it may well be that someday they will all be satisfactorily explained in terms of, for example, the propagation and interactions of EM fields.

But there remains a highly special *intrinsic* feature of light whose explanation must be found along some other path. This intrinsic feature is *luminance*, and it is what is responsible for the "original visibility" that is unique to light. Unlike all of the extrinsic (that is, physical) features of light listed above, luminance is unique in being epistemically accessible only from "the visual point of view."[8]

(I) We can illustrate and reinforce the contrast just drawn with a thought experiment about a physicist named Mary who is completely blind, but comes to know everything physical there is to know about EM waves, about their internal structure and their causal behavior. And yet, because she is blind and thus has no access at all to "the visual point of view," she cannot know about, she must remain ignorant of, the special intrinsic feature of light—luminance—which is accessible from that point of view alone. Evidently, even complete knowledge of the physical facts must still leave her ignorant of the nature of luminance. Luminance must therefore be, in some way, *non*physical.[9]

(J) As just illustrated, any possible physicalist story about the structure and causal functions of EM waves must still leave open an "explanatory gap" between the physical processes and luminance. In particular, it leaves unanswered the following question: Why should mutually-inducing electric and magnetic fields (for example) oscil-

---

[7] Cf. *ibid*, pp. 1, 89-93, 124-26; also Searle, "The Mystery of Consciousness: Part II," pp. 55-56.

[8] Cf. Chalmers, "The Puzzle of Conscious Experience," *Scientific American*, CCLXXIII, 6 (December 1995): 80–86, here pp. 81-82.

[9] Cf. Jackson, "Epiphenomenal Qualia," *Philosophical Quarterly*, XXXII, 127 (1982): 127-36, here p. 130; and Chalmers, "The Puzzle of Conscious Experience," pp. 81-82.

lating at a million billion Hertz and propagating at 300,000 km/sec ever give rise to the intrinsic feature of *luminance*? After all, we can easily imagine a universe that is filled with oscillating EM fields propagating back and forth all over the place, a universe that is nonetheless utterly *dark*, because it is devoid of the additional feature of luminance. We need to know how, when, and why oscillating EM fields *cause* the ontologically distinct feature of intrinsic luminance. Until we understand *that* mysterious causal relation, we shall never understand the ground and real nature of light.[10]

III. CRITICAL COMMENTARY

Concerning (A). As an exercise in term introduction ("original" visibility, and so on), this is strictly harmless, perhaps. But it falsely elevates an extremely peripheral feature of light—namely, its capacity to stimulate the idiosyncratic rods and cones of terrestrial animals—into a deep and presumptively defining feature of light. This is thrice problematic. First, it is arbitrarily selective. Second, it is strictly false that *only* light will stimulate rods and cones (charged particles of suitable energy will also do it, though at some cost to the retina). And third, infrared and ultraviolet light is quite *in*visible to terrestrial eyes. Our eyes evolved to exploit a narrow window of EM transparency in the earth's idiosyncratic atmosphere and oceans. Nothing of ontological importance need correspond to what makes our rods and cones sing.

Concerning (B). The dubious distinction legislated in (A) is here deployed to consign all physical phenomena to a class (things with merely derivative visibility) that *excludes* the phenomenon of light. This division certainly appeals to our default stereotype of a physical object (a tree, or a stone, has merely derivative visibility), but it begs the question against the research program of physicalism, because some unfamiliar physical things may indeed have original visibility, our common-sense expectations notwithstanding. As it turns out, EM waves with a wavelength between .4 and .7 μm are capable of stimulating the retina all by themselves, and thus have original visibility as defined in (A). The argument of (B) is thus a question-begging exploitation of superficial stereotypes and EM ignorance.

Concerning (C). The crucial premise of this argument (premise 3) may seem highly plausible to those who have a common-sense prototype of forces and who are ignorant of the details of EM theory, but it plainly begs the central question against physicalism. (Premise 3 is

---

[10] Cf. Chalmers, The Puzzle of Conscious Experience," pp. 82-83; also Searle, "The Mystery of Consciousness: Part II," pp. 55-56.

the direct denial of the basic physicalist claim.) Moreover, it is false. As mentioned in the preceding paragraph, EM waves of suitable wavelength *are* sufficient for original visibility. The "Luminous Room" thought experiment, concerning the oscillating bar magnet in the pitch-black parlor, is designed specifically to make premise 3 plausible, but that prejudicial story illegitimately exploits the fact that some forms of EM radiation have wavelengths that are simply too long to *interact* effectively with the rods and cones of terrestrial retinas (see figure 2 below). The darkened parlor may *look* to be devoid of light, but, thanks to the oscillating magnet, a very weak form of light is there regardless.

Concerning (D). While superficially plausible, perhaps, this argument refuses to take into account the many ways in which we can be mistaken or misled about the character of the light entering our eyes (for example, the light from a cinema screen appears continuous, but is really discontinuous at 36 frames/sec; the light of an incandescent automobile headlight, while really yellowish, looks white at night; and so forth). Its brief plausibility is a reflection of nothing more than our unfamiliarity with how light is perceptually apprehended and with how that intricate process can occasionally produce false perceptual beliefs. It is a reflection of our own ignorance, rather than of any unique ontological status had by light.

Concerning (E). This argument is sheer question-begging assertion rather than instructive argument. Whether objective properties of light such as spectral redness or spectral blueness are identical with, or distinct from, specific wavelengths of EM radiation is precisely what is at issue. And in this case, it has been plain for a century that these properties are identical. It is also plain that spectral redness, spectral blueness, and their various causal properties—their refractive and absorptive behavior, their velocity and interference effects—are positively *explained*, rather than impotently "left out," by their smooth reduction to EM features.

The point about what an EM vocabulary can or cannot "convey" about certain perceptual properties is a distinct point (and a red herring) to be dealt with below in "*Concerning* (I)."

Concerning (F). This argument also would be found plausible by someone still imprisoned by prescientific prototypes of light. *Invisible light* may well be a conceptual impossibility against the assumptions of the story just told, but we now know better. Indeed, we have learned that *most* light is invisible—and not just "shallowly" invisible, but permanently beyond human visual apprehension (see figure 2). Once again, we find ignorance being paraded as positive knowledge.

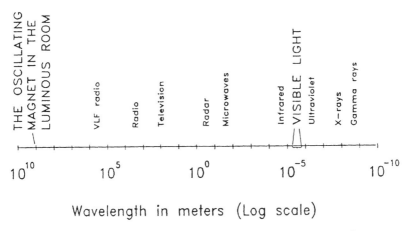

Wavelength in meters (Log scale)

Figure 2. The place of visible light in the extended EM spectrum, and, at an energy-level fourteen orders of magnitude lower, the place of the oscillating magnet's almost undetectable output.

Concerning (G). This summary attempts to find a proper place in nature for the phenomenon touted as ontologically distinct and physically irreducible in arguments (A)–(F). The place suggested is that of a nonphysical *causal* consequence of certain special but purely physical events.

Such a move threatens to violate well-established laws concerning the conservation of both energy and momentum, at least if light is presumed to have any causal powers of its own. But we need not enter into these matters here, for as the critical commentary to this point shows, there is no significant motivation for any such antireductionist research program in the first place. And in the second place, the proper place in nature of light has already been made clear: it has been smoothly and systematically reduced to EM waves.

Concerning (H). Light is here conceded to have a wide variety of physical features—its so-called "extrinsic" or "structural/functional" features—to which some sort of physical explanation is deemed appropriate. But light is also assigned an allegedly special or "intrinsic" feature, a feature that is epistemically accessible through vision, but not through the "structural/functional" stories to which current physical science (alas) is limited.

Once again, our prescientific noninferential epistemic access (namely, vision) to certain entirely physical properties is portrayed as a unique window onto an ontologically special domain. And to com-

pound the felony further, the potential reach of physical explanation is restricted, by arbitrary fiat at the outset (rather than by any
empirical failures revealed during the course of ongoing research),
so as inevitably to fall short of the so-called "intrinsic" features within
the "special" domain at issue.

The "hard problem" is thus *made* transcendently hard at the outset
by presumptive and question-begging fiat, rather than by any substantive considerations. As EM theory has taught us, *there is no* "hard
problem" here at all, and no defensible ontological distinction between intrinsic and extrinsic features. "Luminance"—if we concede
the integrity of this notion at all—is just the normal and entirely
physical capacity of EM waves to excite our own rods and cones (and
to induce chemical changes in photographic film, to free electrons
in a television camera, and so forth).

Concerning (I). This "knowledge" argument equivocates on 'knows
about'. It elevates two distinct modes of epistemic access to light
into a false dichotomy of distinct phenomena thereby accessed—
physical features by scientific description, and a special range of nonphysical features by normal human vision. But for light, at least, we
know perfectly well that there is only one thing here rather than two,
only one class of objective features rather than two.

What Blind Mary is missing is one common *form* of knowledge
about light: she lacks perceptual/discriminative knowledge of
light. And yet, people who have such knowledge are accessing the
very same features of reality that she is obliged to access in other
ways. The difference lies in the manner of the knowing, not in the
nature of the thing(s) known. It is true that no amount of propositional knowledge of light will ever *constitute* the visual apprehension of light, but that is entirely to be expected. They are different
forms of knowledge; they operate with different representational
"palettes" inside Mary's brain. But they both represent, each in
their own distinct way, one and the same entirely physical thing:
light.

Our contemporary scientific knowledge about light aside, one can
see immediately that the crucial divergence here is merely epistemic
rather than ontological (as the argument pretends). For while it is
indeed true that Blind Mary does not know what it is like to *see* spec-
tral-red light, it is equally true, and for exactly the same reasons, that
she does not know what it is like to *see* EM waves at .65 μm. The
deficit here evidently lies with Mary and her epistemic failings, not
with EM waves and their ontological shortcomings vis-à-vis light. For
Mary would continue to have her deficit even if light *were* (as it is)

identical with EM waves. Her deficit, therefore, can hardly weigh against that identity.

Concerning (J). This "open question" argument begs the question in favor of the ontological distinctness of "luminance," and then insists on our providing a *causal* account of how EM waves might produce it. This gets everything backward. We no longer have need for an account of how EM waves might "cause" the various phenomena associated with light, because the systematic reconstruction of optical phenomena within EM theory leads us to believe that light is simply identical with EM waves, and that the assembled properties of light are identical with, rather than caused by, the corresponding properties of EM waves.

The conceivability of a dark universe filled with EM waves shows only that the various cross-theoretic identities motivated by the EM reduction are, as they should be, contingent rather than necessary identities. It should also be pointed out that such an "open question" argument will be maximally appealing to one who is minimally instructed in EM theory. This is because the more one learns about EM waves, about their effects on matter in general and on our eyes in particular, the *harder* it becomes to imagine a consistent scenario in which a universe abuzz with EM waves of all wavelengths remains dark even so. Here, as in so many of the earlier arguments, the audience's presumed ignorance is once more a lubricant that smooths the path of a worthless argument.

This concludes my attempt to construct, and to deflate, a systematic analog for the family of arguments currently so influential in the philosophy of mind. My point, of course, is that the family of arguments on which they are modeled is just as empty of real virtue.

IV. A FINAL NAGEL/SEARLE ARGUMENT FOR IRREDUCIBILITY

A question will inevitably arise over the fairness of the global analogy deployed above. In particular, it will be complained that the global analogy is faulty in placing the *objective* properties of "original visibility" and "luminance" in the role played by the *subjective* properties of original intentionality and inner qualia in the arguments under attack.

The analogy deployed does indeed proceed in precisely this fashion, but this assimilation is the central point of the exercise. It should at least give us pause that the original family of arguments can be collectively and successfully mirrored in a ten-dimensional analogy that deliberately and self-consciously concerns "objective" features. After all, if the analog arguments are at all compelling—and to the electromagnetically uninformed, they will be—then the essential appeal of both families of arguments presumably derives from something other than the unique status of the "subjective."

Second, there is no mystery about what drives the plausibility of the analog arguments. It is the ignorance-fueled appeal of the idea that the epistemic modality of vision is or might be a unique window onto an ontologically distinct class of properties. But in the case of light it is also plain, at least in retrospect, that nothing substantive motivates that repeated insistence. We have to wonder if the same failure might be true of the original family of arguments. After all, and whatever else it might be, introspection *is* an epistemic modality, or perhaps a family of them. And while it may have its own quirks and distinguishing profile, it is entirely unclear whether it, alone among all of our epistemic modalities, constitutes a window onto a unique *ontological* domain of nonphysical properties. None of our other epistemic modalities has any such distinction: they all access some aspect or other of the purely physical world. Why should introspection be any different?

Searle has a further argument, unaddressed to this point, whose burden is to illustrate the ontological cleft he sees between the domain of "outer sense" and the domain of "inner sense," as Immanuel Kant called them. Searle's argument here appeals, uncharacteristically, to the history of science. The argument originally appeared, very briefly, in Thomas Nagel,[11] but Searle has more recently developed it in detail.

Premise (1). We must draw a distinction between the real and objective properties of objects and the contingent subjective effects those properties happen to have on the conscious processes of humans. For example, objective heat (molecular KE) is one thing; the subjective feeling of warmth in humans *produced by* objective heat is quite another.

Premise (2). The scientific reduction of observable phenomena typically ignores or "carves off" their contingent subjective effects on the conscious processes of humans, and reduces only the nonsubjective aspects of the phenomena. (For example, kinetic theory successfully reduces objective heat to molecular KE, but leaves its *subjective conscious effects* on humans aside. EM theory successfully reduces objective spectral colors to different wavelengths of EM radiation, but leaves their *subjective conscious effects* on humans aside. And so forth.)

Premise (3). When we attempt to provide a physicalistic reduction of those subjective conscious effects themselves, we must realize that

---

[11] "What Is It Like to Be a Bat?" *Philosophical Review*, LXXXIII, 4 (1974): 435-50.

here we *cannot* "carve off" their subjective-effects-on-us from their objective properties, and reduce only the latter, because it is precisely those subjective-effects-on-us which we wish to understand. Here, inside the mind, there is no longer any meaningful or defensible distinction between the "objective" and the "subjective" which would allow us to repeat the pattern of reduction described above. The subjective phenomena are exclusively and *essentially* subjective. Any alleged "reduction" would simply leave out what is essential to their nature.

Therefore, mental phenomena are irreducible to physical phenomena. The proper pattern of a physicalist reduction (an "objective"-to-"objective" mapping) uniquely precludes any reduction of the subjective.[12]

What is going on here? Simply this. The Nagel/Searle argument treats a contingent, minor, and remediable feature (of a handful of historical examples of reductions) as if it were a necessary, central, and permanent feature of any possible physicalistic reduction. Specifically, the merely contingent feature that is paraded as essential is the feature: *leaves aside the-effects-on-human-consciousness* (the "C-effects," for short). The argument then points out that this "essential" feature of physicalistic reduction precludes any such reduction in the unique case of C-effects themselves, since "leaving the C-effects aside" is here not an option.

It is indeed true that historical property reductions pay little or no attention to, or provide us with little or no insight into, the C-effects of the various phenomena being reduced. Searle and Nagel seem antecedently convinced that this historical fact is the inevitable reflection of an ontological gulf already fixed between "objective" phenomena and "subjective" phenomena.

That is one (distant) possibility. But there is an obvious alternative explanation of why physicalistic reductions so regularly leave out any account of the human C-effects of the phenomena being reduced, as the historical reduction of heat to molecular energy made no attempt to account for the subjective sensation of warmth, or as the historical reduction of light to EM waves made no attempt to account for the subjective sensation of redness.

The obvious alternative explanation is that *such C-effects are the proper province of a distinct science,* a science such as cognitive neurobiology or computational neuroscience. Searle is wrongly demanding that the kinetic theory of heat do, all by itself, something that clearly

---

[12] Cf. *ibid.*, p. 437; also Searle, *The Rediscovery of Mind*, pp. 116-24.

requires, in addition, an adequate theory of the brain. The fact is, during the late nineteenth century, we were too ignorant about neurobiology for the kinetic theory to suggest any worthwhile hypotheses about the human C-effects of molecular energy. It is no surprise, then, that physicists simply walked past that arcane problem, if it ever occurred to them to address it in the first place. The same is true for the EM theory of light and the problem of our subjective sensations of redness.

Accordingly, this incidental "leaving-aside" need have no metaphysical or ontological significance. This deflationary view is further encouraged by the fact that physicalistic reductions such as the kinetic theory also "leave aside" any explanatory account of millions of other phenomena, so there is no automatic reason to find any special significance in its ignoring of human C-effects in particular. If I may give several examples, historical reductions of heat typically leave aside any attempt to account for:

> heat's effect on Antarctic anchovy production
> heat's effect on bluebird-egg cholesterol levels
> heat's effect on pneumonial infections
> heat's effect on the Gross National Product of Peru
> heat's effect on the rotting of vegetable matter
> heat's effect on the conscious states of humans
> (this list is extendible indefinitely)

The great reductions of classical and modern physics *typically* leave out any account of heat's (or light's, or sound's) effect on all of these things, and of millions more, because no reduction all by itself can presume to account for the ever-more-distant causal effects of its proprietary phenomena as they are progressively articulated into all possible causal domains. There are far too many domains, and causal understanding of the phenomena within those other domains will typically require the resources of further theories in addition to the theory that achieves the local reduction at issue.

It is in no way noteworthy or ontologically significant, then, that the kinetic theory of heat, all by itself, provides no account of any of the arcane phenomena listed above, nor of millions of others as well. In particular, it is neither noteworthy nor ontologically significant that the kinetic theory of heat provides no account of the human conscious response to heat. This marginal and idiosyncratic phenomenon has no more ontological significance than any of the other arcane phenomena just listed. And they all require the resources of theories beyond the kinetic theory of heat to address them adequately. Specifically, heat's effect on

anchovy production needs ecology
egg cholesterol levels needs metabolic chemistry
pneumonia needs immunology and bacteriology
Peru's Gross National Product needs biology and economics
vegetable rotting needs bacteriology and cell chemistry
human conscious experience needs cognitive neurobiology

My counterclaim, then, against Nagel and Searle, is that it is *not* an essential feature of physicalistic reductions that they always "leave aside" human C-effects, or any of the many other effects cited. It is a merely contingent and wholly explicable fact that historical reductions have so far done so. It is not an essential pattern that all physicalistic reductions are doomed-by-nature to follow, nor is it a self-imposed definitional stipulation on what counts as a reduction, as Searle[13] at one point inexplicably suggests. Once we begin to address human C-effects with some appropriately focused science—as neuronal vector-coding theories are already doing, with striking success[14]—then that earlier "pattern" will be well and truly broken. For that pattern reflected only our own scientific ignorance, not some ontological division in nature.

In sum, human conscious experience has no quicksilver history of darting off to one side each time our reductive scientific thumb has tried to pin it down. There have *been* no significant reductive attempts at that target, not, at least, within the grand historical reductions of physics and chemistry. Instead, the phenomena of human conscious experience have quite properly been waiting, patiently and at the sidelines, for the maturation of the only theory that has any realistic hope of providing such a reductive account, namely, an adequate theory of the brain. If and when *that* approach has been fully tried, and proves a failure, *then*, perhaps, it will be time to insist on nonphysical approaches.

---

[13] *The Rediscovery of Mind*, pp. 124, 112-16. Though here is not the place to mount a systematic criticism, it must be said that Searle's 1992 sketch of the nature and varieties of reduction muddies far more than it clarifies. First, it wrongly assimilates ontological reduction to ontological elimination. Second, there simply is no further category or "half-way house"—Searle's so-called "causal reduction"—distinct from ontological reduction. And third, as we just saw, the account attempts to stipulate the closure of certain empirically open questions. To a neutral philosopher of science, Searle's account will appear more as a reflection of his peculiar intuitions in the philosophy of mind rather than as an independently motivated attempt to account for the full range of cases throughout the history of science.

[14] See Austen Clark, *Sensory Qualities* (New York: Oxford, 1994).

Both the appeal to ignorance and the question-begging nature of the Nagel-Searle argument become finally vivid if one plays at constructing a series of parallel arguments to "establish" the physicalistic irreducibility of whatever arcane, complex, and puzzling phenomenon one might choose to consider (something from the preceding list, for example). Simply note that historical reductions of various important phenomena have invariably left that particular phenomenon aside as an unaddressed mystery; pretend that this is an essential pattern, a reflection of an antecedent metaphysical division, or the result of some appropriately exclusive definition of "reduction"; note that said leave-aside pattern (surprise!) precludes any similar reduction of *exactly* the phenomenon at issue; and you are home free. You will then have performed for us the same empty service that Nagel and Searle have performed.

V. SOME DIAGNOSTIC REMARKS ON QUALIA

There is a chronic temptation among philosophers to assign a special epistemological, semantical, or ontological status to those features or properties which form the "discriminational simples" within each of our several sensory or epistemic modalities, such as brightness and colors in the case of vision, sweetness and sourness in the case of taste, and so on. These are the features of the world where one is unable to *say* how it is that one discriminates one such feature from another; one simply can. As well, one is unable to *say* how the meaning of 'red' differs from the meaning of 'green'; one simply has to point to appropriate exemplars.

Such discriminational simples are typically contrasted with properties, such as "being a horse," where one can usually articulate the more elemental constituting features that make up the type in question: size, shapes, configuration, color, texture, and so forth, which more elemental features lead us stepwise back toward the discriminational simples.

Too much has been made of these "simples," for the existence of such discriminable but inarticulable features is entirely inevitable. Such features must exist, if only to prevent an infinite regress of features discriminated by constituting subfeatures discriminated by constituting *sub*-subfeatures, and so on.[15] And their existence is inevitable even on wholly physicalist conceptions of cognition. It simply cannot be the case that *all* conscious feature discriminations are made on the basis of distinct conscious (sub)feature discriminations.

---

[15] See Mary Hesse, "Is There an Independent Observation Language?" in Robert Colodny, ed., *The Nature and Function of Scientific Theories* (Pittsburgh: University Press, 1970), pp. 35-77.

Given any person at any time, there must be some set of features whose spontaneous or noninferential discrimination is currently basic for that person, a set of features whose discrimination does not depend on the conscious discrimination of any more elemental perceptual features. In short, there must be something that counts, for that person, as a set of inarticulable qualia.

Accordingly, we should not be tempted to find anything physically irreducible or ontologically special about such inarticulable features. They need reflect nothing more than the current and perhaps changeable limits of the person's capacity for epistemic and semantic articulation, the current limits, that is, of the person's *knowledge* of the world's fine structure and his own epistemic access to it. Most importantly, there is no reason to expect that the current limits of the typical person's knowledge must mark the boundary of a distinct ontological domain. This is just as true, note, for the epistemic modalities that underwrite (what we loosely call) "introspection" as it is for the epistemic modalities of vision, taste, and audition.

And yet, philosophers have regularly been tempted here, some beyond redemption. Bishop Berkeley rejected the identification of sound with atmospheric compression waves; William Blake and Johann Wolfgang Goethe rejected the identification of light with Isaac Newton's ballistic particles; and Nagel, Jackson, Searle, and Chalmers reject the proposed reduction of inner qualia to physical states of the brain.

There is an important factor here that may help to explain why such features have so frequently been held to be beyond the reach of any physicalist reduction. Specifically, any reduction succeeds by reconstructing, within the resources of the new theory, the antecedently known nature, structure, and causal properties of the target phenomena. That is what intertheoretic reduction is. But if the target phenomena, such as sensory qualia, are features whose internal structure (if any) we are currently unable to articulate, and whose causal properties (if any) are largely unknown to us, then the target phenomena will inevitably seem to offer *the minimum purchase possible* for any aspirant reducing theory. They will display no structure worth reconstructing. They will present themselves as smooth-walled mystery. They will appear to be irreducible to any "structural/functional" theory from conventional science.

But the appearance of seamless simplicity need reflect nothing more than our own ignorance, an ignorance, we should note, that already holds promise of repair. In sum, we should not be too quickly impressed by qualia, whether outer or inner. If cognitive creatures exist at all, then the existence of inarticulable qualia is inevitable, even in a purely physical universe.

If ultimately they are physical, then inner qualia ought to be epistemically accessible from more than just the first-person or "subjective" point of view; they ought to be accessible as well from one or more "objective" points of view, via some appropriate instruments that scan brain activity, for example.

Some will continue to find this implausible on its face. That is mainly because the terms 'objective' and 'subjective' are commonly used in mutually *exclusive* contrast. But the default implication of mutual exclusivity may well be inappropriate in precisely the case at issue. After all, we know that the two epistemic modalities of vision and touch, for example, are not mutually exclusive in the phenomena that they access—one can both see and feel the shape of an object, see and feel that the sun is out, see and feel that rain is falling, and so forth. Why should it be impossible a priori that the epistemic modality we call "introspection" have some similar overlap with one or more of our other epistemic modalities?

Indeed, such overlap appears actual, even by the standards of common sense. One can tell by introspection that one's own bladder is full, but an ultrasound image will tell anyone the same thing. One can tell by introspection that and where one's retinal cells are photo-fatigued (we call it an "after image"), but that too is accessible by nonsubjective means. One can tell by introspection that the cochlear cells of one's inner ear are firing randomly (the condition is called "tinnitus"), but others can access their behavior instrumentally. There are, of course, thousands more such examples.

It would seem, then, that the "subjective" and the "objective" are not mutually exclusive after all. In at least some cases, one and the same (physical) state can be known both subjectively *and* objectively, from both the first-person perspective *and* the third-person perspective. Further, it would seem that the extent and location of the overlap is somewhat fluid, and that it varies as a function of how much background knowledge, conceptual sophistication, and recognitional skill the person has acquired. The process is called "coming to understand explicitly what was hitherto inarticulate," and it is entirely to be encouraged. The more epistemic modalities we can bring to bear on any puzzling phenomenon, the deeper our understanding will become. To insist, in *advance* of real understanding, that a given phenomenon is locked forever within its own epistemic box serves only to block the very research that might dissolve such a prejudicial conception.

VI. A FINAL POINT ABOUT LIGHT

In closing, let me return to the opening family of arguments concerning the irreducibility of light. Someone may remark that, with light, I have used an example that is antithetical to my own reductive inclinations in the philosophy of mind. For while light reduces cleanly to EM waves, light is still famous for having escaped the various *mechanical* reductions (ballistic particle theories, elastic media theories) that everyone in the nineteenth century expected. And it is still famous for having thus emerged as one incarnation of a fundamental and nonmechanical aspect of reality: electromagnetism.

This is quite true, and more than a little instructive. But in the present context it is also instructive (1) that while nonmechanical, light remains an entirely physical phenomenon, and, (2) more importantly, that the modestly special status that light eventually discovered had *absolutely nothing to do* with any of the considerations urged in the family of antireductive arguments in my opening parody. Light's nonmechanical status emerged primarily as a consequence of Special Relativity, as a consequence of the unity of space-time and the impossibility of a universal elastic aether. It was not a consequence or reflection of any of the arguments offered above. It is ironic that, even though light did turn out, unexpectedly, to be a rather special kind of physical phenomenon, the parody-arguments (A)–(J) did nothing whatever to herald it, and they are, after the fact, quite irrelevant to it.

The parallel lesson about mental states is that, even if conscious phenomena are ontologically special in some way, roughly analogous to the case of light, there is no reason to think that the arguments of Searle, Jackson, and Chalmers do anything to illustrate or establish it. Those arguments are no more instructive about the ultimate nature of mental phenomena than arguments (A)–(J) are instructive about the ultimate nature of light.

PAUL M. CHURCHLAND

University of California/San Diego

BEHAVIORAL AND BRAIN SCIENCES (1990) 13, 585–642

Consciousness, explanatory inversion, and cognitive science

John R. Searle

Department of Philosophy, University of California, Berkeley, CA 94720
Electronic mail: searle@cogsci.berkeley.edu

Abstract: Cognitive science typically postulates unconscious mental phenomena, computational or otherwise, to explain cognitive capacities. The mental phenomena in question are supposed to be inaccessible in principle to consciousness. I try to show that this is a mistake, because all unconscious intentionality must be accessible in principle to consciousness; we have no notion of intrinsic intentionality except in terms of its accessibility to consciousness. I call this claim the "Connection Principle." The argument for it proceeds in six steps. The essential point is that intrinsic intentionality has aspectual shape: Our mental representations represent the world under specific aspects, and these aspectual features are essential to a mental state's being the state that it is.

Once we recognize the Connection Principle, we see that it is necessary to perform an inversion on the explanatory models of cognitive science, an inversion analogous to the one evolutionary biology imposes on preDarwinian animistic modes of explanation. In place of the original intentionalistic explanations we have a combination of hardware and functional explanations. This radically alters the structure of explanation, because instead of a mental representation (such as a rule) causing the pattern of behavior it represents (such as rule-governed behavior), there is a neurophysiological cause of a pattern (such as a pattern of behavior), and the pattern plays a functional role in the life of the organism. What we mistakenly thought were descriptions of underlying mental principles in, for example, theories of vision and language were in fact descriptions of functional aspects of systems, which will have to be explained by underlying neurophysiological mechanisms. In such cases, what looks like mentalistic psychology is sometimes better construed as speculative neurophysiology. The moral is that the big mistake in cognitive science is not the overestimation of the computer metaphor (though that is indeed a mistake) but the neglect of consciousness.

Keywords: brain function; cognition; computation; consciousness; explanation; Freud; grammar; intentionality; introspection; language; perception; unconscious processes

1. Introduction.

Ten years ago in this journal I published an article (Searle 1980a; 1980b) criticizing what I call "strong AI," the view that for a system to have mental states it is sufficient for the system to implement the right sort of program with the right inputs and outputs. Strong AI is rather easy to refute; the basic argument can be summarized in one sentence: A system, me for example, could implement a program for understanding Chinese, for example, without understanding any Chinese at all. This idea, when developed, became known as the Chinese Room Argument.

There is a lot more to be said, and I have tried to say some of it. The debate has raged on for ten years and still continues. Though many interesting points have emerged, I think the original argument is quite decisive. In my view, "weak AI," the use of computers to model or simulate mental processes, is untouched by this argument. Weak AI continues and, in its connectionist incarnation, at least, it is flourishing. But strong AI is now primarily of historical interest, though it survives as a sociological phenomenon.

At the time I wrote that article, I thought the major mistake we were making in cognitive science was to think that the mind is a computer program implemented in the hardware of the brain. I now believe the underlying mistake is much deeper: We have neglected the cen-

trality of consciousness to the study of the mind. In this article I will have nothing more to say about the Chinese Room Argument. I assume that it refutes strong AI, but nothing that follows depends on that assumption.

If you come to cognitive science, psychology, or the philosophy of mind with an innocent eye, the first thing that strikes you is how little serious attention is paid to consciousness. Few people in cognitive science think that the study of the mind is essentially or in large part a matter of studying conscious phenomena; consciousness is rather a "problem," a difficulty that functionalist or computationalist theories must somehow deal with. Now, how did we get into this mess? How can we have neglected the most important feature of the mind in those disciplines that are officially dedicated to its study? There are complicated historical reasons for this, but the basic reason is that since Descartes, we have, for the most part, thought that consciousness was not an appropriate subject for a serious science or scientific philosophy of mind. As recently as a few years ago, if one raised the subject of consciousness in cognitive science discussions, it was generally regarded as a form of bad taste, and graduate students, who are always attuned to the social mores of their disciplines, would roll their eyes at the ceiling and assume expressions of mild disgust.

When consciousness is no longer regarded as a suitable topic for scientific discussion, then something else must take its place and in this case that is obviously the unconscious. The idea is that there are unconscious processes, these account for our cognitive capacities, and the task of cognitive science is to lay bare their structure. But what exactly is the notion of the unconscious used in these discussions? Since Freud (1915) we have grown so used to talking about the unconscious that we have ceased to regard it as problematic. The naive notion of the unconscious that we have inherited from Freud is that unconscious mental states are just the same as conscious mental states only minus the consciousness. But what exactly is that supposed to mean? Pretheoretically, I believe most people follow Freud in thinking that an unconscious mental state has exactly the same shape it has when conscious. Naively, we tend to think of unconscious mental states like furniture stored in the dark attic of the mind or like fish deep beneath the surface of the sea. The furniture and the fish have the same shape when invisible that they have when visible; it is just that it is impossible to see them in their unconscious form. Furthermore, somewhere between Freud and Chomsky an important shift took place: Freud apparently regarded unconscious mental states as at least potentially conscious. They may be, for one reason or another, too deeply repressed for the patient to bring them to consciousness without professional assistance, but there is nothing in principle inaccessible to consciousness about the Freudian unconscious. When we get to more recent writers, however, it turns out that many of the mental processes which their theories postulate are in principle inaccessible to consciousness.[1] And it is but a short step from the belief that we are dealing with unconscious mental processes to the belief that these processes are computational, and to the belief that the mind is a computer program operating in the hardware or wetware of the brain.

But, to repeat, what exactly is the conception of unconscious mental processes which is supposed to account for, say, visual information processing or language understanding? How, for example, are we to distinguish between unconscious mental phenomena in the brain and those nonconscious phenomena in the brain which are not mental at all, but are just blind, brute, neurophysiological states and processes? How do we distinguish, for example, between my unconscious belief that Denver is the capital of Colorado when I am not thinking about it and the nonconscious myelination of my axons? Both are features of my brain, but one is in some sense mental and the other is not. The answer to this question is by no means obvious.

2. The Connection Principle

This paper has two aims. First I want to show that we have no notion of an unconscious mental state except in terms of its accessibility to consciousness. Second, I will attempt to lay bare some of the implications of this thesis for the study of the mind.

The first task is to demonstrate what I will call the "Connection Principle."[2] It can be stated in a preliminary fashion as follows: The ascription of an unconscious intentional phenomenon to a system implies that the phenomenon is in principle accessible to consciousness. I leave some of this deliberately vague, particularly what is meant by "in principle," but I will try to transform this vagueness into clarity in the course of discussion. To

substantiate this claim, I will also have to explore the notion of an unconscious mental state.

Before launching into the argument, I need to remind the reader of the distinctions between ontology, causation and epistemology. For any phenomenon, but for biological phenomena especially, we need to know:

1. What is its mode of existence? (ontology)
2. What does it do? (causation)
3. How do we find out about it? (epistemology)

So, for example, if we were examining the heart, the answer to our three questions would be (1) the heart is a large piece of muscle tissue located in the chest cavity (ontology), (2) the heart functions to pump blood throughout the body (causation), and (3) we find out about the heart indirectly through such methods as using stethoscopes, cardiograms and taking pulse, and directly by opening up the chest cavity and looking at the heart (epistemology). These distinctions also apply to both conscious and unconscious mental states.

The argument for the Connection Principle is in six steps.

Step 1. *There is a distinction between intrinsic and as-if intentionality.*

We often make metaphorical attributions of intentionality to systems where the system does not literally have that intentional state or, indeed, any mental life at all. I can, for example, say of my lawn that it is thirsty, just as I can say of myself that I am thirsty. But it is obvious that my lawn has no mental states whatever. When I say that it is thirsty, this is simply a metaphorical way of describing its capacity to absorb water. Whereas when I say "I am thirsty" I am describing an intrinsic mental state in me. We often make such metaphorical attributions of "as-if" intentionality to artifacts: We say such things as, "The carburetor of the car *knows* how rich to make the mixture."; "The thermostat on the wall *perceives* changes in the temperature."; "The calculator *follows rules* of arithmetic when it does addition,", and "The Little Engine That Could is *trying very hard* to make it up the mountain." None of these various attributions is meant to be taken literally, however. To mark this distinction, I say of my thirst that it is a form of intrinsic intentionality, but I say of the attribution of thirst to my lawn that it is only a metaphorical or *as-if* attribution. *As-if*, strictly speaking, is not a type of intentionality, but a type of attribution. (For an extended discussion of this issue see Searle 1984b.)

I have seen efforts to deny this distinction, but it is very hard to take them seriously. If you think there is no principled difference, you might consider the following from the journal *Pharmacology*:

Once the food is past the crico-pharyngus sphincter, its movement is almost entirely involuntary except for the final expulsion of feces during defecation. *The gastrointestinal tract is a highly intelligent organ that senses* not only the presence of food in the lumen but also its chemical composition, quantity, viscosity and adjusts the rate of propulsion and mixing by producing appropriate patterns of contractions. *Due to its highly developed decision making ability,* the gut wall comprised of the smooth muscle layers, the neuronal structures and paracrine-endocrine cells *is often called the gut brain.* (Sarna & Otterson 1988, p. 8, my italics)[3]

This is clearly a case of as-if intentionality in the "gut

brain." Now does anyone think there is no principled difference between the gut brain and the brain brain? I have heard it said that both sorts of cases are the same; that it is all a matter of taking an "intentional stance" toward a system. But just try in real life to suppose that the "perception" and the "decision making" of the gut brain are no different from the real brain.

This example reveals, among other things, that any attempt to deny the distinction between intrinsic and as-if intentionality faces a general *reductio ad absurdum*. If you deny the distinction it turns out that everything in the universe has intentionality. Everything in the universe follows laws of nature, and for that reason everything behaves with a certain degree of regularity, and for that reason everything behaves *as if* it were following a rule, trying to carry out a certain project, acting in accordance with certain desires, and so on. For example, suppose I drop a stone. The stone *tries* to reach the center of the earth, because it *wants* to reach the center of the earth, and in so doing it *follows the rule* $S = \frac{1}{2} g t^2$. The price of denying the distinction between intrinsic and as-if intentionality, in short, is absurdity, because it makes everything in the universe mental.

No doubt there are marginal cases. About grasshoppers or fleas, for example, we may not be quite sure what to say. And no doubt, even in some human cases we might be puzzled as to whether we should take the ascription of intentionality literally or metaphorically. But marginal cases do not alter the distinction between those facts corresponding to ascriptions of intrinsic intentionality and those corresponding to as-if, metaphorical ascriptions of intentionality. There is nothing harmful, misleading, or philosophically mistaken about as-if metaphorical ascriptions. The only mistake is to take them literally.

Step 2. *Intrinsic intentional states, whether conscious or unconscious, always have aspectual shapes.*

I am introducing the term of art, "aspectual shape," to mark a universal feature of intentionality. It can be explained as follows: Whenever we perceive anything or think about anything, it is always under some aspects and not others that we perceive or think about that thing. These aspectual features are essential to the intentional state; they are part of what makes it the mental state that it is. Aspectual shape is most obvious in the case of conscious perceptions; think of seeing a car, for example. When you see a car it is not simply a matter of an object being registered by your perceptual apparatus; rather, you actually have a conscious experience of the object from a certain point of view and with certain features. You see the car as having a certain shape, as having a certain color, and so forth. And what is true of conscious perceptions is true of intentional states generally. A man may believe, for example, that the star in the sky is the Morning Star without believing that it is the Evening Star. A man may, for example, want to drink a glass of water without wanting to drink a glass of H_2O. There is strictly an infinite number of true descriptions of the Evening Star and of a glass of water, but something is believed or desired about them only under certain aspects and not under others. Every belief and every desire, and indeed every intentional phenomenon, has an aspectual shape.

Notice also that the aspectual shape must matter to the agent. It is, for example, from the agent's point of view

that he can want water without wanting H_2O. In the case of conscious thoughts, the way the aspectual shape matters is that it constitutes the way the agent thinks about or experiences a subject matter: I can think about my thirst for a drink of water without thinking at all about its chemical composition. I can think of it *as* water without thinking of it *as* H_2O.

It is reasonably clear how this works for conscious thoughts and experiences, but how does it work for unconscious mental states? One way to get at our question is to ask what fact about an unconscious mental state makes it have the particular aspectual shape that it has, that is, what fact about it makes it the mental state that it is?

Step 3. *The aspectual feature cannot be exhaustively or completely characterized solely in terms of third person, behavioral, or even neurophysiological predicates. None of these is sufficient to give an exhaustive account of aspectual shape.*

Behavioral evidence concerning the existence of mental states, including even evidence concerning the causation of a person's behavior, no matter how complete, always leaves the aspectual character of intentional states underdetermined. There will always be an inferential gulf between the behavioral *epistemic* grounds for the presence of the aspect and the *ontology* of the aspect itself.

A person may indeed exhibit water-seeking behavior, but any water-seeking behavior will also be H_2O-seeking behavior. So there is no way the behavior, construed without reference to a mental component, can constitute wanting water rather than wanting H_2O. Notice that it is not enough to suggest that we might get the person to respond affirmatively to the question, "Do you want water?" and negatively to the question, "Do you want H_2O?" because the affirmative and negative responses are themselves insufficient to fix the aspectual shape under which the person interprets the question and the answer. There is no way just from the behavior to determine whether the person means by "H_2O" what I mean by "H_2O" and whether the person means by "water" what I mean by "water." No set of behavioral facts can constitute the fact that the person represents what he wants under one aspect and not under the other. This is not an epistemic point.

It is equally true, though less obvious, that no set of neurophysiological facts under neurophysiological descriptions constitutes aspectual facts. Even if we had a perfect science of the brain, and even if such a perfect science of the brain allowed us to put our brain-o-scope on the person's skull and see that he wanted water but not H_2O, all the same there would still be an inference; we would still have to have some lawlike connection that would enable us to infer from our observations of neural architecture and neuron firings that they were realizations of the desire for water and not of the desire for H_2O.

Since the neurophysiological facts are always causally sufficient for any set of mental facts,[4] someone with perfect causal knowledge might be able to make the inference from the neurophysiological to the intentional at least in those few cases where there is a lawlike connection between the facts specified in neural terms and the facts specified in intentional terms. But even in

141

these cases, if there are any, there is still an *inference*, and the specification of the neurophysiological in neurophysiological terms is not yet a specification of the intentional.

Step 4. *But the ontology of unconscious mental states, at the time they are unconscious, consists entirely in the existence of purely neurophysiological phenomena.*

Imagine that a man is in a sound dreamless sleep. Now, while he is in such a state it is true to say of him that he has a number of unconscious mental states. For example, he believes that Denver is the capital of Colorado, Washington is the capital of the United States, and so on; but *what fact about him makes it the case that he has these unconscious beliefs?* Well, the only facts that could exist while he is completely unconscious are neurophysiological facts. The only things going on in his unconscious brain are sequences of neurophysiological events occurring in neuronal architectures. At the time when the states are totally unconscious there is simply nothing there except neurophysiological states and processes.

Now we seem to have a contradiction, however: The ontology of unconscious intentionality consists entirely in third person, objective, neurophysiological phenomena, but all the same the states have an aspectual shape that cannot be constituted by such facts, because there is no aspectual shape at the level of neurons and synapses.

I believe there is only one solution to this puzzle. The apparent contradiction is resolved by pointing out that:

Step 5. *The notion of an unconscious intentional state is the notion of a state that is a possible conscious thought or experience.*

There are plenty of unconscious mental phenomena, but to the extent that they are genuinely *intentional* they must in some sense preserve their aspectual shape even when unconscious; but the only sense we can give to the notion that they preserve their aspectual shape when unconscious is that they are possible contents of consciousness.

This is our first main conclusion. But this answer to our first question immediately gives rise to another question: What is meant by "possible" in the previous two sentences? After all, it might be quite *impossible* for the state to occur consciously, because of brain lesion, repression, or other causes. So in what sense exactly must it be a possible content of a thought or experience? This question leads to our next conclusion, which is really a further explanation of Step 5, and is implied by 4 and 5 together:

Step 6. *The ontology of the unconscious consists in objective features of the brain capable of causing subjective conscious thoughts.*

When we describe something as an unconscious intentional state we are characterizing an objective *ontology* in virtue of its *causal* capacity to produce consciousness. But the existence of these causal features is consistent with the fact that in any given case their causal powers may be blocked by some other interfering causes, such as psychological repression or brain damage.

The possibility of interference by various forms of pathology does not alter the fact that any unconscious intentional state is the sort of thing that is in principle accessible to consciousness. It may be unconscious not only in the sense that it does not *happen* to be conscious then and there, but also in the sense that for one reason or another the agent simply *could not* bring it to consciousness; but it must be the *sort of thing* that can be brought to consciousness because its ontology is that of a neurophysiology characterized in terms of its capacity to cause consciousness.

Paradoxically, the naive mentalism of my view of the mind leads to a kind of dispositional analysis of unconscious mental phenomena; only it is not a disposition to behavior, but a disposition – if that is really the right word – to conscious thoughts, including conscious thoughts manifested in behavior. This is paradoxical, even ironic, because the notion of a dispositional account of the mental was introduced precisely to get rid of the appeal to consciousness. I am, in effect, trying to turn this tradition on its head by arguing that unconscious beliefs are indeed dispositional states of the brain, but they are dispositions to produce conscious thoughts and conscious behavior. This sort of dispositional ascription of causal capacities is quite familiar to us from common sense. When, for example, we say of a substance that it is bleach or poison we are ascribing to a chemical ontology a dispositional causal capacity to produce certain effects. Similarly, when we say of the man who is unconscious that he believes that Bush is president we are ascribing to a neurobiological ontology the dispositional causal capacity to produce conscious effects with specific aspectual shapes. The concept of unconscious intentionality is thus that of a *latency* relative to its *manifestation* in consciousness.

To summarize: The argument for the connection principle was somewhat complex but its underlying thrust was quite simple. Just ask yourself what fact about the world is supposed to correspond to your claims. Now when you make a claim about unconscious intentionality there are no facts that bear on the case except neurophysiological facts. There is nothing else there except neurophysiological states and processes describable in neurophysiological terms. But intentional states, conscious or unconscious, have aspectual shapes, and there is no aspectual shape at the level of the neurons. So the only fact about the neurophysiological structures that corresponds to the ascription of intrinsic aspectual shape is the fact that the system has the causal capacity to produce conscious states and processes where those specific aspectual shapes are manifest.

The overall picture that emerges is this: There is nothing going on in my brain but neurophysiological processes, some conscious, some unconscious. Of the unconscious neurophysiological processes, some are mental and some are not. The difference between them is not in consciousness, because, by hypothesis, neither is conscious; the difference is that the mental processes are candidates for consciousness, because they are capable of causing conscious states. But that's all. All my mental life is lodged in the brain. But what in my brain is my "mental life"? It is just two things: conscious states and those neurophysiological states and processes that – given the right circumstances – are capable of generating conscious states. Let's call those states that are in principle accessible to consciousness "shallow unconscious" and those inaccessible even in principle "deep unconscious." Our first conclusion is that there are no deep unconscious intentional states.

142

3. The inversion of the explanation

I believe that the Connection Principle has some quite striking consequences. To anticipate a bit, I will argue that many of our explanations in cognitive science lack the explanatory force we thought they had. To rescue what can be salvaged from them we will have to perform an inversion on their logical structure analogous to the inversion that Darwinian models of biological explanation forced on the old teleological biology which preceded Darwin. In our skulls there is just the brain with all of its intricacy, and consciousness with all its color and variety. The brain produces the conscious states that are occurring in you and me right now, and it has the capacity to produce lots more which are not occurring. But that is it. Where the mind is concerned that is the end of the story. There are brute, blind neurophysiological processes and there is consciousness; but there is nothing else. If we are looking for phenomena which are intrinsically intentional but inaccessible in principle to consciousness there is nothing there: no rule following, no mental information processing, no unconscious inferences, no mental models, no primal sketches, no 2½D images, no three-dimensional descriptions, no language of thought and no universal grammar. In what follows, I will argue that the entire cognitivist story which postulates all these inaccessible mental phenomena is based on a preDarwinian conception of the function of the brain.

Consider the case of plants and the consequences of the Darwinian revolution for the explanatory apparatus we use to account for plant behavior. Prior to Darwin it was common to anthropomorphize plant behavior and say things such as the plant turns its leaves toward the sun in order to aid in its survival. The plant *wants* to survive and flourish, and in order *to do so* it follows the sun. On this preDarwinian conception there was supposed to be a level of intentionality in the behavior of the plant. But this level of supposed intentionality has been replaced by two other levels of explanation, a "hardware" level and a "functional" level. At the hardware level we have discovered that the actual movements of the plant's leaves in following the sun are caused by the secretion of a specific hormone, auxin. Variable secretions of auxin are quite sufficient to account for the plant's behavior, without any extra hypothesis of intentionality. Notice furthermore that this behavior plays a crucial role in the plant's survival, and at the functional level we can say things such as that the light-seeking behavior of the plant functions to help the plant survive and reproduce.

The original intentionalistic explanation of the plant's behavior turned out to be false, but it was not just false. If we get rid of the intentionality and invert the order of the explanation, the intentionalistic claim emerges as trying to say something true. To be sure that what happened is absolutely clear, I want to show how in replacing the original intentionalistic explanation by a combination of the mechanical hardware explanation and a functional explanation, we are inverting the explanatory structure of the original intentionalistic explanation:

(a) The original intentionalistic explanation: *Because it wants to survive*, the plant turns its leaves toward the sun or *In order to survive* the plant turns its leaves toward the sun.

(b) The mechanical hardware explanation: Variable secretions of auxin cause plants to turn their leaves toward the sun.

(c) The functional explanation: Plants that are going to turn their leaves toward the sun anyway *are more likely to survive than plants that do not*.

In (a) the form of the explanation is teleological and the *representation* of the goal, survival, functions as the *cause* of the behavior, turning toward the sun. But in (c) the teleology is eliminated and the behavior that now, by (b), has a mechanical explanation, causes the brute fact of survival, which is now no longer a goal but just an effect that happens.

The moral I will later draw from the entire discussion can now be stated at least in a preliminary form: *Where nonconscious processes are concerned we are still anthropomorphizing the brain in a way that we were anthropomorphizing plants before the Darwinian revolution.* It is easy to see why we make the mistake of anthropomorphizing the brain; it is after all the home of anthropos. But to ascribe a vast array of intentional phenomena to a system where the conditions on that ascription are being violated is still a mistake. Just as the plant has no intentional states because it does not meet the conditions for having intentional states, so those brain processes which are in principle inaccessible to consciousness have no intentionality, because they do not meet the conditions for having intentionality. The ascriptions of intentionality we make to processes in the brain which are in principle inaccessible to consciousness are either metaphorical, like metaphorical ascriptions of mental states to plants, or they are false, as our ascriptions to plants would be false if we tried to take them literally. Notice, however, that they are not *just* false; they are trying to say something true, and to get at what is true in them we have to do the same inversion of the explanation in cognitive science that we did for plant biology.

To work out this thesis in detail we will have to consider some specific cases. I will start with theories of perception and then proceed to theories of language in order to show what a cognitive science which respects the facts of the brain and the facts of consciousness might look like.

Irvin Rock concludes his excellent book on perception (Rock 1984) with the following observations:

Although perception is autonomous with respect to such higher mental faculties as are exhibited in conscious thought and in the use of conscious knowledge, I would still argue that it is intelligent. By calling perception "intelligent" I mean to say that it is based on such thoughtlike mental processes as description, inference, and problem solving, although these processes are rapid-fire, unconscious, and nonverbal . . . "Inference" implies that certain perceptual properties are computed from given sensory information using unconsciously known rules. For example, perceived size is inferred from the object's visual angle, its perceived distance, and the law of geometrical optics relating the visual angle to object distance. (p.234)

But now let us apply this thesis to the explanation of the Ponzo illusion as an obvious example.

Though the two parallel lines are equal in length, the

143

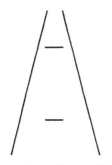

Figure 1. The Ponzo illusion. The upper of the two equal and parallel lines is generally seen as longer.

top line looks longer. Why? According to the standard explanation, the agent is unconsciously following two rules and making two unconscious inferences. The first rule is that converging lines from lower to higher in the visual field imply greater distance in the direction of the convergence and the second is that objects that occupy equal portions of the retinal image vary in perceived size depending on perceived distance from the observer (Emmert's Law). On this account, the agent unconsciously infers that the top parallel line is further away because of its position in relation to the converging lines; second, he infers that the top line is larger because it is further away. Thus there are two rules and two unconscious inferences, none of whose operations are accessible to consciousness, even in principle. It should be pointed out that this explanation is controversial and there are lots of objections to it (see Rock 1984, pp. 156ff). But the point here is that the *form* of the explanation is not challenged and that is what I am challenging now. I am interested in this type of explanation, not just in the details of this example.

There is no way that this type of explanation can be made consistent with the Connection Principle. You can see this if you ask yourself which facts in the brain are supposed to correspond to the ascription of all these unconscious mental processes. We know that there are conscious visual experiences and we know that these are caused by brain processes, but where is the additional mental level supposed to be in this case? Indeed, in this example it is very hard to interpret it literally at all without a homunculus; we are postulating logical operations performed over retinal images, but who is supposed to be performing these operations? Close inspection reveals that in its very form this explanation is anthropomorphizing the nonconscious processes in the brain in the same way that the preDarwinian explanations of plant behavior anthropomorphized the nonconscious operations of the plant.

The problem is not, as is sometimes claimed, that we lack sufficient empirical evidence for the postulation of mental processes which are in principle inaccessible to consciousness; rather, it is not at all clear what the postulation is supposed to mean. We cannot make it coherent with what we know about the nature of mental states and the operation of the brain. We think, in our pathetic ignorance about brain functioning, that some day an advanced brain science will locate all these uncon-

scious intelligent processes for us. But you only have to imagine the details of a perfect science of the brain to see that even if we had such a science there could be no place in it for the postulation of such processes. A perfect science of the brain would be stated in neurophysiological (i.e., hardware) vocabulary. There would be several hardware levels of description, and, as with the plant, there would also be functional levels of description. These functional levels would identify those features of the hardware that we find interesting in the same way that our functional descriptions of the plant identify those hardware operations in which we take an interest. But just as the plant knows nothing of survival, so the nonconscious operations of the brain know nothing of inference, rule following, or size and distance judgments. We attribute these functions to the hardware relative to our interests, but there are no additional mental facts involved in the functional attributions.

The crucial difference between the brain on the one hand and the plant on the other is this: The brain has an intrinsically mental level of description because at any given point it is causing actual conscious events and is capable of causing further conscious events. Because the brain has both conscious and unconscious mental states we are also inclined to suppose that in the brain there are mental states which are intrinsically inaccessible to consciousness. But this thesis is incoherent, and we need to make the same inversion in these sorts of explanations that we made in the explanations of the plant's behavior. Instead of saying, "We see the top line as larger because we are unconsciously following two rules and making two inferences," we should say: "We consciously see the top line as farther away and larger." Period. End of the intentionalistic story.

As with the plant there is also a functional story and a (largely unknown) mechanical hardware story. The brain functions in such a way that lines converging above appear to be going away from us in the direction of the convergence and objects which produce retinal images of the same size will appear to vary in size if they are perceived to be at different distances away from us. *But there is no mental content whatever at this functional level.* In such cases, the system functions to cause certain sorts of conscious intentionality, but the causing is not itself intentional. And the point, to repeat, is not that the ascription of deep unconscious intentionality is insufficiently supported by empirical evidence: It is incoherent, in the sense that it cannot be made to cohere with what we already know to be the case.

"Well," you might say, "the distinction does not really make much difference to cognitive science. We continue to say what we have always said and do what we have always done; we simply substitute the word 'functional' for the word 'mental' in these cases. And this is a substitution many of us have been doing unconsciously anyway, since many of us tend to use these words interchangeably."

I think the claim I am making does have important implications for cognitive science research, because by inverting the order of explanation we get a different account of cause-and-effect relations, and in so doing we radically alter the structure of psychological explanation. In what follows, I have two aims: I want to develop the original claim that cognitive science requires an inversion

144

of the explanation comparable to the inversion achieved by evolutionary biology and I want to show some of the consequences that this inversion would have for the conduct of our research.

I believe the mistake persists largely because we lack hardware explanations of the auxin type. I want to explain the inversion in a case where we have something like a hardware explanation. Anyone who has seen home movies taken from a moving car is struck by how much more the world jumps around in the movie than it does in real life. Why? Imagine that you are driving on a bumpy road. You consciously keep your eyes fixed on the road and the other traffic even though the car and its contents, including your body, are bouncing around. In addition to your conscious efforts to keep your eye on the road something else is happening unconsciously; your eyeballs are constantly moving inside their sockets in such a way as to help you continue to focus on the road. You can try the experiment right now by simple focusing on the page in front of you and shaking your head from side to side and up and down.

Now in the car case it is tempting to think we are following an unconscious rule, as a first approximation: Move the eyeballs in the eye sockets relative to the rest of the head in such a way as to keep vision focused on the intended object. Notice that the predictions of this rule are nontrivial. Another way to do it would have been to keep the eyes fixed in their sockets and move the head, and in fact some birds keep retinal stability in this way. (If an owl could drive, this is how he would have to do it, since his eyeballs are fixed.) So we have two levels of intentionality:

A conscious intention: Keep your visual attention on the road.

A deep unconscious rule: Make eyeball movements in relation to the eye sockets that are equal and opposite to head movements in order to keep the retinal image stable.

In this case the result is conscious, though the means for achieving it are unconscious. But the unconscious aspect has all the earmarks of intelligent behavior. It is complex, flexible, goal-directed; it involves information processing and has a potentially infinite generative capacity. That is, the system takes in information about body movements and prints out instructions for eyeball movements, and there is no limit on the number of possible combinations of eyeball movements the system can generate. Furthermore, the system can learn because the rule can be systematically modified by putting magnifying or miniaturizing spectacles on the agent. And without much difficulty one could tell any standard cognitive science story about the unconscious behavior: a story about information processing, the language of thought, and computer programs, just to mention obvious examples. I leave it to readers as a five finger exercise to work out the story according to their favorite cognitive science paradigm.

The problem, however, is that all these stories are false. What actually happens is that fluid movements in the semicircular canals of the inner ear trigger a sequence of neuron firings that enter the brain over the eighth cranial nerve. These signals follow two parallel pathways, one of which can "learn" and one of which cannot. The pathways are in the brain stem and cerebellum and they transform the initial input signals to provide motor output "commands," via motorneurons that connect to the eye muscles and cause eyeball movements. The whole system contains feedback mechanisms for error correction. It is called the vestibular ocular reflex (VOR). (Lisberger 1988; Lisberger & Pavelko 1988). The actual hardware mechanism of the VOR has no more intentionality or intelligence than the movement of the plant's leaves due to the secretion of auxin. The appearance of an unconscious rule being followed, unconscious information processing, and so on is an optical illusion. All the intentional ascriptions are *as-if*. So here is how the inversion of the explanation goes. Instead of saying:

Intentional: To keep my retinal image stable and thus improve my vision while my head is moving, I follow the deep unconscious rule of eyeball movement.

We should say:

Hardware: When I look at an object while my head is moving, the hardware mechanism of the VOR moves my eyeballs.

Functional: The VOR movement keeps the retinal image stable and this helps to improve my vision.

Now why is this shift so important? In any scientific explanation, among other things, we are trying to say exactly what causes what. In the traditional cognitive science paradigms there is supposed to be a deep unconscious mental cause and it is supposed to produce a desired effect, such as perceptual judgments, or grammatical sentences. But the inversion eliminates this mental cause altogether. There is nothing there except a brute physical mechanism that produces a brute physical effect. These mechanisms and effects are describable at different levels but none of them so far is mental. The apparatus of the VOR functions to improve visual efficiency, but the only intentionality is the conscious perception of the object. All the rest of the work is done by the brute physical mechanism of the VOR. So the inversion radically alters the ontology of cognitive science explanation by eliminating a whole level of deep unconscious psychological causes. The normative element that was supposed to be inside the system in virtue of its psychological content now comes back in when a conscious agent outside the mechanism makes judgments about its *functioning*. To clarify this last point I have to say more about functional explanations.

4. The logic of functional explanations

It might appear that I am proposing that, unproblematically, there are three different levels of explanation – hardware, functional, and intentional; and that where deep unconscious processes are concerned we should simply substitute hardware and functional explanations for intentional ones. But, in fact, the situation is a bit more complicated than that. Where functional explanations are concerned the metaphor of levels is somewhat misleading, because it suggests that there is a separate functional level different from the causal levels. That is not true. The so called functional level is not a separate level at all, but simply one of the causal levels described in terms of our interests. Where artifacts and biological individuals are concerned, our interests are so obvious that they may seem inevitable and the functional level

145

may seem intrinsic to the system. After all, who could deny, for example, that the heart *functions* to pump blood? But remember that when we say the heart functions to pump blood the only facts in question are that the heart does, in fact, pump blood; that fact is important to us, and it is causally related to a lot of other facts that also are important to us, such as the fact that the pumping of blood is necessary for staying alive. If the only things that interested us about the heart were that it made a thumping noise or that it exerted gravitational attraction on the moon we would have a completely different conception of its "functioning" and correspondingly of, for example, heart disease. To put the point bluntly, in addition to its various causal relations the heart does not have any functions. When we speak of its functions we are talking about those of its causal relations to which we attach some *normative* importance. So the elimination of the deep unconscious level marks two major changes: It gets rid of a whole level of psychological causation and it shifts the normative component outside the mechanism to the eye of the beholder of the mechanism. Notice, for example, the normative vocabulary that Lisberger uses to characterize the function of the VOR. "The function of the VOR is to stabilize retinal images by generating smooth eye movements that are equal and opposite to each head movement." Furthermore, "an accurate VOR is important because we require stable retinal images for good vision" (Lisberger 1988, pp. 728–29).

The intentional level, on the other hand, differs from nonintentional functional levels. Though both are causal, the causal features of intrinsic intentionality combine the causal with the normative. Intentional phenomena such as rule following and acting on desires and beliefs are genuinely causal phenomena; but as intentional phenomena they are essentially related to such normative phenomena as truth and falsity, success and failure, consistency, rationality, illusion, and conditions of satisfaction generally (See Searle 1983, especially Chapter 5, for an extended discussion). In short, the actual facts of intentionality contain normative elements; but where functional explanations are concerned the only *facts* are brute, blind physical facts and the only norms are in us and exist only from our point of view.

The abandonment of the belief in a large class of mental phenomena which are in principle inaccessible to consciousness would therefore result in treating the brain as an organ like any other. Like any other organ the brain has a functional level – indeed many functional levels – of description; and like any other organ it *can be described as if* it were doing "information processing" and implementing any number of computer programs. But the truly special feature of the brain, the feature that makes it the organ of the mental, is its capacity to cause and sustain conscious thoughts, experiences, actions, memories, and so forth.

The notion of an unconscious mental *process* and the correlated notion of the principles of unconscious mental processes are also sources of confusion. If we think of a conscious process which is *purely* mental we might think of something like humming a tune soundlessly to oneself in one's head. Here there is clearly a process and it has a mental content. But there is also a sense of *mental process* where it does not mean *process with mental content*, but rather *process by which mental phenomena are related*.

Processes in this second sense may or may not have a mental content. For example, in the old associationist psychology there was supposed to be a process by which the perception of A reminds me of B, and that process works on the principle of resemblance. If one sees A, and A resembles B, then one will have a tendency to form an image of B. In this case the process by which one goes from the perception of A to the image of B does not necessarily involve any additional mental content at all. There is supposed to be a principle on which the process works, namely, resemblance, but the existence of the process according to the principle does not imply that there has to be any further mental content other than the perception of A and the thought of B or the thought of B as resembling A. In particular, it does not imply that when one sees A and is reminded of B one follows a rule whose content requires that if one sees A, and A resembles B, then one should think of B. In short, *a process by which mental contents are related need not have any mental content at all in addition to that of the relata;* even though, of course, our theoretical talk and thoughts of that principle will have a content referring to the principle. Now, this distinction is going to prove important, because many of the discussions in cognitive science move from the claim that there are processes which are *mental* in the sense of causing conscious phenomena (the processes in the brain which produce visual experiences, for example) to the claim that those processes are mental processes in the sense of having mental content, information, inference, and so on. The nonconscious processes in the brain that cause visual experiences are certainly mental in one sense, but they have no mental content at all and thus in that sense are not mental processes.

To make this distinction clear let us distinguish between those processes, such as rule following, which have a mental content that functions causally in the production of behavior, and those processes which do not have a mental content but associate mental contents with input stimuli, output behavior, and other mental contents. The latter class I will call "association patterns." If, for example, whenever I eat too much pizza I get a stomach ache, there is definitely an association pattern, but no rule following. I do not follow a rule: when you eat too much pizza get a stomach ache; it just happens that way.

5. Some consequences: Universal grammar, association patterns, and connectionism

It is characteristic of intentionalistic explanations of human and animal behavior that *patterns* in the behavior are explained by the fact that the agent has a representation of that very pattern or a representation logically related to that very pattern in its intentional apparatus, and that representation functions causally in the production of the pattern of behavior. Thus, we say that people in Britain drive on the left because they follow the rule: Drive on the left; and that they do not drive on the right because they follow that same rule. The intentional content functions causally in producing the behavior it represents. There are two immediate qualifications. First, the intentional content of the rule does not produce the behavior all by itself. Nobody, for example, goes for a drive just to be following the rule; and nobody talks just

146

for the sake of following the rules of English. And second, the rules, principles, and so forth may be unconscious and, for all practical purposes, they are often unavailable to consciousness even though, as we have seen, if there really are such rules they must be, at least in principle, accessible to consciousness.

A typical strategy in cognitive science has been to try to discover complex patterns such as those found in perception or language and then to postulate combinations of mental representations which will explain the pattern in the appropriate way. Where there is no conscious or shallow unconscious representation we postulate a deep unconscious mental representation. Epistemically, the existence of the patterns is taken as evidence for the existence of the representations. Causally, the existence of the representations is supposed to explain the existence of the patterns. But both the epistemic and the causal claims presuppose that the ontology of deep unconscious rules is perfectly in order as it stands. I have tried to challenge the ontology of deep unconscious rules, and if that challenge is successful the epistemic and the causal claims collapse together. Epistemically, both the plant and the VOR exhibit systematic patterns, but that provides no evidence at all for the existence of deep unconscious rules; an obvious point in the case of the plant, less obvious but still true in the case of vision. Causally, the pattern of behavior plays a functional role in the overall behavior of the system, but the representation of the pattern in our theory does not identify a deep unconscious representation that plays a causal role in the production of the pattern of behavior, because there is no such deep unconscious representation. Again, this is an obvious point in the case of the plant, less obvious but still true in the case of vision.

Now, with this apparatus in hand, let us turn to a discussion of the status of the alleged rules of "universal grammar." I concentrate my attention on universal grammar, because grammars of particular languages, like French or English, whatever else they contain, obviously contain a large number of rules that are accessible to consciousness. The traditional argument for the existence of universal grammar can be stated quite simply: The fact that all normal children can readily acquire the language of the community in which they grow up without special instruction and on the basis of very imperfect and degenerate stimuli, and further that children can learn certain sorts of languages, exemplified by natural human languages, but cannot learn all sorts of other logically possible language systems, provides overwhelming evidence that normal children contain in some unknown way in their brains a special "language acquisition device," and *this language acquisition device consists at least in part of a set of deep unconscious rules.*

With the exception of the last italicized clause, I agree entirely with the foregoing argument for a language acquisition device. The only problem is with the postulation of deep unconscious rules. That postulation is inconsistent with the Connection Principle. It is not surprising that there has been a great deal of discussion about the sorts of evidence that one might have for the existence of these rules. And these discussions are always inconclusive, because the hypothesis is empty.

Years ago, I raised epistemic doubts about Chomsky's confident attribution of deep unconscious rules and sug-

gested that any such attribution would require evidence that the specific content, the specific aspectual shape, of the rule was playing a causal role in the production of the behavior in question (Searle 1976). I claimed that simply predicting the right patterns would not be enough to justify the claim that we are following deep unconscious rules; in addition, we would need evidence that the rule was "causally efficacious" in the production of the pattern. With certain qualifications, Chomsky accepts the requirements. Since we are agreed on these requirements, it might be worth spelling them out:

1. The use of the word *rule* is not important. The phenomenon in question could be a principle, or a parameter, or a constraint, and so on. The point, however, is that it is at a level of intrinsic intentionality. For both Chomsky and me, it is not merely a matter of the system behaving *as if* it were following a rule. There must be a difference between the role of rules in the language faculty and, for example, the role of "rules" in the behavior of plants and planets.

2. *Behavior* is not at issue, either. Understanding sentences, intuitions of grammaticality, and manifestations of linguistic competence in general, are what we are referring to by the use of the short-hand term "behavior." There is no behaviorism implicit in the use of this term and no confusion between competence and performance.

3. Neither of us supposes that all of the behavior (in the relevant sense) is caused by the rules (in the relevant sense). The point, however, is that in the best causal explanation of the phenomena, the rules "enter into" (Chomsky's phrase) the theory that gives the explanation.

Now, with these constraints in mind, what exactly was Chomsky's answer to the objection?

Suppose that our most successful mode of explanation and description attributes to Jones an initial and attained state including certain rules (principles with parameters fixed or rules of other sorts) and explains Jones's behavior in these terms; that is, the rules form a central part of the best account of his use and understanding of language and are directly and crucially invoked in explaining it in the best theory we can devise. . . .

I cannot see that anything is involved in attributing causal efficacy to rules beyond the claim that these rules are constituent elements of the states postulated in an explanatory theory of behavior and enter into our best account of this behavior (Chomsky 1986, pp. 252–53).

In the same connection, Chomsky also quotes Demopoulos and Matthews (1983):

As Demopoulos and Matthews (1983) observe, "the apparent theoretical indispensibility of appeals to grammatically characterized internal states in the explanation of linguistic behavior is surely the best sort of reason for attributing to these states [and, we may add, to their relevant constituent elements] a causal role in the production of behavior" (Chomsky 1986, p. 257).

So the idea is this: The claim that the rules are causally efficacious is justified by the fact that the rules are constituent elements of the states postulated by the best causal theory of the behavior. The objection that I want to make to this account should by now be obvious: In stating that the "best theory" requires the postulation of deep unconscious rules of universal grammar, all three authors

147

are presupposing that the postulation of such rules is perfectly legitimate to begin with. Once we cast doubt on the legitimacy of that assumption, however, then it looks like the "best theory" might just as well treat the evidence as association patterns that are not produced by mental representations that in some way reflect those patterns, but are produced by neurophysiological structures that need have no resemblance to the patterns at all. The hardware produces patterns of association, in the sense defined above, but the patterns of association play no causal role in the production of the patterns of behavior; they just are those patterns of behavior.

Specifically, the evidence for universal grammar is much more simply accounted for by the following hypothesis: There is, indeed, a language acquisition device innate in human brains, and this language acquisition device constrains the form of languages that human beings can learn. There is, thus, a hardware level of explanation in terms of the structure of the device, and there is a functional level of explanation, describing which sorts of languages can be acquired by the human infant in the application of this mechanism. Now, no further predictive or explanatory power is added by saying that there is in addition a level of deep unconscious rules of universal grammar, and indeed, I have tried to suggest that that postulation is incoherent anyway. For example, suppose children can only learn languages that contain some specific formal property F. Now that is evidence that the language acquisition device makes it possible to learn F languages and not possible to learn non-F languages. But that is it. There is no further evidence that the child has a deep unconscious rule representing F, and no sense has been given to that supposition anyway.

The situation is exactly analogous to the following: Humans are able to perceive colors only within certain range of spectrum. Without formal training, they can see blue and red, for example, but they cannot see infrared or ultraviolet. This is overwhelming evidence that they have a "vision faculty" that constrains what sorts of colors they can see. But now, is this because they are following the deep unconscious rules "if it is infrared, don't see it" or "if it is blue, it is ok to see it"? To my knowledge, no argument has ever been presented to show that the rules of universal *linguistic* grammar have any different status from the rules of universal *visual* grammar. Now ask yourself why exactly you are unwilling to say that there are such rules of universal visual grammar? After all, the evidence is just as good as, indeed it is identical in form with, the evidence for the rules of universal linguistic grammar. The answer, I believe, is that it is quite obvious to us from everything else we know that there is no such mental level. There is simply a hardware mechanism that functions in a certain way and not others. I am suggesting here that there is no difference between the status of deep unconscious universal visual grammar and deep unconscious universal linguistic grammar: Both are nonexistent.

Notice that it is not enough to rescue the cognitive science paradigm to say that we can simply decide to treat the attribution of rules and principles as as-if intentionality; because as-if intentional states, not being real, have no causal powers whatever. They explain nothing. The problem with as-if intentionality is not merely that it

is ubiquitous – which it is – but its identification does not give a causal explanation, it simply restates the problem which the attribution of real intentionality is supposed to solve. Let us see how this point applies in the present instance. We tried to explain the facts of language acquisition by postulating rules of universal grammar. If true, this would be a genuine causal explanation of language acquisition. But now suppose we abandon this form of explanation and say simply that the child acts *as if* he were following rules, but of course he is not really doing so. If we say that, we no longer have an explanation. The cause is now left open. We have converted a psychological explanation into speculative neurophysiology.

Now, if I am right we have been making some stunning mistakes. Why? I believe it is in part because we have been supposing that if the input to the system is meaningful and the output is meaningful then all the processes in between must be meaningful as well. And certainly there are many meaningful processes in cognition. But where we are unable to find meaningful conscious processes we postulate meaningful unconscious processes, even deep unconscious processes. And when challenged we invoke that most powerful of philosophical arguments: "What else could it be?" "How else could it work?" Deep unconscious rules satisfy our urge for meaning, and besides, what other theory is there? Any theory is better than none at all. Once we make these mistakes our theories of the deep unconscious are off and running. But it is simply false to assume that the meaningfulness of the input and output implies a set of meaningful processes in between; and it is a violation of the Connection Principle to postulate in principle inaccessible unconscious processes.

One of the unexpected consequences of this whole investigation is that I have quite inadvertently arrived at a defense – if that is the right word – of connectionism. [See Hanson & Burr: "What Connectionist Models Learn: Learning and Representation in Connectionist Networks" *BBS* 13(3) 1990.] Among their other merits, at least some connectionist models show how a system might convert a meaningful input into a meaningful output without any rules, principles, inferences, or other sorts of meaningful phenomena in between. This is not to say that existing connectionist models are correct – perhaps they are all wrong. But they are not all obviously false or incoherent in the way that the traditional cognitivist models which violate the Connection Principle are.

6. Two objections

I want to discuss two objections. One I thought of myself, though several other people[3] also gave me different versions of it; one is due to Ned Block (personal communication).

First objection: Suppose we had a perfect science of the brain. Suppose, for example, that we could put our brain-o-scope on someone's skull and see that they wanted water. Now suppose that the "I-want-water" configuration in the brain was universal. People want water if and only if they have that configuration. This is a total sci-fi fantasy, of course, but let's pretend. Now let's suppose that we found a subsection of the population that had exactly that configuration but could not "in principle"

148

bring any desire for water to consciousness. They engage in water-seeking behavior but they are unable "in principle" to become conscious of the desire for water. There is nothing pathological about them; that is just the way their brains are constructed. Now if this is possible – and why not? – then we have found a counterexample to the Connection Principle, because we have found an example of an unconscious desire for water which it is "in principle" impossible to bring to consciousness.

I like the example, but I do not think it is a counterexample. Characteristically in the sciences we define surface phenomena in terms of their microcauses; we can define red in terms of wavelengths of a certain number of nanometers, for example. If we had a perfect science of the brain of the sort imagined we would certainly identify mental states by their microcauses in the neurophysiology of the brain. But – and this is the crucial point – the redefinition works as an identification of an unconscious mental phenomenon only to the extent that we continue to suppose that the unconscious neurophysiology is still, so to speak, tracking the right conscious mental phenomenon with the right aspectual shape. So the difficulty is with the use of the expression "in principle." In the imagined case, the "I-want-water" neurophysiology is indeed capable of causing the conscious experience. It was only on that supposition that we got the example going in the first place. The cases we have imagined are simply cases where there is a blockage of some sort. They are like Weiskrantz's (1982) "blindsight" examples, only without the pathology. [See also Campion, Catto & Smith: "Is Blindsight an Effect of Scattered Light, Spared Cortex, and Near-Threshold Vision?" *BBS* 6(3) 1983.] But there is nothing "in principle" inaccessible to consciousness about the phenomena in question, and that is why it is not a counterexample to the Connection Principle.

Second objection (due to Ned Block): The argument has the consequence that there could not be a totally unconscious intentional zombie. But why could there not be? If such a thing is possible – and why not? – then the Connection Principle entails a false proposition and is therefore false.

Actually, there could not be an intentional zombie and Quine's famous argument for the indeterminacy of translation (Quine 1960, Chapter 2) has inadvertently supplied us with the proof: For a zombie, unlike a conscious agent, there simply is no fact of the matter as to exactly which aspectual shapes its alleged intentional states have. Suppose we built a "water-seeking" zombie. Now, what fact about the zombie makes it the case that he, she or it is seeking the stuff under the aspect "water" and not under the aspect "H_2O"? Notice that it would not be enough to answer this question to claim that we could program the zombie to say, "I sure do want water, but I do not want any H_2O" because that only forces the question back a step: What fact about the zombie makes it the case that by "water" it means what we mean by "water" and by "H_2O" it means what we mean by "H_2O"? And even if we complicate its behavior to try to answer this question there will always be alternative ways of interpreting its verbal behavior which are consistent with all the facts about verbal behavior but which give inconsistent attributions of meaning and intentionality to the zombie. And, as Quine has shown in laborious detail, the problem

is not that we could not know for sure that the zombie meant, say, "rabbit" as opposed to "stage in the life history of a rabbit," or "water" as opposed to "H_2O"; there is simply no fact of the matter at all about which the zombie meant. But where there is no fact of the matter about aspectual shape there is no aspectual shape, and where there is no aspectual shape there is no intentionality. Quine, we might say, has a theory of meaning appropriate for verbally-behaving zombies. But we are not zombies and our utterances do, on occasion at least, have determinate meanings with determinate aspectual shapes, just as our intentional states often have determinate intentional contents with determinate aspectual shapes (Searle 1987). But all of that presupposes consciousness.

7. The unconscious

What is left of the unconscious? I said earlier that our naive pretheoretical notion of the unconscious was like the notion of objects which, like the fish in the sea or furniture in the dark attic of the mind, keep their shapes even when unconscious. But now we can see that this notion is mistaken. It confuses the conscious form of an intentional state with a causal capacity to cause the state in that form, that is, it confuses the latency with its manifestation. It is as if we thought the bottle of poison on the shelf had to be poisoning something all the time in order really to be poison.

The final conclusion I want to draw from this discussion is that we have no unified notion of the unconscious. There are at least four different notions.

First, there are as-if metaphorical attributions of intentionality to the brain which are not to be taken literally. For example, we might say that the medulla wants to keep us alive so it keeps us breathing even while we are asleep.

Second, there are Freudian cases of shallow unconscious desires, beliefs, and so forth. It is best to think of these as cases of repressed consciousness, because they are always bubbling to the surface, though often in a disguised form. In its logical behavior the Freudian notion of the unconscious is quite unlike the cognitive science notion in the crucial respect that Freudian unconscious mental states are always potentially conscious. (I wish I had more time to discuss these in detail, but they are irrelevant to the main point of this dispute.)

Third, there are the (relatively) unproblematic cases of shallow unconscious mental phenomena which just do not happen to form the content of my consciousness at any given point in time. Thus, most of my beliefs, desires, worries, and memories are not present to my consciousness at any given moment such as the present one. Nonetheless, they are all *potentially* conscious in the sense I have explained (if I understand him correctly, these are what Freud meant by the "preconscious" as opposed to the "unconscious" [Freud 1949]).

Fourth, there is supposed to be a class of deep unconscious mental intentional phenomena which are not only unconscious but which are in principle inaccessible to consciousness. These, I have argued, do not exist. Not only is there no evidence for their existence, but the postulation of their existence violates a logical constraint on the notion of intentionality.

8. Conclusion

This discussion is the upshot of the application of two principles. Always ask yourself: What do you know for sure? and: What facts are supposed to correspond to the claims you are making? Now, as far as the inside of the skull is concerned, we know for sure that there is a brain and that at least sometimes it is conscious. With respect to those two facts, if we apply the second principle to the discipline of cognitive science, we get the results I have tried to present.

ACKNOWLEDGMENT
I am indebted to a very large number of people for helpful comments and criticisms on the topics discussed in this article. I cannot thank all of them, but several deserve special mention: indeed many of these patiently worked through entire drafts and made detailed comments. I am especially grateful to David Armstrong, Ned Block, Francis Crick, Hubert Dreyfus, Vinod Goel, Stevan Harnad, Marti Hearst, Elisabeth Lloyd, Kirk Ludwig, Irvin Rock, Dagmar Searle, Nathalie van Bockstaele, and Richard Wolheim.

NOTES
1. Chomsky, Noam (1976): "Human action can be understood only on the assumption that first-order capacities and families of dispositions to behave involve the use of cognitive structures that express systems of (unconscious) knowledge, belief, expectation, evaluation, judgment, and the like. At least, so it seems to me (p. 24). These systems may be unconscious for the most part and even beyond the reach of conscious introspection" (p. 35).

Among the elements that are beyond the reach of conscious introspection is "universal grammar" and Chomsky says: "Let us define universal grammar (UG) as the system of principles, conditions, and rules that are elements or properties of all human languages not merely by accident but by necessity – of course, I mean biological, not logical, necessity" (p. 29).

2. The argument here is a condensed version of a much longer development in Searle (1989). I have tried to keep its basic structure intact; I apologize for a certain amount of repetition.

3. I am indebted to Dan Rudermann for calling my attention to this article.

4. For these purposes I am contrasting "neurophysiological" and "mental," but in my view of mind/body relations, the mental simply is neurophysiological at a higher level (see Searle 1984a). I contrast mental and neurophysiological as one might contrast humans and animals without thereby implying that the first class is not included in the second. There is no dualism implicit in my use of this contrast.

5. Specifically, David Armstrong, Alison Gopnik, and Pat Hayes.

Open Peer Commentary

Commentaries submitted by the qualified professional readership of this journal will be considered for publication in a later issue as Continuing Commentary on this article. Integrative overviews and syntheses are especially encouraged.

Consciousness and accessibility

Ned Block
Department of Linguistics and Philosophy, Massachusetts Institute of Technology, Cambridge, MA 02139
Electronic mail: block@cogito.mit.edu

Searle's Connection Principle says that unconscious mental states must be in principle accessible to consciousness. If deep unconscious rules, representations, states, and processes are not in principle accessible to consciousness, then they are not mental. I don't think that many in the cognitive science community care whether these phenomena are *mental* or not; the important point is that they are representational. But since Searle's argument applies as well to representationality as to mentality, we can move on to the real issues.

What does Searle mean by "accessibility in principle?" One of the real issues of which I speak is what "in principle" comes to. (Another, to be discussed later, is what consciousness is.) Searle clarifies his stance by describing people (I will call them the Less Conscious People) who have a desire for water, despite there being a "blockage" that prevents this desire from having any disposition to become conscious. What makes the Less Conscious People's desire in principle accessible to consciousness? The answer, as I read Searle, is that the Less Conscious People have the same brain state that is the "I want water" configuration in us. We have the "I want water" brain state just in case we want water, and we satisfy the Connection Principle since we can become conscious of our desire for water. So the presence of the "I want water" brain state in them justifies ascribing the desire for water to them. Though they are unable to become conscious of their desire for water, we should think of their desire as in principle accessible to consciousness because its brain state gives rise to awareness of the desire for water in us. As Searle notes, similar reasoning would justify ascribing visual knowledge to blind-sighted patients.[1]

Once Searle's point is set out in this way, it becomes clear that it is susceptible to a straightforward objection. For we can imagine a species of More Conscious People who bear the same relation to us that we bear to the Less Conscious People. The point calls for a concrete example.

Consider the dialect of English in which there is a difference between the pronunciation of the "ng" in "finger" and in "singer." In the dialect I have in mind, the "ng" in "finger" might be said to be hard, whereas the "ng" in "singer" is soft. (Actually, the "g" is deleted in "singer," and the "n" is velar.) In this dialect there is a rule that at least *describes* the phenomenon. For our purposes, we can take it as: Pronounce the "ng" as soft in "nger" words derived from "ng" verbs – otherwise hard (see Chomsky & Halle 1968, pp. 85–87; see Halle 1990 for other examples). One bit of weak evidence in favor of the hypothesis that such an internally represented rule (note, incidentally, that like many other cognitivists, I say "internally," not "mentally") actually *governs* our behavior is that this hypothesis predicts certain facts about the pronunciation of new words. If you tell a member of the dialect group in question that "to bling" is to look under tables, asking what you call one who blings, they say "blinger" with a soft "g." This result rules out the hypothesis that "nger" word pronunciation is simply a matter of a memorized list, and it also rules out certain alternative hypotheses of rules governing behavior. Nonetheless, I concede that there is no strong evidence for the hypothesis that an internal representation of the mentioned rule governs our behavior. But we need not tarry over this matter, since Searle's quarrel is with the very idea of the "deep unconscious," not with the strength of the empirical evidence.

To return to the point, we can now imagine a species of people, the aforementioned More Conscious People, some of whom speak the dialect of English just mentioned, and are also conscious of using the "nger" rule mentioned. Let us further suppose that in the More Conscious People, the use of this rule is coextensive with a certain brain state, call it the applying-the-"nger"-rule brain state. To complete the analogy, let us now suppose that you and I also have the applying-the-"nger"-rule brain state just in case we belong to the dialect group that makes the mentioned distinction between "singer" and "finger." Here is the punch line: The very reason that Searle gives for postulating blockage in the Less Conscious People applies to us. We have a blockage that keeps us from becoming conscious of our application of the "nger" rule. Since a similar story can be told for any "deep unconscious" phenomenon, this point can be used to legitimize any of the cognitivist's favorite rules, representations, states, or processes. Thus Searle's clarification of his notion of in principle accessibility undermines his overall claim against the deep unconscious.

What does Searle mean by "Conscious?" One way of stating Searle's argument for the Connection Principle is this: Mentality requires aspectual shape, but there is no matter of fact about aspectual shape without (potential) consciousness; hence mentality requires (potential) consciousness. The main line of cognitivist reply should be to challenge the second premise, arguing for a different theory of aspectual shape, namely, a language of thought theory. This line of reply will be no surprise to Searle. I prefer to follow a less traveled path, taking seriously Searle's point that consciousness is a neglected topic.

The word "conscious" is notoriously ambiguous. In one sense of the term, for a state to be conscious there must be something "it is like" to have it. This is the sense that figures in the famous inverted spectrum hypothesis: Perhaps, what it is like for me to look at things that we agree are red is the same as what it is like for you to look at things that we agree are green.

There are many other senses of "consciousness, "including Jaynes's (1977) "internal soliloquy" sense in which consciousness was actually discovered by the ancient Greeks.[2] There is one sense of 'consciousness' that is particularly relevant for our concerns, one in which a state is conscious to the extent that it is accessible to reasoning and reporting processes. In connection with other states, it finds expression in speech. Something like this sense is the one that is most often meant when cognitive science tries to deal in a substantive way with consciousness, and it is for this reason that consciousness is often thought of in cognitive science as a species of *attention* (see Posner 1978, Chapter 6, for example).

Now there is some reason to take Searle's notion of consciousness to be the last sense, the accessibility sense. It is only in this sense that the phenomena postulated by Freud and by cognitive scientists are *clearly and obviously* unconscious. However, in this last sense of "conscious," Searle's Connection Principle is implausible and the argument for it is question-begging. (I am assuming here that Searle will tighten his notion of "in principle accessibility" to avoid the conclusion of the last section that all of the "deep unconscious" is in principle accessible.) If consciousness is simply a matter of access to reasoning and reporting processes, then two states could be exactly alike in all intrinsic properties, yet differ in that one is situated so that reasoning and reporting processes can get at it, whereas the other is not. Yet the state that is badly situated with respect to reasoning and reporting processes might be well situated with respect to other modules of the mind, and may thus have an important effect on what we think and do. Searle would have to say that the first (well situated) state is mental, whereas the second is not. But who would care about such a conception of mentality? This is why I say that according to the present sense of "conscious," the Connection Principle is implausible.

Recall that Searle's argument for the Connection Principle involves the premise that there is no matter of fact about aspectual shape without consciousness. But what reason would there be to believe this premise if all that consciousness comes to is a relation to reasoning and reporting processes? Two states

might have exactly the same aspectual shape despite the fact that one can be detected by certain mechanisms and the other can be detected only by different mechanisms.

Further, for the access sense of "conscious," the metaphor of the fish that Searle rejects is quite appropriate. Just as the same type of fish can be below or above the water, the same type of mental state (with the same aspectual shape) can be either accessible or inaccessible to mechanisms of reasoning and reporting. If Searle's argument is to get off the ground, he must take consciousness to be an intrinsic property of a conscious state, not a relational property.

It is time to move to the obvious candidate, the "what it is like" sense. Understanding Searle this way, we enter deep and muddy waters where Searle cannot so easily be refuted, but where he cannot so easily make his case either. I think we can see that his argument depends on a point of view that cognitive scientists should not accept, however. An immediate problem for Searle with this sense of consciousness is this: How does Searle *know* that there is nothing it is like to have the rules and representations he objects to? That is, how does he know that what he calls the "deep unconscious" *really is* unconscious in the what-it-is-like sense? Indeed, how does he know that there is nothing it is like to have Freudian unconscious desires? (Recall that in the present sense of "unconscious" there is nothing it is like to be in any unconscious state.) Our reasoning and reporting mechanisms do not have direct information about these states, so how are we to know whether there is anything it is like to have them? Suppose you drive to your office, finding when you arrive that you have been on "automatic pilot," and recall nothing of the trip. Perhaps your decisions en route were not available to reasoning and reporting processes, but that does not show that there was nothing it was like to, say, see a red light and decide to stop. Or consider a "deep unconscious" case. If subjects wear headphones in which different programs are played to different ears, they can obey instructions to attend to one of the programs. When so doing, they can report accurately on the content of the attended program, but can report only superficial features of the unattended program, for example, whether it was a male or a female voice. Nonetheless, information on the unattended program has the effect of favoring one of two possible readings of ambiguous sentences presented in the attended program. (See Lackner & Garrett 1973.) Does Searle know for sure that there is nothing it is like to understand the contents of the unattended program? Let us be clear about who has the burden of proof. Anyone who wants to reorient cognitive science on the ground that the rules, representations, states, and processes of which it speaks are things that there is nothing it is like to have must show this.

The underlying issue here depends on a deep division between Searle and the viewpoint of most of cognitive science. Cognitive science tends to regard the mind as a collection of semiautonomous agencies – modules – whose processes are often "informationally encapsulated," and thus inaccessible to other modules (see Chomsky 1986; Fodor 1983; Gazzaniga 1985; and Pylyshyn 1984) Though as Searle says, cognitivists rarely talk about consciousness (and to be sure, many cognitivists – Dennett, Harman, and Rey, for example – explicitly reject the what-it-is-like sense), the cognitivist point of view is one according to which it is perfectly possible that that there could be something it is like for one module to be in a certain state, yet that this should be unknown to other modules, including those that control reasoning and reporting. Searle will no doubt disagree with this picture, but his conclusion nonetheless depends on a view of the organization of the mind that is itself at issue between him and cognitivists.

Suppose Searle manages to refute the point of the last paragraph by showing that there is nothing it is like to be in states that are unavailable to reasoning and reporting. That is, suppose that the access sense and the what-it-is-like sense of 'conscious' apply to exactly the same things. Still, the issue arises as to which is *primary*. Searle will no doubt say that our states are accessible to reasoning and reporting mechanisms precisely when and because there is something it is like to have them. But how could he know that this is the way things are, rather than the reverse, that is, there being something that it is like to have a state is a byproduct of accessibility of the state to reasoning and reporting mechanisms. If the latter is true, once again the metaphor of the fish would be right. For an unconscious thought would have its aspectual shape, whether or not reasoning and reporting processes can detect it; it would be only when and because they detect it that there would be anything it is like to have the thought.

The upshot is this: If Searle is using the access sense of "consciousness," his argument doesn't get to first base. If, as is more likely, he intends the what-it-is-like sense, his argument depends on assumptions about issues that the cognitivist is bound to regard as deeply unsettled empirical questions.

NOTES
1. It is worth mentioning that it is easy to arrange experimental situations in which normal people act like blind-sighted patients in that they give behavioral indications of possessing information that they *say* they do not have. See the discussion below of the Lackner & Garrett (1973) experiment. Thus in one respect we are not very different from the Less Conscious People.
2. See Block, 1981, for a critique of Jaynes (1977), and Dennett, 1986, for a ringing defense of the importance of Jaynes's notion of consciousness.

Intention itself will disappear when its mechanisms are known

Bruce Bridgeman

Professor of Psychology, Clark Kerr Hall, University of California, Santa Cruz, Ca 95064
Electronic mail: *psy160@c.ucsc.edu*

The problem of mentalistic explanation is both more and less than meets the eye. It is less because some of Searle's as-if examples were never meant to be taken literally. The problem is more serious than Searle implies, however, because intentional language for brain processes is always metaphorical; intention is a result, not a process or state.

Searle gives a particularly clear explanation of the contrast between intentionalistic and mechanical/functional explanations with his examples of anthropomorphosed plants and the successful mechanical/functional explanation of the VOR (vestibular ocular reflex). But he exaggerates in assigning intentionality wherever cognitive scientists use intentionalistic language. Searle admits that "We often make . . . metaphorical attributions of as-if intentionality to artifacts," but he maintains that the as-if character is lost when descriptions of brain processes are concerned. Not so – in fact, the contrast between functional and as-if explanation is quite explicit in the neurosciences, and is taught to students of physiological psychology.

A recent textbook (with which I am particularly familiar) makes this clear in words that almost echo Searle's:

Biologists often speak, in a kind of verbal shorthand, as though useful traits were evolved purposefully, using statements such as: Fish evolved complex motor systems to coordinate their quick swimming movements." This is the intentionalistic statement. The textbook goes on: "What they really mean, though, is that the fish that by chance happened to have a few more neurons in the motor parts of their brains (Searle's mechanical hardware explanation) . . . survived in greater numbers and had more offspring than those that happened to have fewer neurons or less effectively organized ones (Searle's functional explanation). . . . The shorthand of purposeful language will sometimes be used in this book, though the more biologically

valid interpretation should always stand behind it. (Bridgeman 1988, p. 10, parenthetical comments added)

Thus purposeful language in neuroscience explanations should always be taken metaphorically, a colorful and compact means of exposition that can always be unpacked to the two-step mechanical/functional argument by the informed student.

In this context, Rock (1984) receives somewhat of a "bum rap" from Searle. The processes of perception work as if they were intelligent, and Rock makes clear even in the quoted passage that the intelligence describes processes in brains, not conscious insights. The perceptual parts of the brain merely process information in ways that in other contexts are interpreted as intelligent.

The second part of my argument, that the intentionalistic explanation is more of a problem than Searle makes it out to be, comes from a closer look at the processes that Searle is still willing to describe in intentionalistic terms, that is, those explicitly identified as conscious.

Perhaps the primary problem with the concept of unconscious mental state is not the "unconscious" or the "mental," but the assumption of a static "state." The state. of course not Searle's invention, is a problem because it derives ultimately from introspection and nothing else. At the start of his target article, Searle questions the relatively small role of consciousness in cognitive science, noting that mention of the term was completely suppressed in respectable circles until recently. This was not always so, however – psychology as a separate discipline was in fact founded on the basis of introspection, or careful examination of the contents of consciousness. Implicit in this effort was the assumption that mental life was indeed accessible to consciousness. The assumption turned out to be false. Freud formalized the insight that some aspects of brain function were unconscious, and neurophysiology has revealed more and more nonconscious processing in the brain. Everywhere we look, nonconscious processes such as early vision, parsing of sensory tasks to different cortical areas even within a modality, or coding of different sorts of memory dominate the brain.

Psychology has been understandably wary of returning to consciousness, and has done so only with an array of new techniques. But it is already clear that the role of consciousness in mental life is very small, almost frighteningly so. The aspects of mental life that require consciousness have turned out to be a relatively minor fraction of the business of the brain, and we must consider consciousness to be a brain system like any other, with particular functions and properties. It looms large only in our introspections.

More specifically, whenever we examine an aspect of what seems to be conscious it turns out to be made of simpler parts. The process of seeing, for instance, is made of a great cascade of neural processing based on a welter of relatively simple algorithms. What had seemed like visual intelligence turns out to be only processing after all, when we look empirically at how it works. (A similar fate has befallen seemingly intelligent AI efforts.) Wherever we look for intentionality, we find only neurons, as Searle laments. We can predict that intentionality will evaporate in the twenty-first century, as certainty did in the twentieth.

My final comment is on the problem of aspectual shape. The resolution, seen from psychology, is that just the fact that a conscious manifestation has aspectual shape is no reason that the memories on which it is based should have any such property. Why not construct the aspectual shape during the process of bringing memory from storage? With apologies to Searle, the computer analogy applies here, in a kind of lilies-of-the-field argument – if the humble computer has a given capacity or property, why not us? The information stored on my computer's disc has no margins, lines or paragraphs; it looks nothing like the form it will have when it is displayed on my terminal. When it is needed, the bare-bones disc information is recoded, formatted, and made accessible to an editor, high-level language, or other system program. Even the forms of the letters are added upon recall, from a sparse ASCII code. When I look for the commentary in the machine, I find only bits; when I look for intentionality in the brain, I find only neurons.

To carry the analogy to the brain, we need not expect aspectual shape at the level of neurons. The brain stores fragments of what we remember; and recall to consciousness is literally a remembering, assembling the members of the memory from hints and assumptions. The process is obvious in some amnesics, where events that never occurred are constructed to hide memory deficits. But the same process takes place, usually more successfully, in all of us. In short, memory is always an informed confabulation. The neurological fact of recoding means that aspectual shape is not the problem that it had seemed to be.

Conscious mental episodes and skill acquisition

Richard A. Carlson

Department of Psychology, The Pennsylvania State University, University Park, PA 16802
Electronic mail: *cvy@psuvm.bitnet*

I agree with several of Searle's major points – that it is a mistake for cognitive science to postulate mental states that are in principle unconscious, that taking metaphorical ("as-if") ascriptions of intentionality literally leads to an illusion of explanatory force, and that contemporary thinking about the mind is characterized by gross neglect of consciousness. Something like the Connection Principle could provide an important constraint on cognitive theory. Searle seems pessimistic about the symbol-processing paradigm, but offers little guidance for cognitive scientists who would like to accept the Connection Principle.

I will try to make two points in this commentary. First, the Connection Principle implies a model for psychological explanation that has been followed in some contemporary work in cognitive psychology. Furthermore, at least some current work in the symbol-processing paradigm can be construed in a way consistent with this model. Second, ambiguities concerning the notion of "in principle" accessibility to consciousness must be resolved if the Connection Principle is to provide a useful constraint on cognitive theory. In particular, difficult issues about the causal status of unconscious mental states must be addressed.

Conscious mental episodes. Searle's "dispositional" view of unconscious mental states implies that these states participate in intentional causal relations only when they are conscious. At the psychological level, then, the theorist's task is to describe relations among conscious states: "mental episodes consist of conscious states yielded from other conscious states by nonconscious operations – inherently nonconscious as brain processes that are discernible only as the relations among those states" (Dulany, Carlson & Dewey 1985, p. 30). There is nothing methodologically extraordinary about this approach. It is widely accepted that our descriptions of "stimuli" and "responses" in experiments are intentional, and standard tools of experimentation, measurement, and modeling can be applied to describe mental episodes conceived as conscious states and their relations (Dulany 1984). The relations among conscious states (Carlson & Dulany 1988; Dulany 1968) and between conscious states and behavior (Carlson & Dulany 1985; Dulany, Carlson & Dewey 1984) are highly systematic, when those states are appropriately assessed.

Many theories in cognitive psychology can be construed on this model. For example, some "mental logic" theories of deductive reasoning (e.g., Rips 1983) could be construed as describing relations among conscious states whose representational contents constitute the premises and conclusions of de-

ductive arguments. On this view, the rules of mental logic are not themselves mental states but descriptions of nonconscious operations relating mental states. It is the individual who has the conscious states, not some mysterious homunculus who performs the logical operations.

Cognitive scientists concerned with mental activities such as problem solving have relied on verbal protocols that are naturally interpreted as data about conscious states (e.g., Ericsson & Simon 1983). Production system models of problem solving (e.g., Anderson 1983) can also be construed as theoretical statements about the nonconscious relations among conscious states. The psychological causation in these models occurs in a working memory that might be identified with consciousness, and individual productions might be construed as hypotheses about relations between conscious states and particular nonconscious dispositions to produce other conscious states or actions.

Accepting the Connection Principle might change our understanding of these theories, but it does not provide a basis for rejecting symbol processing in favor of connectionism.

Skill acquisition and the unconscious. Searle is apparently content with the Freudian notion of unconscious mental states, because these states may in principle be brought to consciousness. On my reading of Freud, however, unconscious mental states may participate in intentional causal relations without becoming conscious. My purpose in raising this point is not to debate interpretations of Freud, but to point out an important ambiguity in the idea of states that are "in principle" accessible to consciousness. The question is whether the kind of unconscious mental causality suggested by Freud is ruled out by the Connection Principle. Rather than discuss phenomena as controversial – and rarely discussed by cognitive scientists – as those considered by Freud, consider the phenomenon of skill acquisition.

A standard observation about the acquisition of skill is that the involvement of consciousness in performance diminishes with increasing skill (e.g., Anderson 1982; Searle 1983, p. 100). We might say that the intentionality of performance "collapses" – a performance that once required a whole sequence of intentional states may after sufficient practice require (at the level of conscious states) only a representation of the overall goal. Thus, for example, it seems unlikely that the British drivers Searle mentions (part 5, paragraph 1) consciously consult a rule about driving on the left – at least, skilled British drivers are probably thinking about their destinations, not such basic "rules of the road."

Why is this a problem? We might say that the mental state whose content is the rule "drive on the left" no longer actually becomes conscious, but can function causally without violating the Connection Principle because it is "in principle" accessible to consciousness. This interpretation, however, has two problems. First, it does not seem to fit with the empirical facts – experts often cannot report (i.e., bring to consciousness) any rules governing their skilled performances. Second, Searle's dispositional account of unconscious mental states seems to rule out this kind of causal functioning. An alternative interpretation might be that what was once a conscious rule-following performance is now a preintentional background skill (Searle 1983, Chapter 5). If so, however, the states that participate in causing the performance must lack aspectual shape – since all and only intentional states have aspectual shape. It will be hard to account for skilled performances, however, without granting aspectual shape to nonconscious processes postulated in accounts of skilled performance. In this sense, it seems that some nonconscious states may be mental. This may not be a counterexample to the Connection Principle, but it suggests that the notions of in-principle accessibility to consciousness and intentional causation must be spelled out in greater detail before the model of psychological explanation sketched above can be applied to such problematic cases as skilled performance.

Accessibility "in principle"

Noam Chomsky
Massachusetts Institute of Technology, Cambridge, MA 02139
Electronic mail: *chomsky@psyche.mit.edu*

Consider the patient P.S., studied by Marshall and Halligan (1988). Presented simultaneously with two line drawings of a house, in one of which the left side is on fire, and asked where she would prefer to live, P.S. selected the house that was not burning. Suppose P.S. also gives the obvious reasons. Then, uncontroversially, P.S. has a mental representation of a house on fire and uses it in reaching her decision. In the case in question, however, P.S. had sustained right cerebral damage and judged the drawings to be identical, regarding the choice question as silly "because they're the same" but reliably selecting the nonburning house when forced. Marshall and Halligan conclude that "her preference judgments clearly indicated covert knowledge" of the flames on one side of the house; P.S. had a mental representation, but lacked conscious access to it. The authors might proceed to construct a theory of rules and representations to account for such facts as these; perhaps the theory would turn out to be independent of the cerebral damage, the two cases differing only in access to consciousness. This is all straightforward science, taking the mental simply as "neurophysiological at a higher level" (Searle), that is, as one of the many levels at which true explanatory theory can be constructed: in terms of atoms, molecules, cells, neurons, computational systems, and so on.

Now consider Jones and Smith, both of whom can calculate square roots. Suppose that behavioral and physiological evidence leads us to conclude that Jones does it by algorithm J, and Smith by the different algorithm S. Suppose that Jones and Smith also inform us, upon introspection, that this is what they are doing. Then uncontroversially, they are following rules, have the relevant mental states, and so on.

Suppose, however, that Jones reports (honestly) that he is using S. The evidence is now more complex. The correct conclusion might be that Jones's mental states and processes are as before, but that the "inner eye" of consciousness is giving a false picture.

Consider now Brown, who also calculates square roots but with no conscious access, like P.S. Suppose the relevant evidence indicates that he is using S, not J. What is missing, then, is only the (notoriously weak) evidence from introspection. For the natural scientist, Brown has the same mental states and processes as Smith, not Jones, but the inner eye focuses differently.

Let us turn now to Mary, who is presented with the expressions:

(1) John is too clever to catch
(2) John is too clever to be caught
(3) John is easy to catch
(4) John is easy to be caught

She tells us that (1) and (2) are true if John is so clever that one cannot catch him, and that (3) – but not (4) – is true if it is easy to catch John; the obvious analogy fails for (4), which she furthermore regards as somehow deviant. Suppose we have as much behavioral and physiological evidence as we like to support such conclusions about Mary. Proceeding in the manner of the sciences, we will seek an explanatory theory, in whatever turn out to be the appropriate terms. A rather successful theory (a version of universal grammar, UG) explains these facts by attributing to Mary a computational system involving representations with empty categories of various kinds, along with principles of control, case, theta-marking, and others, all of these notions spelled out with some precision and applying to examples of many other kinds as well, in English and other languages (and leaving many questions unanswered). We thus attribute to Mary mental states and processes to which she happens to have no access.

Searle maintains that it is not clear what "the postulation of mental processes which are in principle inaccessible to consciousness . . . is supposed to mean." He is quite right; the reason is that his phrase "in principle" is hopelessly obscure. Deleting it, all is clear enough, and we can proceed in the normal manner of rational inquiry, determining the true theory of rules and representations as best we can, and relating it to other characterizations of the brain in terms of cells, and so forth (including neurophysiology if it proves to be relevant, as generally assumed, largely on faith).

The crucial notion, then, is Searle's "in principle." His "Connection Principle" (CP) requires accessibility *in principle* for attribution of mental states and processes. It is his task, then, to explain this mysterious concept and inform us how it applies in such cases as those just illustrated. These problems do not arise for the natural scientist, who invokes no such notion here – wisely. Note that although Searle's discussion proceeds in the material mode, and presents the CP as the conclusion of an argument, in fact, it amounts to little more than a recommendation for the use of the term "mental," which we might understand if he would explain his notion "in principle." Call the notions that meet this condition, whatever it is, "S(earle)-intentionality," "S-mental," and so on.

Take P.S. Does she have access "in principle" to the line drawings? If not, then we reject the CP and the concept of S-mental as pointless. In section 6, Searle attempts to defend the CP from such considerations (which he calls potential "counterexamples," but misleadingly, since terminological stipulations do not have counterexamples). In such cases, he says, P.S. has "a blockage of some sort," but there is no inaccessibility "in principle." Thus S-mental states can be attributed.

Consider now Brown. If he lacks access "in principle," then the CP, again, is a pointless terminological stipulation. So let us assume that in this case, too, Brown has "a blockage"; thus Brown has access "in principle" to the algorithm S, not J. Suppose that Jones reports (honestly) that he is following S; then to save the CP we could again say that "in principle" his inner eye could have given a true picture, had it not been somehow out of kilter. Note again that no such pointless questions arise from the naturalistic perspective.

Suppose that Brown belongs to a different species. Would he then have access "in principle"? If not, then again we abandon the CP and S-intentionality. To save the CP, then, let us suppose that this too is a case of Searlean "blockage": Brown does have access to J "in principle."

Now turn to Mary. The point of Searle's paper is that there is a serious if not lethal problem when one pursues the path indicated earlier, which is in fact the normal path of the natural sciences. This is so because Mary lacks access to representations with empty categories, to the principles of case theory, and so on. And apparently this is to be understood not as "blockage," but as "inaccessibility in principle." That is, there is no possible species that would have such introspective access. Whatever the status of this claim, it need delay us no further.

There is, then, no reason to pay any attention to the CP. It is about as interesting as the proposal that (E-)mental states can be properly attributed only to someone who can report them in English. We now have many new facts, of as much interest as the facts regarding S-intentionality provided by the CP. For example, Mary will have E-mental states, but Marie will not – a puzzle.

Dispensing with all of this, we can proceed, as in normal science, to regard the mental as the molecular, cellular, and so forth, at a "higher level." The criteria for existence will be understood, as usual, in terms of role within the best explanatory theory. In particular, we need not trouble ourselves with the mysterious notion of "in principle" and the pointless puzzles to which it gives rise.

Searle claims that there is a simpler explanation for such evidence as (1)–(4) than the one provided in terms of UG (with empty categories, case theory, etc.), namely, there is "a hardware level of interpretation." By the same logic, we can eliminate the strange and complex constructions of quantum theory and molecular biology: it is much simpler to say that the phenomena are explained by unknown properties of unknown mechanisms.

Searle also offers an alleged reductio of the explanation of (1)–(4) in the manner indicated. The latter, he says, is "exactly analogous" to what he calls sarcastically "the deep unconscious rule" (R): "if it is infrared, don't see it." There is an analogy, but not the one Searle gives. His (R) is analogous to the rule: "if it is (4), don't understand it and regard it as deviant." (R) is thus an odd formulation of one of the facts to be explained, but contrary to what Searle assumes, it is in no way analogous to the proposed explanation of these facts. The difference between the principles of UG and Searle's (R) is the difference between significance and vacuity, a rather considerable one.

There are numerous other problems in Searle's account, but putting them aside, the "cognitive science paradigm," as he terms it, is in no danger from these considerations. For many years, Searle has been attempting to provide a critique of the approach that I and others have taken to UG. In response, I have repeatedly argued that his objections reduce ultimately to a demand for conscious accessibility, one that is arbitrary and pointless, despite its distinguished ancestry. His latest effort is of value, I think, in making it more clear than before that this is exactly what is at stake, and that it makes no sense at all. Putting aside questions about Searle's account of the history of the matter (see Chomsky 1980a, Chapter 6) there is no doubt that the demand for conscious accessibility has deep roots. Searle's paper provides another indication that it is a fundamental error, and should be abandoned.

Aspects and algorithms

Andy Clark

School of Cognitive and Computing Sciences, University of Sussex, Brighton, BN1 9QH, England
Electronic mail: *andycl@syma.sussex.ac.uk*

Ultimately, I suppose, there is only physics. But our scientific world-view posits a set of overlaid structures; the chemical, the social, even the computational. Searle offers a timely, elegant, and disturbing assault on one of those layers – the citadel of the information-processing unconscious. His goal is to show that information-processing descriptions (except insofar as they pick out states which are in principle accessible to consciousness) have no intrinsic place in such a hierarchy of levels. His main tool is a requirement of aspectual fixity (i.e., the world must be presented under a particular description, e.g., the glass of water presented *as* water and not as H_2O). The only way of getting such aspectual fixity, he argues, is by appeal to the nature of our conscious awareness.

Searle may be right about aspectual fixity. Indeed, the whole notion of an aspect seems bound up with that of a conscious point of view. But if this is to be an argument against the ontology of information-processing psychology, Searle must show not just that we lack (for the unconscious cases) a notion of *aspectual* fixity, but also that we lack a notion of *algorithmic* fixity. This I believe he fails to do.

Searle's argument depends on demonstrating that there is (in the case of in principle unconscious "mental contents") no logical space available between neurophysiological descriptions of the brain, on the one hand, and functional descriptions which are merely a reflection of the particular interests of conscious theorists on the other. But (*pace* Searle) I believe we have a robust and respectable notion of just such a space. I shall try to say why.

Searle's idea is that we must treat putative cases of non-conscious information processing either as gestures at the actual neurophysiology or as statements about functional role. He is careful to add, however, that the functional description is not to be seen as a legitimate level of a scientific ontology (unlike, say, chemistry). Instead it is "simply one of the causal levels described in terms of our interests"). Thus, when we say that the heart functions to pump blood, all we are doing (according to Searle) is highlighting those of its (brutal, physical· causal powers "to which *we* attach some normative importance" section 4, p. 12, my emphasis).

But this downgrading of the idea of a genuine functional level of description is deeply unconvincing. To see why, we need to consider the way (a) a story about historical origins and b ways of differentiating algorithms, can combine to yield a robust and fully realistic notion of the nature of *selecting a particular algorithm*.

The historical dimension is straightforward (see, for example, Millikan 1986). It is surely true that the heart has the function it has quite independently of my attaching normative importance to that function (pumping blood). For that physical structure would not exist were it not for its ability to perform that function – it has been selected (by a nonconscious mother nature) *because of* its functional role.

Consider now the idea of a cognitive module whose task is (let's suppose) the computation of shape from shading in a low-level vision system. We can argue, in ways directly analogous to the above, that the functional description picks out a proper, full-blown level of scientific description. For the structure was selected *because of* its ability to compute shape from shading. It is at this point that the reality of algorithms must be recognized. For there will be many different ways of computing shape from shading. The theorist will now want to discover *which* algorithm nature as exploited. There are various ways to do this. Different algorithms will yield different relative response times for different classes of problems. And different pathologies will be possible according to exactly which information-processing strategy is in use. (Virtually all of psycholinguistics and much of cognitive neuropsychology are devoted to such fine-grained differentiation of algorithms). Nothing in the target article looks to undermine this conception of the functional and algorithmic fixity of particular information-processing descriptions – a fixity achieved independently of the conscious accessibility of its contents and the idiosyncratic normative interests of the human theorist.

A second worry is that the crucial notion of what is consciously accessible "in principle" is left dangerously vague, so vague, in fact, that it runs the risk of becoming trivial. Thus we are told that any genuinely contentful but unconscious mental state is one which is at least "the *sort of* thing that can be brought to consciousness." But in response to (legitimate) worries about pathologies and "blockages" (see, e.g., sect. 6, para. 2 and 3) Searle is forced to unpack this in terms of a notion of the "in principle" as opposed to the "actual" accessibility of the contents. But this is problematic. Consider the following thought-experiment. Suppose (I do not for a moment believe it to be so simple!) that conscious awareness was caused by the development of an extra mental module which monitors and recodes the activity of other modules. Suppose, furthermore, that humans have and hamsters lack this module. But it turns out to be biologically feasible to "add on" a consciousness module to the hamster so that it can monitor, integrate, and recode its own lower level information processing. Does this show that the hamster actually had those (unconscious) mental states all along? Were those states (in Searle's terms) in principle consciously accessible? If so, it seems as if there may be no limit, after all, on the extent of our unconscious mental contents! Searle might reply that imagining such extensions to the hardware is illegitimate. But suppose we discover, in the human pathological cases, that what is causing the "blockage" is, precisely, damage

155

to such a module. Then how can we justify allowing determinate unconscious content where the creature has a damaged from birth) recoding module yet disallowing it to creatures who simply lack the module? Indeed we may, if we wish, suppose that the damage just is the absence of the module from birth! What this case suggests is that the notion of in-principle accessibility to consciousness, once it is weakened to deal with the cases which Searle addresses in section 6, may become too liberal to be of use. Perhaps there is a way of adding hardware to a cactus so that it becomes aware of the way its own growth is stimulated by sunlight! Even so, this had better not imply that the cactus, here and now, has just such latent, but genuine, intentional states.

In sum, the stress on aspectual fixity looks doubly blighted. It is blighted because the crucial notion of in-principle accessibility of some state to conscious (aspect-fixing) awareness is ill-specified. And it is blighted because such fixity may in any case be a red herring as regards information-processing psychology. For the content-using descriptions of such a discipline require only the cheaper algorithmic fixity which can be purchased by a combination of causal history (the device was selected because of its information-processing abilities) and instantaneous structure (it can be seen, for example, to implement a *specific* shape from shading algorithm).

The ability versus intentionality aspects of unconscious mental processes

Maria Czyzewska,[a] Thomas Hill,[b] and Pawel Lewicki[b]
[a]*Department of Psychology, Southwest Texas State University, University Station, San Marcos, TX 78666 and* [b]*Department of Psychology, University of Tulsa, Tulsa, OK 74104-3189*

This commentary will be short because its main point is rather straightforward. In his target article, Searle argues strongly against the general notion of unconscious mental processes as typically postulated by cognitive science and maintains that this notion is "a mistake." His major argument is that unconscious mental processes imply some kind of intentionality, whereas, as he argues, we cannot invoke the notion of "intrinsic intentionality" without reference to consciousness. Thus, according to his position, unconscious mental processes "cannot exist."

The major problem with Searle's position is that he appears to understand the notion of unconscious processes in a manner that is based more on the lay (and "everyday language") tradition than on contemporary cognitive research on mental processes, where these terms have the status of hypothetical constructs in an explanatory system aimed at generating empirically verifiable hypotheses. Searle's analysis of unconsciousness stresses mostly the aspect of the "intentionality" which is not only irrelevant to (or even inconsistent with) what cognitive investigators of those processes have studied, but it also appears conceptually erroneous, at least in not being instrumental (because the concept of intentionality seems very difficult to operationalize for cognitive research).

The distinction between conscious and unconscious mental processes in cognitive research is based not on the motivational (or psychodynamic) concept of "intentionality" but rather on the informational concept of "ability," that is, whether specific information *can* be accessed at specific levels of processing (e.g., the "ability" to report verbally on information or the functional "accessibility" of specific information). There are quite trivial (and difficult to refute) reasons for maintaining the current (cognitive) notion of unconsciousness as defined in terms of informational access. The most important of these is the abundant empirical evidence for the existence of qualitative differences between the functional status of information that (a) can

be accessed by the individual in a consciously controlled manner (e.g., reported, identified, explicitly used in reasoning) and information that (b) cannot be accessed in this way but is still stored in memory and can systematically affect cognitive processes Hill et al. 1989). [See Holender: "Semantic Activation Without Conscious Identification in Dichotic Listening Parafoveal Vision, and Visual Masking: A Survey and Appraisal" *BBS* 9(1) 1986.]

Searle's arguments might be useful in explicating some common features of the "metaconceptual" or philosophical nature of conscious and unconscious processes. They do not appear to provide evidence for any flaws in the current distinction between these two categories of processes (as understood in cognitive psychology), however, because this distinction is simply justified by the empirical data themselves. Even if one could imagine convincing linguistic or other formal arguments against using the specific term *conscious/unconscious* to denote the empirical differences mentioned above, there would still be a need to make up a term for those differences because their functional and conceptual importance for the entire field of cognitive science is firmly established Lewicki & Hill 1989). To convince cognitive researchers otherwise would require demonstrating that the huge body of empirical data in support of this distinction is somehow invalid.

We acknowledge the importance of philosophical reflection on even those issues and questions that can be ultimately resolved only by empirical research e.g., the specificity of unconscious as compared to conscious mental processes). The contribution of such reflections, however, should be measured by the extent to which they provide guidance for empirical research (i.e., its eventual specific functionality for that research). It appears that in order to offer such guidance, Searle's Connection Principle is in need of further elaboration.

ACKNOWLEDGMENT
Writing this commentary was supported by the grant MH-42715 from the National Institute of Mental Health.

Language and the deep unconscious mind: Aspectualities of the theory of syntax

B. Elan Dresher[a] and Norbert Hornstein[b]
[a]*University of Toronto, Toronto, Ontario, Canada M5S1A1 and* [b]*University of Maryland College Park, MD 20742*
Electronic mail: dresher@utorepas.bitnet and 212129@umdd.bitnet

Searle focuses on three questions to which we "need to know" the answers if we are to understand a biological phenomenon:
(1) What is its mode of existence?
(2) What does it do?
(3) How do we find out about it?
He leaves out the one type of question that most scientists spend their time on:
(4) What is its fine structure?
To use Searle's example: What sort of pump is the heart? This type of question is of particular concern to us, for it is in trying to answer (4) that linguists have been led to postulate rich innate representational structures which are not accessible to consciousness. These are precisely the sorts of entities that Searle believes to border on the incoherent.

To help see how such entities bear on Searle's claims, it is worth considering a sample account in a bit of detail. Consider the data in (5):
(5) a. The men like each other
b. *Each other like the men
c. The men expect each other to win
d. *The men expect Mary to kiss each other

156

These data raise the following question: why do speakers of English consider sentences (5a) and (5c) to be fully acceptable while (5b) and (5d) are not? One account proceeds as follows:

(6) a. Each other is an anaphor.
   b. An anaphor must be locally bound by its antecedent.
   c. For one element to bind another it must enjoy a certain structural prominence relative to the bindee: for our purposes, it must be higher in the tree.
   d. Local binding also requires a certain kind of structural proximity: for cases like (5), there may be no intervening subject between the binder and the bindee.

The structural descriptions of the examples in (5) are given in (7):

(7) a. $[_S]_{NP}$ The men]$_i$ $[_{VP}$ like $[_{NP}$ each other]$_i]]$
   b. $[_S$ $[_{NP}$ each other]$_i$ $[_{VP}$ like $[_{NP}$ the men]$_i]]$
   c. $[_S$ $[_{NP}$ the men]$_i$ $[_{VP}$ expect $[_S$ $[_{NP}$ each other]$_i$ to win]]$
   d. $[_S$ $[_{NP}$ the men]$_i$ $[_{VP}$ expect $[_S$ $[_{NP}$ Mary]$_j$ to kiss $[_{NP}$ each other]$_i]]$

Sentences (5b) and (5d) are judged to be unacceptable because their structural descriptions are ungrammatical. In both, the anaphor *each other* lacks a local binder: in (7b), its only potential binder *the men* is lower in the tree than *each other* is, and so may not bind it by (6c); in (7d), a subject *Mary* intervenes, preventing local binding by (6d). The ill-formedness of structures (7b) and (7d) accounts for the unacceptability of (5b) and (5d). By contrast, (7a) and (7c) are well-formed. In both, *the men* is higher in the tree than *each other* and no subject intervenes, and so *the men* can locally bind *each other* as required.

This sort of explanation is typical of what one finds in syntactic theory. The details differ but the format is always the same. What we aim to account for is a speaker's judgements. Our account presupposes (a) that the speaker categorizes certain expressions as anaphors; (b) that the speaker mentally constructs hierarchically structured phrase markers which must satisfy various prominence relations and locality restrictions; and (c) that a speaker's acceptability judgements are (at least partially) a function of the well-formedness of the sentential representations that the speaker constructs and examines. Observe that (a) and (b) suppose that a speaker analyses utterances under a description: they have aspectual shape. It is relative to these types of descriptions which the speaker is tacitly assumed to construct that the account of the judgements in (5) proceeds.

So it appears that we have a standard form of explanation, exploited in linguistics, which attributes unconscious aspectual representations to speakers. It is aspectual in that it asserts that part of what native speakers do is to represent utterances in a certain way. It is clearly not accessible to consciousness. Finally, let us propose that the principles (6b–d) are innate, a part of universal grammar (UG). In short, this is just the sort of account that Searle considers to be a "stunning mistake." For if the explanation is not in terms of "mental content" in his narrow sense (that is, in principle accessible to consciousness), it must, he argues, be neurophysiological, or else it is empty.

The explanation we offered is not neurophysiological. It is cast in more abstract terms, necessarily, as we currently know nothing about the brain side of these issues. Of course, our ignorance of how the brain instantiates the principles of (6) is not, in Searle's view, an excuse for continuing to hold a "pre-Darwinian conception of the function of the brain." To avoid error, let us then recast our account along the lines he suggests. Instead of:

(8) When I hear the sentence, "Each other like the men," it strikes me as odd because I follow the deep unconscious rules of (6).

We should say (sect. 3, para. 18):

(9) When I hear the sentence, "Each other like the men," it strikes me as odd because the hardware mechanism of the brain produces this effect in me.

Apart from adhering to Searle's categories, the statement in (9) explains nothing at all, unless it somehow incorporates the rules of (6). For example, we could tack onto the end of (9): ". . . the hardware mechanism produces this effect in me by instantiating the rules of (6)." This change, however, is entirely terminological, at least at present. It has no consequences whatever for research into UG or the neurobiology of language.

Searle suggests that when we ever discover what the hardware mechanism is, we may find that the "neurophysiological structures . . . need have no resemblance to the [mental] patterns at all." In our case, the relevant hardware mechanisms that instantiate (6) might have nothing in them that corresponds to the UG constructs we have posited, such as "anaphor," "local binding," and so on. These are very weak claims and at present there is no reason to endorse Searle's scepticism. Furthermore, Searle's examples from vision suggest that what we are looking for in the domain of language are not really analogous. Though we do not need a rule, "If it is infrared, don't see it," what receptors will stand in for, "If it is a sentence with an anaphor not locally bound, don't say it (find it odd, have trouble processing it, and so on)"?

Searle's other objections also do not apply in our case. The hypothesis in (8) is not empty; it has empirical consequences. Nor does there appear to be any confusion of mechanical and functional levels; it is not teleological, so the example of the sun-seeking plant and auxin is not relevant here.

The capacities we are attributing to the speaker involve a typical data-processing task that we could easily program a computer to do. Nevertheless, this sort of account does not fall prey to Searle's Chinese Room Argument, whatever its merits. If successful, the Chinese Room Argument at most shows that data-processing accounts are not sufficient to explain all of human cognitive competence. It does not show that data-processing might not be necessary. Nor does it show that some parts of human cognition might not be fully accounted for in data-processing terms.

Different people are interested in different sorts of things. Searle's model does not seem to have room for the kinds of questions that many researchers in linguistics or vision think about. These people are aiming to adumbrate the fine structure of particular capacities. Currently, the best stories we have require postulating internal representations that are inaccessible to consciousness. Of course, one might with Searle accept the facts and simply refuse to countenance the question of *how* the facts are to be explained: But that is the question that most interests some of us.

Searle's Freudian slip

Hubert L. Dreyfus
University of California, Berkeley, CA 94720
Electronic mail: *dreyfus@cogsci.berkeley.edu*

Searle tells us that "when we describe something as an unconscious intentional state we are characterizing an objective *ontology* in virtue of its *causal* capacity to produce consciousness. . . . [This] leads to a kind of dispositional analysis of unconscious mental phenomena; only it is not a disposition to behavior, but a disposition . . . to conscious thoughts, including conscious thoughts manifested in behavior. . . . The concept of unconscious intentionality is thus that of a *latency* relative to its *manifestation* in consciousness." (sect. 2, para. 31). Thus, unconscious intentional states, for example, unconscious beliefs, are really just brain states, but they can legitimately be considered mental states because they have the same neural structure as brain states which, if not blocked in some way, would be conscious. They have no intentionality as mere brain states, but they do have latent intentionality.

This is a plausible account of how we can speak of unconscious

157

beliefs, but this view has strange consequences when generalized to unconsciously motivated actions. I'll use an example Searle and I discussed in his seminar. Suppose that at a family dinner Bill "accidentally" spills a glass of water into his brother Bob's lap because, as Bill's therapist tells him later, he has an unconscious desire to annoy Bob. The explanation of this behavior requires not just *latent* beliefs such as Bill's long-standing belief that Bob stole his mother's affection, but also actually occurring beliefs, such as that the accident will upset Bob, and actually occurring desires, such as the desire that Bob be upset. It also requires what Searle calls a "prior intention" – to upset Bob by means of spilling the water – and an "intention in action" which "governs" Bill's bodily movements, so that these movements are a case of spilling water, not of moving H_2O around. The issue is one of aspectual shape. According to Searle, unconscious brain states cannot determine aspectual shape and yet the example requires that the intention in action have an actual, not just a potential aspectual shape. Otherwise, one cannot say what Bill is doing or, indeed, that he is doing anything at all.

The same of course holds for the standard cases such as desiring one's mother. When this desire leads to a slip or a dream it is under the aspectual shape of desire for one's *mother*, not one's aunt's sister. There is no mention of one's aunt in the slip and she does not appear in the dream. In fact, the Freudian unconscious is full of beliefs and desires that are not merely *capable* of affecting behavior under some aspectual shape should they become conscious; they affect behavior under some aspectual shape precisely while remaining unconscious.

If Freud is right about the unconscious then there seem to be only two ways out; let us call them the instrumentalist and the minimalist ways. The instrumentalist, like Dennett, could claim that all we need in order to make sense of unconsciously motivated behavior is as-if intentionality. [See Dennett: "Intentional Systems in Cognitive Ethology: The Panglossian Paradigm Defended" *BBS* 3 1985.] Searle might add that the attribution of as-if intentionality in such cases is justified only when the behavior is caused by a brain state that is potentially conscious. This would mean that in the unconscious case the same brain state that normally caused a bodily movement by way of instantiating an intention in action caused the same bodily movement but without there being an intention in action. But it seems strange that one brain state could produce the same movement in these two entirely different ways. Why should it be just as if there is unconscious aspectual shape unless there *is* unconscious aspectual shape?

If, to avoid this problem, one says that all unconscious motives are a tiny bit conscious – that Bill had a dim sense that he wanted to annoy his brother by his action – we have to face the fact that Bill was not able to detect any such intention and will vehemently deny that he had any dim or marginal awareness of it. If we nonetheless insist that he must have been dimly conscious of his unconscious motivation, then we might as well say that all unconscious processes are dimly conscious – a view that neither Searle nor his opponents want to endorse.

If Searle accepts the Freudian unconscious, it seems he is committed to the view that it is in principle possible for there to be a world in which people have mental states which cause them to behave, talk, and perceive (or at least do something that we must make sense of as behaving, talking, and perceiving) without any actual consciousness being involved. The only difference between this view and cognitivism would be that Searle insists that to count as a mental state a brain state must be at least potentially conscious, while the cognitivists deny this. But as long as all agree that we do not have to *be* conscious of our beliefs and desires for them to be aspectually efficacious, it does not seem important whether we are potentially conscious of them. If brain states can function causally just like intentional states, independently of consciousness, that is all the cognitivist needs.

Consciousness as physiological self-organizing process

Walter J. Freeman
Department of Molecular and Cell Biology, University of California, Berkeley CA 94720
Electronic mail: *wfreeman@garnet.berkeley.edu*

Searle's main goal is an explanation of the causes and causal operations of consciousness, reinstating it as a legitimate topic of philosophical discourse. His key premise for this effort is reflected in his statement that in "any scientific explanation . . . we are trying to say exactly what causes what." In my opinion, this view is incorrect. A scientific description, whether by a physical, medical, or social scientist, is a set of relationships among a number of measured variables, and an explanation is a set of equations that interrelates them in symbolic form. To recall a truism, scientists ask how, not why; an apple falls according to the law of the inverse square, it is not pulled by gravity. They do this not to avoid teleological wish fulfillment but to keep to their business. Judges, journalists, lawyers, and philosophers require nonscientific explanations in terms of causes. A pathologist must say: Subject A died with disease B; and a coroner must say: Subject A died of disease B. Coroners have legal responsibility to assign causes of death. Theirs are societal and not scientific judgments.

We all postulate causal relationships at the ontological level, which comprises anatomical and physiological aspects of the brain, and at the intentional level, which includes mental and psychological aspects of the brain. Certain patterns of activity of nerve cells in the brain cause related patterns of muscle activity, and certain thoughts cause interrelated types of behavior. What is not clear is what Searle achieves by his claim that patterns of neural activity cause thoughts, and that thoughts in turn cause patterns of neural activity. If these two aspects of brain function coexist as two facets of an entity, like a coin and its monetary value, then there is no causation between them. If they are interactive, then Searle has the same problem as that dealt with by Descartes and by Eccles (1970): How does the soul grasp the reins of the pineal gland to control the flow of spinal fluid through the central aqueduct? How at the quantal level does it control release of vesicles at synapses? Dualism enters not with multiple aspects of brains but with postulates of causation between the aspects at different hierarchical levels.

I think it may be useful to raise the question why it is that a differential equation, which makes no intrinsic causal statement, should be satisfactory as an explanation for scientists, whereas philosophers need something more. Why do people feel that they understand an event or a process if they can conceive it in terms of cause and effect? Why is it so *satisfying* to be able to do this? An answer to this question can be formulated, I believe, from an understanding of brain mechanisms of consciousness, which come from analyses of brain activity using differential equations as descriptors of relations and not of causes (Freeman, in press).

One of the first insights to come forth in the applications of nonlinear dynamics to brain function is the property of emergent order. A large and complex system such as the brain, when left to itself with adequate food supplies and facilities for waste disposal, tends to develop more order rather than less, contrary to what is supposedly mandated by the second law of thermodynamics. Immediately we have a problem. What causes the newly emergent order in a self-organizing system? According to classical doctrine, nothing can cause itself; yet complex brain activity patterns appear and evolve as we watch, and we have no explanation at biochemical or psychological levels for what causes them. I suggest that we experience these brain activity patterns as consciousness.

A second insight is that perceptual patterns are not imposed on the brain by external stimuli. Perception begins as an

158

emergent activity within the brain, which leads to an invitation for the admission of stimuli. An "aspectual shape" (Searle's term) begins as an emergent pattern in the limbic system, which is projected through the brain in two modes. One mode constitutes a motor command to carry through a searching action by placing the exteroceptors optimally in respect to past inflow of sensations. The other mode constitutes neural messages to all of the sensory systems to prepare them for the consequences of the anticipated action. This mode is subsumed under the process of reafference. The sensory systems then receive sensory inflows and respond to them in ways that are determined largely by stored experience and by the modulations imposed by re-afference. All these new activity patterns are reflexively transmitted and modulated through serial stages back into the limbic system, where they are integrated with the proprioceptive and interoceptive consequences of the recent action, not as intended but as executed. The updated limbic activity pattern provides the basis for the next motor command and its reafferent messages. I suggest that we experience this cycle as "I act . . . I feel," and use it as a metaphor of "cause . . . effect" that we project into the world outside our brains – "It causes . . . its effect."

A third insight is that these processes evolve in a sequence of frames in time like the freeze-shift action of a motion picture. Each new frame is the creation of an activity pattern by the several areas of cerebral cortex. I suggest that we may experience as a thought each frame as it arises newly from the old by self-organization, and that it is created, not retrieved or resurrected from a "memory store" in the manner of operation of a digital or analog computer memory. Brains are less like libraries than nurseries and farms.

These processes can be described experimentally in terms of sequential relationships of activity patterns in different parts of the brain (Freeman 1975). They can also be described metaphorically in terms of cause and effect, using a distinction posed by Searle between the "intrinsic" and the "as-if." In the immense complexities of multiple time-varying and distributed nonlinear feedback systems we cannot know the causes.

Here lies the issue. If we seek to model or replicate a system, we must explain it heuristically in terms of cause and effect. If we seek to predict, control and understand it, then we must describe its inner relationships. The greatest advances in science of the preceding two centuries – the cell doctrine of Virchow, the Darwinian concept of evolution, the field concept of Faraday with "action at a distance," relativity and quantum theory – all have central cores of noncausal relationships, onto which users have grafted explanatory causal matrices. Some of these have been unsuccessful, such as the "ether" that was supposedly needed to "explain" gravitational action at a distance, or the "hidden variables" to "explain" causal conundrums in quantum mechanics. Some are questionable, such as the concept of "survival of the fittest" to explain causally the statistical relationships of natural selection. Others are partially successful, such as the germ theory of disease, which works well for pneumonia but not for schizophrenia.

These causes are add-ons extrinsic to the theories. We do not need the causes in order to use the theories but to humanize them. Hence the problem of the causes of consciousness will not be explained away. It will become ever more pressing when machine intelligence is eventually achieved by computer and cognitive scientists, and the Machine Liberation Front inherits the mantle of responsibility from animal rights activists.

Grammar and consciousness

Robert Freidin
Department of Philosophy/Program in Linguistics, Princeton University, Princeton, NJ 08544
Electronic mail: *bob@clarity.princeton.edu*

Searle's Connection Principle (henceforth CP) would prohibit the postulation of rules and principles of universal grammar (UG) as part of a cognitive model of language because such constructs are inaccessible to consciousness, whereas it would, according to Searle, allow for language-particular rules which are alleged to be accessible to consciousness. In Searle's own words: "grammars of particular languages, like French or English, whatever else they contain, obviously contain a large number of rules that are accessible to consciousness." Since the notion "accessible to consciousness" lies at the heart of the CP, it is worth trying to understand how rules of UG differ from language-particular rules. Although Searle does not mention any explicit examples of the latter, this distinction is spurious for reasons that will be elaborated below. Moreover, if we take the notion "accessible to consciousness" as a criterion for the legitimacy of cognitive models then we are obliged to reject the rich and substantive discussions of language and cognition in linguistics, philosophy, and psychology during the past 35 years as illegitimate – not a step that one would take without strong justification, including some idea of the benefits to be gained.

The distinction between language-particular rules and rules of UG that Searle wants to make on the basis of accessibility to consciousness appears to be spurious in one of two ways: Either both kinds of rules are inaccessible to consciousness, or both are accessible. To see how this works, let us compare a language-particular rule of English with a principle of UG as applied to English.

As a rule of English, consider the prohibition against split infinitives.[1] An explicit formulation of this rule requires a grammatical analysis of structures. Thus we must somehow distinguish the infinitival "to" from the preposition "to," which is identical both phonologically and orthographically. Exactly how this rule operates will depend on a structural analysis of the infinitival "to" and the verbal form that follows (i.e., either a so-called main verb or an aspectual auxiliary verb (perfective "have" or progressive "be"). The crucial point is that these structural analyses are inaccessible to consciousness. Conscious introspection tells us nothing at all about the correctness of the categories we apply in grammatical analysis. Furthermore, the operation of the rule in mental computation is also inaccessible to consciousness.

What is accessible to consciousness with regard to language is that some sentence X has this and/or that interpretation, or that some sentence Y is deviant. As an explicit example, consider the sentence in (1).

(1) A report by three senators in favor of that proposal just appeared.

A native English speaker will recognize that (1) has two different interpretations: one where the phrase "in favor of that proposal" modifies "senators," and the other where this phrase modifies "report." Presented with (2), our English speaker will be conscious of the fact that only one of the two interpretations of (1) is possible for (2) – namely, the one where "in favor of that proposal" modifies "report."

(2) A report by three senators just appeared in favor of that proposal.

The interpretation of (2) where "in favor of that proposal" modifies "senators" either won't be recognized or the sentence under that interpretation will be regarded as deviant. If we substitute the relative clause "who know John" for "in favor of that proposal" in (1–2), giving (3–4), our English speaker will recognize that the resulting sentences are not analogous to (1–2).

(3) A report by three senators who know John just appeared.

(4) A report by three senators just appeared who know John.

(3) allows only one interpretation where "who know John" modifies "senators" and (4) will be recognized as deviant.

UG provides an explanation for our English speaker's conscious knowledge concerning these sentences. By hypothesis, we take these examples to have certain structural representations (which are in fact inaccessible to consciousness). On the basis of these structural representations we can make the relevant distinctions. Thus (1) has two interpretations because there are two distinct structural representations available given the phrase structure of English and certain general principles of interpretation that apply to phrase structure. The fact that (2) has only one interpretation follows from the structural analysis of (2) in conjunction with the UG principle of *subjacency*. This principle says that the connection between a phrase and its source position in underlying structure cannot cross more than one "bounding category." Assuming that the underlying source position of "in favor of that proposal" in (2) is in the subject noun phrase as in (1), and that NP is a bounding category for English, then the interpretation of (2) on the nonexistent (or deviant) reading would involve crossing two NPs in violation of subjacency. This is indicated in (5) where the source position is given as a trace which is coindexed with the phrase that has moved.

(5) [$_{NP}$ A report by [$_{NP}$ 3 senators e$_i$]] just appeared [$_i$ in favor of that proposal]

Subjacency automatically generalizes to (4) since the relative clause could only appropriately modify "senators." This principle of UG extends to other constructions and is also subject to some parametric variation across languages (see Freidin & Quicoli 1989).

Thus UG in the form of subjacency provides an explanation for some interesting facts about our conscious knowledge of English, though what we are actually conscious of is rather limited. Given the CP on Searle's account we are not allowed to postulate principles of UG or structural analyses of sentences as part of the mental representation of language since these are inaccessible to consciousness. Rather, these facts that English speakers are conscious of can only be explained as "neurophysiology." If we limit explanation of language and cognition to the level of neurophysiology, we will of course get nowhere since virtually nothing is known about language at that level. It's clear that a great deal of insight into the problems and issues in the study of language and cognition is lost by adopting the CP, and Searle's account so far offers nothing in its place.

ACKNOWLEDGMENT
I am indebted to Carlos Otero for his very helpful comments on this material.

NOTE
1. Assume for the sake of the discussion that the rule is valid. It really doesn't matter what "rule of English" we apply here or whether it is of the form "never do this" as opposed to "always do this." As an instance of the latter we could have considered the subject/verb agreement rule which requires that the tensed verb or auxiliary of a finite clause agree in number and person with the subject. Note that an explicit characterization of what constitutes the subject involves the postulation of quite abstract structures, structures which are inaccessible to consciousness.

Unconscious mental processes

Clark Glymour

Philosophy Department, Carnegie-Mellon University, Pittsburgh, PA 15213
Electronic mail: *c909 + @ andrew.cmu.edu*

Searle's formal argument is something like this:
(i) All mental states are "aspectual."
(ii) No neurological descriptions involve intentionality.
(iii) No neurological states are "aspectual."

(iv) All unconscious states – in contrast to conscious states – are purely neurological states.

Conclusion: No unconscious states are mental states.

I won't dispute (i), although I'm not fully clear about its sense. But the inference from (i) and (ii) to (iii) equivocates between facts and facts under descriptions. Perhaps the argument could be rephrased to avoid the difficulty, but the essential thing is that insofar as (iii) and (iv) are clear, no one is in any position to know that they are true.

Of course, explicit arguments, let alone reconstructions of them, aren't usually the whole story. Searle's idea must be more than that unconscious states and processes aren't mental – else cognitive scientists could properly be unconcerned, and simply respond that they are describing the computational architecture underneath and around what Searle calls "mental." For the same reason, Searle's idea must be more than that unconscious systems don't *follow rules*, in the sense that you or I follow a rule when doing long division, because cognitive scientists could (and should) remain unimpressed. Unconscious computational systems, they should hold, *instantiate* rules, they don't follow them, the special thing about stored program computers is that they can store a representation of the programs they instantiate, and storing such a representation (and not some other) literally causes the processes in them that instantiate the program.

I read Searle's target article more seriously as an attack on the very idea of "computational architecture" and programs as an explanation of human behavior: On my guess as to how to read him, Searle claims to have an argument that there could be no such thing as a stored program computer that produces computations that truly explain its behavior and that is not an artifact. Altogether, the view seems to be roughly as follows: Suppose that Mt. Rushmore were carved out by the wind and for some reason stood unobserved until recently. Until it was seen by conscious eyes Mt. Rushmore would not depict American presidents (or more obscurely, would depict them only potentially). There would be certain structural similarities between Mt. Rushmore and George Washington, but they wouldn't constitute depiction. Similarly, if the wind carved out "George Washington didn't sleep here," the indentations wouldn't *say* anything until someone read them. An intentional relation, depiction, denotation, or signification requires a third term, consciousness. So, first structures that are beneath consciousness – universal grammar and so on – are not intentional. But following Freud's teacher, Brentano, intentionality is the essential mark of the mental. Now consider "following rules." A falling body instantiates the law of gravity, but it doesn't *follow* a rule in doing so. Following a rule requires consciousness. So, second, unconscious systems can't follow rules. But neither can they instantiate rules they represent unconsciously. For by the first result, no physical aspect of an unconscious, nonartifact could state or represent a rule, so, third, unconscious systems that are not artifacts cannot store representations of rules they instantiate. For all of these reasons, there can be no such thing as a system that computes unconsciously according to a stored program, and so no such hypothesis can explain human behavior.

But just why should we think that an unconscious nonartifact cannot, of itself, encode or represent a rule? Why could there not be, in Block's phrase, an "intentional zombie"? So far as I can tell Searle's argument is this: In order for something to be a stored program computer there must be a truth to the matter about whether or not in the course of physical changes the system goes through sequences of states that instantiate rules that are represented within the system. For computers that are artifacts there are such matters of fact, determined by the conscious intentions of the designers. If someone consciously follows a rule then there is also a fact to the matter. But for systems that, like the brain, are not artifacts, to have an unconscious computational structure there would have to be a fact of the matter about whether some structure in the brain (the "program," or part of it) denotes a set of input/output sequences

160

or "data structures" of which a process in the brain is an instance. Now denotation is a meaning relation, and from Quine's indeterminacy argument we know that there is in this circumstance no fact of the matter about them.

But Quine's "argument" is nothing more than behaviorist dogma. There is no more to it logically than the assertion that since synonymy and reference are underdetermined by behavior there is therefore (!) no truth to the matter about meaning. Programs are also underdetermined by behavior, but that is an epistemological point, not an ontological proof that there is no fact to the matter about which programs a system embodies.

On my reading, Searle's essay contains only an unsound argument on the following question: Could there be an object that is unconscious, not an artifact, and of which it is true that it is a stored program digital computer, and the fact that it is a stored program digital computer is the correct explanation of some of what it does? His argument aside, what is the answer? Certainly there are imaginable circumstances in which we would have a hard time resisting that conclusion: were we to find a copy of an IBM computer on Venus perhaps, or, as Alison Gopnik suggests (personal communication), were we to find a brain mechanism underlying particular conscious skills that is perfectly reproduced, save for a bit of circuitry characteristic of all conscious processes, in an unconscious process. One might hold (as I take it Searle would) that such inferences are no more than irrational (even if irresistible) anthropomorphism. I think that would be wrong.

Computational predicates given by a statement of a Turing machine program, for example, are complex structural descriptions, not material descriptions. They are much like descriptions such as "is an Abelian group." When applied to a system under a physical (or neurophysiological) description they are essentially Ramsey sentences. For any system specified by material properties, for example, by neurophysiological and physical properties, and for any two such distinct, variable aspects of the environment of the system, it is just a matter of fact whether or not there are definable properties whose extensions for the system realize a Post machine with a specific program for computing one aspect of the environment from another. Predicates such as "is a Post machine" or "is a Turing machine" or "is an URM machine" are similarly structural. They are true of some things and not of others. The phenomenal properties of my chair don't constitute a Universal Turing machine, and that is as much a matter of fact about my chair as that its arms are made of oak.

The problems in the very idea of computational cognitive science are very different from those Searle alleges. They are, first, that we don't understand very clearly what sort of structural predicate "is a programming system" or "is a computing system" may be. We don't, in other words, have useful characterizations of what individuates programming systems, and so we don't know quite what we are saying when we say the brain is such a system. And second, the most fundamental and urgent difficulties of contemporary cognitive science are epistemological, not ontological. They are difficulties of the underdetermination of program and computational architecture by all possible behavioral evidence, not of incoherence. That is reason enough for cognitive psychology to tolerate both modesty and neurophysiology.

Intentionality: Some distinctions

Gilbert Harman

Department of Philosophy, Princeton University, Princeton, NJ 08544
Electronic mail: *ghh@clarity.princeton.edu*

Step 1 of Searle's argument for the Connection Principle, "There is a distinction between intrinsic and as-if intentionality," conflates three distinctions. First, there is the dis-

tinction between intrinsic and relational intentionality. Second, there is the distinction between real and merely as-if intentionality. Third, there is the distinction between the intentional and the mental.

1. The intentional and the mental. The term "intentional" is a technical philosophical term used for any phenomenon that is directed towards an object that may or may not exist. A picture is an intentional phenomenon because it may picture something as a unicorn that does not exist. Since a picture does not have to be a mental picture, something can be intentional without being mental.

Mental phenomena are often intentional. You can think of something that does not exist. You can try to do something that you will not do. Ponce de Leon looked for the Fountain of Youth. The Fountain of Youth was the intentional object of his search: He was looking for something that does not exist.

Whether something can be mental without being intentional is controversial, depending on such things as whether sensations always have intentional objects or whether there are any truly objectless emotions. Harman (1990) argues that the mental is always intentional.

The needs of a plant or animal are intentional phenomena. A lawn may need to have water even if there isn't going to be any water for it. The lawn's need for water is an intentional but of course not a mental phenomenon.

2. Intrinsic versus relation. "Intrinsic" is another technical term in philosophy. A feature of a thing is intrinsic to that thing if the thing has that feature purely by virtue of the way the thing is in itself and apart from its relations to other things. That a person has a stomach may seem to be an intrinsic feature of the person, although as I'll indicate in a moment, that can be disputed. That the person is a father is not an intrinsic feature, because it involves a relation to something else. A person cannot be a father without there being a child that the person is the father of.

Suppose a person has the thought, "Here's a glass of water." Is the content of that thought intrinsic to the person? Putnam (1975) uses his twin-earth argument in favor of a negative conclusion. Whether the person is thinking about water or something else depends not just on what is inside that person's head but also on what sorts of liquid make up the rivers and lakes in that person's world, what the person drinks, and so on. So some real intentionality is not intrinsic.

In fact, it's quite unclear whether anything is really intrinsic to an object. A stomach is defined by its function in digesting food; so, whether an organ is a stomach or not may depend in part on certain functional relations to external objects. The color of an object may depend on relations to potential observers and so not be an intrinsic feature of the object. Whether the mass of an object depends on its relations to other objects is a deep question of physics.

Searle says, "when I say, 'I am thirsty,' I am describing an intrinsic mental state in me." Apart from general worries about intrinsic features of things, Searle's remark is clearly incorrect. The fact that the mental state in question is a state of thirst is clearly not intrinsic to the state. What makes it a state of thirst is in part its relation to dehydration and to drinking. A state that was phenomenologically the same but without such relations would not be thirst.

Furthermore, a person can be thirsty without realizing it. What makes it true that a person is thirsty is not the presence of a conscious mental state in that person but the presence of a functional state of a certain sort.

3. Real versus as-if. When gravity attracts an object it is as if the object needs or wants to go down. That is as-if intentionality. There is no real teleology in this case, no real intentionality. An object cannot be gravitationally affected by something that does not exist.

When a lawn needs water, that is not, as Searle claims, merely as-if intentionality. It's the real thing. The water that the lawn needs may not exist; the lawn still needs water.

161

4. Criticism of the beginning of Searle's argument. Searle's first two sentences are these:

> We often make metaphorical attributions of intentionality to systems where the system literally does not have that intentional state or, indeed, any mental life at all. I can, for example, say of my lawn that it is thirsty, just as I can say of myself that I am thirsty.

There are at least three mistakes in these two sentences. (1) Literally to attribute intentionality to something is not to attribute "mental life" to it. A picture of a unicorn is literally a picture of a unicorn without having any sort of mental life. (2) To attribute thirst to the lawn is not, either literally or metaphorically, to attribute any mental life to the lawn. (3) We can attribute thirst to a person without meaning to say anything about the mental life of that person, since we know that a person can be thirsty without this affecting the person's mental life.

Searle is also mistaken in his further remark, "When I say that it [the lawn] is thirsty, I am simply using a metaphorical way of describing its capacity to absorb water." To describe a lawn as thirsty is to speak teleologically about what the lawn needs. When paper towels or kitchen sponges are described as thirsty, that's clearly to speak metaphorically about their "capacity to absorb water." But it is quite unclear that there is anything metaphorical in the claim that a lawn is thirsty.

Searle's vision of psychology

James Higginbotham

Department of Linguistics and Philosophy, Massachusetts Institute of Technology, Cambridge, MA 02139

Electronic mail: *higgy@cogito.mit.edu*

John Searle's conception of psychological inquiry leaves no room for the ascription of thoughts to an agent that are not at least potentially conscious thoughts of similar agents. This conception rests on substantive assumptions, some of which, naturally, are not defended in the paper itself. Searle's argument proceeds from these assumptions to the denial of the cogency of some standard explanations in cognitive science, examples being the explanation of the Ponzo illusion in terms of unconscious inference, and the explanation of the human capacity to learn language in terms of a prior grasp of the nature of universal grammar.

Searle wants not only to dismiss explanations of these kinds as explanations, but also to account for their appearing to be explanations, and so, I take it, for the existence of disciplines that invoke them. And, finally, he wants to call attention to the empirical phenomena (such as the human capacity to learn language) that constitute the factual core for which the illusory explanations have been invoked.

We may therefore distinguish, besides possible internal criticisms of Searle's position, two sorts of external criticisms. The first would consist in the critical discussion of Searle's substantive assumptions, and the second in the examination of the allegedly illusory explanatory programs. I begin by sketching one example of explanation under such a program.

In linguistics, one typically extends the conception of language as rule-governed beyond what is explicitly or reflectively available to us, so as to include unconscious states of knowledge that guide us in speaking or interpreting. Native speakers of the English language will judge the (a) sentences grammatical and the (b) sentences ungrammatical in the following examples:

(1) (a) I consider him a traitor
    (b) I consider him traitors
(2) (a) I consider them traitors
    (b) I consider them a traitor

The standard explanation of this fact, in the terms of the mentalistic psychology that reflects in Searle's view a "big mistake in cognitive science," is that our judgment is consequent upon our tacit knowledge that predicates in English agree in number with their subjects. The explanation is thus of the same sort as ordinary explanations of judgment based upon knowledge, except that the knowledge is tacit rather than explicit, unconscious rather than conscious. Searle eschews such explanations, although he sees them as perhaps partially rescuable as "speculative neurophysiology." The questions external to Searle's discussion are whether they are cogent, his misgivings aside, and correlatively whether there is any other sort of explanation in the offing.

We may agree from the beginning (and not just for the sake of argument) that our conception of unconscious knowledge is parasitic on that of conscious knowledge, in that we would not invoke unconscious knowledge except in explanations of the type familiar to us from the practical syllogism, and we would not identify a brain state as a mental state except that it could be brought into connection with our mental life, more or less as we commonsensically speak of it. If these criteria are not observed, then we would be using what Searle calls "as-if" intentionality, and it is evident that the explanatory programs we are considering do not intend that. The extended conception of rules of language does indeed make use of genuine (Searle's "intrinsic") intentionality. Suppose now that some of the well-supported rules proposed in the inquiry are rules that cannot appear in the consciousness of the subject. The external critic of Searle asks: Why must that matter? Searle's paper embodies an answer to this question, to which I will return: In the meantime, I note as a point of logic that the criteria I have assumed are needed to justify the ascription of unconscious knowledge can be satisfied even if not every item of unconscious knowledge can be brought to consciousness, and that Searle's misgivings are themselves external to the discipline in question.

In what follows, I want first of all to clear away some points where I think Searle has not stated his own position as well as one might. I will then try to isolate Searle's substantive assumptions, concluding that at least one strong assumption that he needs for deriving his skeptical conclusion is unwarranted. Even granting the assumption, I will argue, there is a specific thesis within the argument of the paper that is stronger than we have reason to accept. Last, I will return to the question of whether Searle has dealt adequately, even within his own perspective, with the disciplines he considers suspect.

First, some clarifications. Searle writes that there is a question how in mentalistic psychology we are to distinguish unconscious mental phenomena from phenomena that are not mental at all: "How do we distinguish, for example, between my *unconscious* belief that Denver is the capital of Colorado when I am not thinking about it and the *nonconscious* myelination of my axons?" But by referring to the first state of the brain as an unconscious *belief*, Searle has already distinguished it as mental. What he means to ask, I take it, is how we are to know which brain states, as physically described, are also mental states. His first answer to this question is that a brain state is mental if it is "capable of generating conscious states." This standard is evidently very generous, since it would admit as mental the states that form the background, for example, of the conscious judgment that (1b) is ungrammatical, and there is no obvious reason why one of these states is not the state of knowing unconsciously that subjects agree in number with their predicates. Later, Searle remarks that the fact that a brain state causes conscious mental states does not have to make it mental in the sense of having a content: It may be, in his terminology, a mere "association pattern." But his discussion to this point gives no reason for *withholding* the ascription of content to an association pattern. What should induce us to withhold it, in Searle's view, is that it is not potentially conscious. But here we should distinguish the question of whether we can merely become conscious that we are following some rule from the question of whether we can comprehend ourselves as following the rule (e.g., as going through the steps of the deduction of the ungrammaticality of [1b]) in the course of coming to form the judgment or other

162

manifestation of its use. I think it must be the impossibility of the latter that would, in Searle's view, disqualify a brain state from having the content that mentalistic psychology would allow it as having; for otherwise it would be sufficient in rebuttal to observe that our unconscious rules are rules that we may know that we come to follow, even if we cannot be aware of ourselves as following them.

Parenthetically, it may be noted that there are a number of ways in which unconscious thoughts may be in principle unconscious that would not lend any support to Searle's position. Consciousness, for example, may be too limited in capacity, or too slow in operation, for it to be feasible to have a creature that follows consciously with proper efficiency and speed the rules that we follow unconsciously (these possibilities have been suggested by several authors, of widely different views on the problems of cognition).

I will assume, then, that the conclusion Searle intends to reach is that, unless we can *become conscious of ourselves* as following rule *r*, we cannot say of ourselves that we are following it, or identify any brain state as the state of following it. To reach this conclusion, Searle deploys first of all a physicalist assumption, or rather a range of such assumptions; these are expressed in the paper in terms of the sufficiency of neurophysiology for mental activity. (Physicalism, of Searle's kind or another, has often been taken for granted in contemporary discussions in the philosophy of mind and psychology, and as such has not been so much a philosophical or scientific thesis as the price of admission to the debate.) My first aim is to sort the assumptions out.

Searle maintains the thesis that the neurophysiological is "causally sufficient" for the mental, and also the thesis that a "perfect science of the brain would be stated in a neurophysiological . . . vocabulary." The first thesis obviously does not imply the second; but neither does the second imply the first unless, indeed, the "perfect science of the brain" would by definition also be the perfect science of our mental life. But Searle holds that the neurophysiological is not constitutive of the mental, so the second thesis must for him be independent – the "perfect science of the brain" is not the perfect science of our mental life, since, according to Searle, supplementary connections between the neurophysiological and the mental are necessary (see the discussion in step 3 of his argument).

One might wonder why Searle is so confident that neurophysiology is the most detailed physical level at which the laws of the activity of the brain are to be found. Perhaps deeper, subatomic phenomena are crucial. But in fact Searle's reference to neurophysiology is inessential. All that is required is that there be some level of physical description *L* such that the properties of our bodies at level *L* are "causally sufficient" for the mental, in whose vocabulary a "perfect science" of the body would be stated, and which is not constitutive of the mental itself. It is evidently the second assumption that wants defending. If there is, in addition to "blind, brute neurophysiological processes," also blind, brute rule following at level *L*, then the first and third assumptions may be true, and the second false, and no argument against the existence of unconscious rules emerges.

The sleeping man knows that Denver is the capital of Colorado, or has the rule of number agreement between subjects and predicates. In Searle's picture, this can only be so because he is disposed to entertain this knowledge consciously (under the right circumstances) when awake. Suppose we have no notion of unconscious mental states of any type except insofar as we know about such states in relation to consciousness. Even so, we do not yet reach Searle's conclusion. Searle's argument requires much more than the manifestation of unconscious thought in the etiology of some conscious behavior, for it requires that the attribution to an agent of an unconscious thought that *p* should be justified by the possible consciousness to that agent or to another similar one of *that very thought*. I do not see that a reason has been given in support of the latter conclusion.

Above I have expressed doubt that Searle's version of physicalism – the causal sufficiency of the neurophysiological for the mental – will support his rejection of unconscious rules, and I have doubted also that the connection between the realm of tacit thought or action, including both the conscious and the unconscious, requires any point-by-point correlation with the realm of potentially conscious thought or action. In conclusion, I turn briefly to Searle's examination of the disciplines that do invoke unconscious thoughts and rules, and the factual core that he does accept.

Searle grants the existence of a language faculty, or a specialized part of us that constitutes our language readiness. He does not grant the description of this faculty as knowledge of universal grammar. In his view we have, at best, a language sensitivity, on a par with color sensitivity. We are able to take in certain languages, but not others, as we are sensitive to certain colors and not to others.

As it stands, Searle's comparison of languages with colors will not do, since we project our languages from linguistic experience, and that activity has no counterpart in color perception. What is left of the projection, Searle says, is that for certain properties *F* we may say that our language readiness "makes it possible to learn *F* languages and not possible to learn non *F* languages." No hint of theory is in sight here; rather, we have a replacement of explanations of the customary sort with sheer description of the facts that they were supposed to explain. It is thus highly misleading for Searle to say that the evidence for universal grammar is "more simply" accounted for in his view, since the alleged account consists merely in stating the empirical facts.

Is there any account of those facts, in fact or on the horizon, that Searle would find in principle acceptable? He suggests that (some) connectionist views are acceptable, but I do not see that his statements about the nature of these views can be correct. He writes that connectionism proposes no "meaningful phenomena" between inputs and outputs. On the contrary, the nodes of connectionist systems are put forth as meaningful, even if their meanings do not express rules. I think Searle needs to be more precise about this.

To summarize, I have suggested that Searle's strong physicalist assumption, that the "perfect science of the brain" would contain no mention of rules, is not well supported, and certainly draws no hint of support from the causal sufficiency of the neurophysiological (or other physical level) for the mental; that even if it is true that unconscious thought is only reasonably and literally to be posited for the explanation of typically conscious activity, it cannot be inferred that each unconscious thought is capable of being conscious; and that Searle has not presented a possible alternative to familiar forms of psychological explanation.

It is obvious that the question of what constitutes "in principle" consciousness is difficult to pin down in any way that would make Searle's thesis more precise, both because consciousness may be too limited a faculty for us to be conscious of very much that we actually think about, and because, as his reply to the first of the objections he considers at the end of his paper makes clear, a sufficient reason for saying that something is "in principle" conscious to us is that some (hypothetical) creature roughly similar to us actually would be conscious of it. Both the objection and Searle's reply, however, fail to consider the possibility that the realm of unconscious thought and rule following may be justified in terms of the resulting overall theory of our rational activity, so that it is by no means required that each unconscious thought have its own bit of distinctive conscious manifestation.

I am not arguing against Searle that the appeal to unconscious thoughts or rule following is a trivial extension of what we do already in the practical syllogism, unworthy of the fuss he makes over it. On the contrary, that appeal needs continual reexamination in the light of the theoretical advances and bottlenecks to which it gives rise, and it needs to be brought into relation to

163

what we know about the makeup of our nervous systems. The difficulty for Searle is with his thesis, or vision of proper psychology, that we already know enough about the physical background to our mental lives that we can in our enlightenment dismiss unconscious knowledge out of hand. Supposing that unconscious knowledge is grounded in the physical properties of the nervous system, we might say, following Searle, that cognitive science is indeed engaged in speculative neurophysiology. Buy why does Searle suppose that explanations in that science cannot both be speculative neurophysiology and be true?

Matter, levels, and consciousness

Jerry R. Hobbs
Samoff Research Institute International, 333 Ravenswood Ave., Menlo Park, CA 94025
Electronic mail: *hobbs@al.sri.com*

Searle's ontology is at once richer and more barren than that of most cognitive scientists. He says repeatedly that matter exists (in particular, neurons), and conscious, intentional experience exists, and that is all. If we accept this, it follows trivially that there are no deeply unconscious rules operating at a symbolic level intermediate between neurophysiology and intelligent activity. There are no such rules because they are neither neurons nor potentially conscious intentional experience. The six carefully argued steps in his demonstration could have been dispensed with entirely. Throughout his long argumentation, however, we find no justification for an ontology so barren.

For example, in step 4 Searle says, "Well, the only facts that could exist while he is completely unconscious are neurophysiological facts. The only things going on in his unconscious brain are sequences of neurophysiological events occurring in neuronal architectures. At the time when the states are totally unconscious there is simply nothing there except neurophysiological states and processes." This is repetition, not argument.

Levels. Such an ontology runs counter to the spirit of the entire scientific enterprise, for Searle apparently would not admit such complex, large-scale, diffuse organizations of matter as geological faults, immune systems, data structures, and nations. There is nothing but matter and conscious experience. To a scientist's arguments that such entities are useful descriptively and that their reality is at least plausible, he would apparently reply in triumph that these intermediate levels can be dispensed with, and for the sake of parsimony, should be.

Science is organized by levels, a strategy that is successful probably because nature is organized by levels. There are at least two ways we can view these levels. First, we can view them as *levels of description.* Nature cannot be usefully described solely in terms of the motions of elementary particles. We have found it convenient to define or hypothesize larger-scale entities and to couch our theories in terms of them. We then try to account for the behavior of these larger entities in terms of the entities provided by the one or two levels down. Thus, chemists seek to understand in quantum theoretic terms why molecules react as they do.

Second, we can view the levels as *levels of organization.* That is, they are not merely convenient fictions that allow our poor, finite minds to understand what is going on. There is something in nature that actually corresponds to these large-scale entities and actually behaves approximately in the manner that our theories describe. The argument for assuming that these things are really out there in the world has often been stated: We should adopt the ontology implied by our most successful theories. The reality of the ontology is the best explanation for the success of the theory. Molecules, cells, tissues and organs,

organisms, herds, and nations are not merely stories we tell. They really do exist.

Levels of description are not necessarily levels of organization. Explanations based on a metaphor of intentional action may once have been useful in plant physiology; they may have corresponded to a level of description. But this did not come to be viewed as a level of organization. The biochemical mechanisms involved turned out to be simple enough that this intermediate level was not required.

Searle seems to lack this notion of entities at different levels. His unwillingness to accept the notion of two different entities operating at two different levels but realized in exactly the same portion of matter is, in fact, the reason for his failure to accept the "systems reply" to the Chinese Room Argument – an argument that is now primarily of historical interest, though of course it still rages on as a sociological phenomenon.

What I take to be the standard view, or hope, in cognitive science today is the following: Intelligent activity is implemented in a symbolic level, which is in turn implemented in a connectionist architecture, which is in turn implemented in neurophysiology. Variations on this view dispense with the symbolic or the connectionist level. (Searle's "discovery" that the symbolic level could be eliminated in favor of a more powerful connectionist level is a rather commonplace view among workers in the field.) The reasons for this strategy are clear. We can observe intelligent activity and we can observe the firing of neurons, but there is no obvious way of linking the two together. So we decompose the problem into three smaller problems. We can formulate theories at the symbolic level that can, at least in a small way so far, explain some aspects of intelligent behavior; here we work from intelligent activity down. We can formulate theories at the connectionist level in terms of elements that behave very much like what we know of the neuron's behavior; here we work from the neuron up. Finally, efforts are being made, with modest success, to implement the key elements of symbolic processing in connectionist architecture. If each of these three efforts were to succeed, we would have the whole picture.

What success there has been at the symbolic level in modeling intelligent activity has been due to hypothesizing operations that are very close to conscious intentional mental operations – idealized, cleaned up versions of them. In the most successful symbolic theories, the processor at the symbolic level does such things as manipulate symbols, follow rules, draw inferences, and plan. (And contrary to what Searle says, this does not require a homunculus. There are no little men in our machines.) Symbolic operations are for the most part based on metaphors drawn from folk psychology, and their success is largely parasitic on the success of folk psychological explanations of intentional behavior. This "anthropomorphizing" is not a "mistake" that Searle has discovered. It is a quite deliberate strategy.

I find incomprehensible Searle's statement that these symbolic operations have no causal powers. When they are written down on paper, of course they have no causal powers. Just like real intentions, they have causal powers if and only if they are implemented in neurophysiology or in electronics or in some other adequate material medium. Moreover, most computer programs implementing this strategy do not merely *behave* as though they were following rules. The rules really are there in the data structures, and the following really is there in the procedures that manipulate the data structures.

The symbolic level is at least a level of description. Whether it will turn out to be a level of organization we simply don't know today, although many cognitive scientists believe it successful enough already to adopt the ontology it implies, and say that symbolic operations really do exist. It may be, as Searle suggests and many connectionists believe, that the symbolic level will wither away as better direct accounts relating neurophysiology and intentional behavior become available. Or it may be that a

164

symbolic level will come to be viewed as being as useful a concept and as real in psychology as tissues and organs are in physiology.

There *is* a notion of levels that Searle makes use of, but this is a quite different notion and should probably be thought of not as levels but as perspectives one can take at any level. There are exactly two of these, and they may be called the *structural* ("hardware") perspective and the *functional* perspective. From a structural perspective, we attempt to decompose the entity into its constituent parts and tell a causal story about how the properties of the entity emerge from the properties of its parts. From a functional perspective, we consider the entity as a whole, undecomposed, and ask how it influences and is influenced by its environment. It is not an intrinsic feature of the functional perspective, as Searle says, that it be related to human interests, although, unsurprisingly, most functional accounts we have bothered to construct are.

Consciousness. I said that Searle's ontology is not only more barren but also richer than that of most cognitive scientists. He believes not merely in the existence of conscious, intentional experience, but in its special, fundamental, explanatory power. It seems to play a role equal to that of matter: Matter can cause, and conscious intentions can cause. Moreover, this is a position that he attributes to cognitive science, in what can only be characterized as a fundamental misunderstanding.

The assumption is so deeply ingrained in what Searle has written that it is not easy to spot. But near the end of Section 5, he says, "*as-if* intentional states, not being real, have no causal powers whatever. *They explain nothing*," and "*as-if* intentionality . . . simply restates the problem *which the attribution of real intentionality is supposed to solve*" (italics mine). It looks very much as if Searle believes that the postulation of deeply unconscious intention-like operations at the symbolic level is an attempt to appropriate the causal, explanatory power that he supposes conscious intentionality to have. But this is no part of the strategy of cognitive science.

It really is true of Freudian psychology that the hypothesis of subconscious intentions is meant to borrow the causal, explanatory power of conscious intentions. In folk psychology, conscious intentions have explanatory power; explanation can often stop there. In the social sciences, where we often take folk psychology as a background science and hence as unproblematic, conscious intentions can have explanatory power; explanation can again stop there. However, cognitive science, in its use of deeply unconscious intentions, is not seeking "intentionalistic *explanations*" in the sense of Section 3. In cognitive science, intentions do not have special explanatory power. Explanation cannot stop there, because intentional behavior is precisely what is to be explained. Conscious intentions have no causal powers except insofar as they are viewed as complex processes implemented, through who knows how many levels, in neurophysiology. They cause only because matter causes and they are implemented in matter.

Searle seems to believe that intentionalistic explanations are valid in psychology when applied to potentially conscious intentions; he accuses cognitive scientists of illegitimately extending them to "deeply unconscious intentions." But in fact cognitive scientists have rejected intentionalistic explanations altogether, and use the hypothesis of deeply unconscious intentions as a way station to a structural, computational explanation. If deeply unconscious rules and representations could not be realized computationally, and hence materially, they would be of no value.

Searle closes his article with the statement, "We know for sure that there is a brain and that at least sometimes it is conscious." Could it be that he believes, since each of us knows for sure that our brain is sometimes conscious, that that fact requires no further – structural – explanation?

Consciousness, in any case, seems a very shaky foundation to build a science on. I know I'm conscious. I can't be sure about other people, but it's a courtesy I'm willing to extend. I would extend it to orangutans as well, but probably not to birds or to any computer program I've ever encountered, and certainly not to amoebas.

Although the intelligent activity that cognitive science seeks to model is normally exhibited by conscious people, there is a sense in which Searle is quite right in saying that the field has largely "neglected" consciousness. There are some aspects of consciousness that can be modeled in symbolic terms, such as focus of attention and knowledge of one's own beliefs. But how the subjective experience of consciousness could emerge from complex arrangements of neurons is a mystery apparently inaccessible to present-day cognitive science. The neglect is not because it is in bad taste to study it, but because there are no very good ideas about it.

I can imagine Searle replying to all this with, well then, the symbolic operations are just *as-if* intentionality, and not intrinsic. I believe it is possible to justify attributing intrinsic intentionality to them on the basis of their place in a large system of similar rules, some deeply unconscious and some quite conscious and frequently verbalized. However, in cognitive science, since intrinsic intentionality confers no extra explanatory power, the issue is not very important.

"Consciousness" is the name of a nonentity[1]

Deborah Hodgkin[a] and Alasdair I. Houston[b]

[a]*Physiology Department, University of Cambridge, Cambridge CB2 3GE, England and *[b]*Department of Zoology, University of Oxford, Oxford OX1 3PS, England*
Electronic mail: *houston@vax.oxford.ac.uk*

Searle's conclusion that an economical, functional neurophysiology can effectively replace the confused conceptions of unconscious processes implementing complex grammatical or perceptual rules, is entirely welcome, basically correct, and a productive general position from which to tackle difficult information-processing problems like language and memory. So welcome is this "Darwinian shift" that it is tempting to suggest: If these are the arguments that will persuade a certain kind of cognitive scientist to stop attributing grammatical rules and so on to unidentified cognitive processes, so be it. One can then simply enjoy the ensuing battle in which Searle deftly deploys his inimitable combination of logic and theatre to disarm all opponents. It is not clear, however, that the arguments embodied in the Connection Principle are either necessary to achieve this final conclusion or do in fact damage the class of cognitive theories at which they are directed. Accepting the Connection Principle may also commit cognitive science to an unduly restrictive set of assumptions, which may be almost as damaging as the pre-Darwinian position on which it seems to be based.

(1) Whereas Searle is right in saying that many theories in cognitive science are guilty of muddling epistemic and causal levels, it is obviously not essential to use the Connection Principle to draw a clear distinction between the formal theoretical specification of a particular subject matter – such as grammar or perspective – and a description of how the brain achieves these feats. The fact that such theories are frequently beguilingly expressed in terms of computations, rules, inferences, and so on, has often deluded cognitive scientists into treating them as descriptions of processes. This tendency is particularly prevalent among those who study language, probably because, as Searle suggests, so little is known about the neural processes underlying language. In the area of vision, although there is still a long way to go, it is possible, at least occasionally, to recognise that some theories are highly unsuit-

165

able as potential models of brain processing (e.g., early models of how to extract shape from shading, Horn 1977), whereas others are more suitable (e.g., Lehky & Sejnowski 1988). Similar distinctions between theory and its behavioural implementation have been drawn in other areas of biology, for example, the distinction between the theory of kin selection and the actual ways animals behave toward their relatives (Dawkins 1979).

2. The principal tenet of Searle's argument is the denial of deep unconscious processes that are in principle inaccessible to consciousness. Searle argues that "both the epistemic and the causal claims presuppose that the ontology of deep unconscious rules is perfectly in order" and that by challenging this ontology "the epistemic and the causal claims collapse together." In fact, many of the offending cognitive theories do not invoke deep unconscious processes of the kind which are in principle inaccessible to consciousness. Precisely because many of them are in fact analytic theories, that is, putative formal specifications of the problem, they tend to be expressed in terms of rules, mathematical theorems, inferences, and so on, which are exactly the sort of things of which we can be conscious. Such rules were attributed to unconscious processes because people knew that formal representations of them did not describe the way they were performing such feats as understanding language, seeing perspective, and so on. They were implausible conscious mechanisms, not impossible ones. In fact, Emmert's Law is precisely the sort of process an interested blind child might consciously learn to use so as to construct a faithful reproduction of Brunelleschi's theory of perspective. Whether we would then say that learning Emmert's Law allows the blind child to understand perspective in the same way that, say, learning echolocation skills allows blind children to build up a richly textured, two and three dimensional world is more complicated. But Searle has demonstrated that simply instantiating a set of rules does not provide a sufficient account of knowing something. So even if we accepted that deep unconscious processes are in principle incoherent, a large number of cognitive theories would remain unscathed and could blithely continue misattributing theoretical terms to cognitive processes.

(3) For his paradigm of unconscious thought, Searle focuses on sleep and dreams and their well-known propensity for false belief. There are other cognitive areas of mental activity, however, where it may be easier to find cognitive processes which are in principle inaccessible to consciousness. One prominent theory of preattentive processes (Treisman 1980; 1988) suggests that at early stages of vision, information about the colour, shape, and movement of objects may be registered independently and in parallel, and only subsequently combined into the familiar conjunctions of coloured, moving shapes of which we are conscious. [See Strong & Whitehead: "A Solution to the Tag Assignment Problem" BBS 12(3) 1989.] (The psychophysical evidence fits in with the increasing neurophysiological evidence for separate cortical visual areas dealing mainly with form [V1] colour [V4] and movement [V5]; see Cowie 1979 and Zeki & Shipp 1988). This suggests that there are cognitive events which can be unconscious in principle precisely because conscious experience is "wired up" to appreciate conjunctions of colour and form. This preattentive parallel registration of information is not a case of something which could be conscious but happens not to be, as Searle treats the examples of blindsight, repression, and the desire for water/H_2O. They are the kind of cognitive events which cannot be conscious in principle, even though they may form essential constituents of some conscious event and its attendant aspectual characteristics. Obviously Searle's argument would lose all force if he allowed the reductio ad absurdum that conscious thought has an infinitely "surreal" imagination in which anything is possible; the key claim in the connection principle is that "any unconscious intentional state is the sort of thing that is in principle accessible to consciousness."

The experimental investigation of preconscious processes in

perception, memory, and language is among the most exciting areas of cognitive science, e.g., Treisman 1985, Treisman & Gormican 1988. The evidence is often indirect, based on inferences from careful measurements, but no more so than electron micrographs, X-ray crystallography or protein chemistry. It is clearly cognitive/mental and not just neurophysiological. It will be interesting to hear Searle's account of whether such preconscious constituents of conscious events are deemed to have aspectual shape and if not, how they achieve their content.

Although the Darwinian shift in biology made it possible to reposition various kinds of functional and causal explanation, it did not thereby solve the problem of the complex relationship between theoretical insights and experimental discovery. Searle covers a wide range of biological examples but this sometimes results in potentially confusing language. In behavioural biology, causation is distinct from function and the Darwinian approach may result in a series of explanations; a functional account relates behaviour to the animal's Darwinian fitness; a computational model might bring this behaviour about in appropriate circumstances and physical instantiation. Unpacking what Searle describes as "several functional layers" may be a much more exacting and interesting process of discovery than he makes it sound.

NOTE
1. Title is taken from William James (1912) Does consciousness exist? Essays in radical empiricism, London.

On doing research on consciousness without being aware of it

Daniel Holender
Laboratoire de Psychologie expérimentale, Université libre de Bruxelles, B-1050 Bruxelles, Belgium
Electronic mail: R07203@BBRBFU01.bitnet

My commentary on Searle's thought-provoking target article is divided into two parts. In the first part, I am basically asking the question of whether the Connection Principle is refutable. In the second I discuss some consequences of the neglect of consciousness in current interpretations of a subclass of empirical data supposed to reflect preperceptual processes.

Refutability of the connection principle. Searle (1980c) concluded his discussion of Chomsky (1980b) by asserting that in exchange for being entitled to postulate rules and representations not accessible to introspection, Chomsky should at least care more about showing their causal role through some indirect effect. Suppose Chomsky has now provided the requisite information. Would Searle concede that mental phenomena that are in principle inaccessible to consciousness have finally been discovered? Or does the present statement that unconscious mental phenomena are in principle accessible to consciousness imply that Searle no longer considers this an empirical issue?

Granting that the evidence for tacit language knowledge may not be satisfactory, what about other recent evidence for dissociations between implicit and explicit knowledge in neuropsychological syndromes (e.g., Schacter et al. 1988), memory (e.g., Schacter 1987), and learning (e.g., Reber 1989)? Most of the extensive recent work suggesting the existence of several kinds of implicit knowledge is based on what Erdelyi (1985; 1986) has called the *dissociation paradigm* of the unconscious. In this paradigm, an indicator of knowledge available to the brain shows a positive effect even though an indicator of the information available to consciousness, such as introspective reporting, does not disclose any such knowledge.

Not all the recent claims for the existence of implicit knowledge are equally challenging for Searle's position because not all forms of implicit knowledge have the same properties. He could rightly argue, as he does for blindsight, that the forms of implicit knowledge resulting from brain damage are "in principle" available to consciousness. [See Campion & Latto: "Is Blindsight an Effect of Scattered Light, Spared Cortex, and Nearthreshold Vision? *BBS* (3) 1985.] Implicit memory may not affect Searle's position either because not all the interpretations of the observed indirect effects rely on unconscious processes that qualify as mental. Implicit learning is more challenging to the Connection Principle, however, because some authors, such as Broadbent (e.g., Hayes & Broadbent 1988), would argue that there is a mode of learning resulting in tacit knowledge that, although completely unavailable to consciousness, shows intrinsic intentionality in playing a causal role in the production of behavior.

The problem is that Searle can still argue that the dissociation between implicit and explicit learning may simply result from a failure to index adequately the availability of information to consciousness, perhaps because of a failure to ask the right question at the right time. There are indeed serious experimental indeterminacies in the dissociation paradigm as it is commonly used (Erdelyi 1986). The two main problems are finding an adequate indicator of consciousness and showing that this indicator has zero sensitivity.

Reingold and Merikle (1988) have argued that both problems can be solved provided the indicator of availability of the information to the brain and the indicator of availability of the same information to consciousness are measured on the same metric. This would enable us to conclude that unconscious knowledge exists simply by finding a greater sensitivity of the former than the latter indicator, even if the latter shows greater than zero sensitivity.

Reingold & Merikle's proposal was cast in the framework of subliminal perception but it can obviously be generalized to the investigation of any form of implicit knowledge by means of the dissociation paradigm. Assume that we find a situation satisfying the auxiliary conditions specific to the form of knowledge under study and demonstrating the existence of unconscious mental contents exerting a causal role in the determination of behavior. If so, should we still consider these mental contents as in principle accessible to consciousness rather than in principle inaccessible to it but devoid of intentionality?

Some consequences of the neglect of consciousness. What follows concerns exclusively Searle's class of "association patterns" mediating between sensory inputs and conscious mental representations. For Searle, these processes are purely neurophysiological and devoid of any mental content. For Fodor (1983; 1985), parts of these processes are modular input systems that violate the Connection Principle in being both computational and cognitively impenetrable.

By analogy with the examples developed in the target article, the following should adequately characterize Searle's position with respect to modular input systems. He probably thinks that there is no lack of empirical evidence for Fodor's postulation of mental processes in principle inaccessible to consciousness. The mistake is rather in the postulation of meaningful unconscious processes where no meaningful conscious ones can be found. Some connectionist models offer a way to characterize the mediation between a meaningful input and a meaningful conscious percept that does not violate the Connection Principle in anthropomorphizing input systems. Hence, this subset of connectionist models may well offer an adequate description of how neurophysiological perceptual processes can generate conscious perceptual contents.

Unlike Searle, I am much less concerned with the ontology of the processes we are supposed to study than with their epistemological status, while sharing his belief that the neglect of consciousness is a mistake. One reason why it is a mistake is that it is misleading. It is misleading because it often makes us overlook the influence of postperceptual decisional processes on

the effects of the variables we are manipulating. Unlike Searle, I do think we *in fact* lack solid evidence for concluding that our current experimental tasks enable us to tap into unconscious processes, whether mental or neurophysiological (Holender 1986; 1987a; 1987b; in press). Although Fodor has adequately characterized the current theoretical biases of a nonnegligible subgroup of cognitive psychologists, the experimental evidence for modular input systems is at best extremely scanty. Unless connectionist models provide a good account of postperceptual decisional processes, they do not fare better in accounting for the majority of our data either.

The foregoing assertions can be illustrated using the paradigmatic example of the shift in the interpretation of the data obtained with the lexical decision task. In this task, one measures the latencies for classifying letter strings into words and nonwords. First, Balota and Chumbley (1984), among others, provided a convincing demonstration that the variations in latency induced by the manipulation of word and nonword properties in no way reflect the lexical access process; they reflect postperceptual decisional processes instead. Second, a similar shift in interpretation has been proposed for the semantic priming effect – the finding that the positive lexical decision is faster when the target word is preceded by a semantically related priming word, compared to a semantically unrelated word. The semantic priming effect has traditionally been accounted for in terms of both a postperceptual decisional component and an unconscious, automatic spreading activational component. de Groot (1985) and Neely et al. (1989), among others, however, have provided strong evidence that the priming effect can be accounted for entirely in terms of postperceptual decisional processes.

Those involved in the research under discussion are confident that a complete shift from a preperceptual to a postperceptual interpretation provides a valid account of most of the effects observed with the lexical decision task, yet most are extremely reluctant to relinquish automatic spreading activation as an explanatory concept. Both de Groot (1985) and Neely (in press) have tried to rescue it by relying on the putative demonstrations of priming by masked unconscious primes.

Unfortunately, previous conclusions about unconscious priming (Holender 1986; 1987a) remain valid for the new experiments published since then: None of the tentative demonstrations of unconscious priming have provided the requisite controls to ensure that the priming word was not consciously identified at the time of presentation. Furthermore, the postperceptual interpretation of semantic priming in the lexical decision alluded to above precludes any priming effect in the absence of a conscious representation of the meaning of the priming word. Indeed, if instead of having lived with a spreading activation metaphor for the last 20 years or so we had been more biased in thinking about the priming effect in terms of postperceptual decisional processes, we would not have called it a "priming" effect in the first place.

The moral of this story is that experimental paradigms die reluctantly and the concepts they generate often survive even longer. Numerous stories similar to that of the priming effect have plagued the modern approach to cognition. The pattern is always the same. We start believing that an experimental task allows us to tap into early preperceptual processes. Later – generally much later – we realize that postperceptual decisional processes can account for our observations. In the meantime, we are in a way doing research on consciousness without being aware of it or, perhaps sometimes without wanting to know.

There is more to consciousness than the elementary decisions needed for responding adequately in the very simple tasks imposed by cognitive psychologists on their subjects. Yet it is now clear that many of the effects on performance of the variables manipulated in these tasks reflect nothing other than postperceptual decisional processes dealing with conscious mental contents. The problem is that once they become aware of this, cognitive psychologists start looking for still other tasks supposed to tap into preperceptual processes, they never seem to get around to studying consciousness in a positive way.

ACKNOWLEDGMENTS
This work has been supported by the Belgian "Fonds de la Recherche Fondamentale Collective" (Convention 2.4562.88) and the Belgian Ministry of Scientific Policy (Action de Recherche concertée "Processus cognitifs dans la lecture" and National Incentive Program for Fundamental Research in Artificial Intelligence.)

Is Searle conscious?

John C. Kulli

Department of Anesthesiology, University of Rochester School of Medicine, Rochester, NY 14642
Electronic mail: jkul/a db1.cc.rochester.edu

As an anesthesiologist, I share Searle's interest in determining the presence or absence of consciousness. In analyzing the problem, I divide the nervous system into two types of neurons, those concerned with consciousness, "C" neurons, and those which take care of unconscious functions. "U" neurons (the use of the word "neuron" in this context is shorthand for "otherwise unspecified subpart of the brain"). The goal of anesthesia is to interfere temporarily with the function of the C neurons without disturbing the U neurons. To do this rationally, I must (1) be able to discriminate reliably between the patient who is conscious and the patient who is unconscious, and (2) identify the C neurons. The general approach to (2) is to observe, with Searle's "brain-o-scope," as it were, the function of the nervous system in patients who are first conscious, then unconscious, then conscious again, in an attempt to identify neurons which are affected by a particular anesthetic agent. Comparisons of the effects of different anesthetics might then allow us to identify, for example, those neurons affected by all anesthetics (Hille 1983). Several candidate brain-o-scope technologies allow us to construct three-dimensional images of brain activity (Rogers et al. 1990). The intractable part of the problem is the detection of the conscious state (Michenfelder 1988).

Has Searle helped me? It is exactly this problem that Searle addresses: What does it mean for someone (or part of someone) to be conscious? Searle describes the criteria he uses to judge the existence of the mental state (he uses the expressions conscious, mental, and intentional interchangeably). To qualify as a mental state, a brain phenomenon must have intentionality, which cannot exist independent of aspectual shape (Searle's step 2). Aspectual shape seems to be similar to Titchener's (1910) context theory of meaning. But aspectual shape, although Searle finds it easy to understand for conscious thoughts, cannot be defined by objective criteria (Searle's step 3); and thus unconscious processes cannot be shown to have it. Therefore, seemingly conscious actions on the part of the brain that are not (in principle) accessible to consciousness are just brute neurophysiology and not mental at all. Examples of this would be the machinery that underlies the Ponzo illusion or the language acquisition device.

Searle's analysis shows a consistent viewpoint throughout: The conscious part of his brain (the C neurons, as it were) are engaged in the interesting task of determining whether Searle's U neurons are themselves conscious. The principle that the C neurons use to rule out consciousness in the U neurons is, unfortunately, so restrictive that it rules out consciousness in any or all of the neurons of any other person. From the point of view of the C neurons, there is, in another person, only brute neurophysiology to be seen. Searle feels that this rules out a full characterization of the aspectual shape; there is no intentionality without aspectual shape; thus the C neurons should conclude that only they are conscious. Returning to the Chinese Room,

Searle, moving the characters around according to the rule book, not understanding Chinese at all, corresponds to the U neurons of the Chinese Room; he is, by his own argument, unconscious. The narrowness of the grounds Searle uses to judge the existence of consciousness is such that the only conclusion is solipsism. I leave the discussion of Searle's conclusion, that lack of unconscious mental events forces us to abandon functional cognitive theories, to other commentators.

Several different possible theories can explain the unconsciousness of the U neurons, which Searle does not distinguish. For example: (1) The U neurons are different from (cruder, simpler, more primitive than?) the C neurons and therefore cannot support consciousness. (2) The U neurons are really conscious but just don't (or can't) communicate with the C neurons (see Pucetti 1981, for an example). (3) The positivist objection: It is not meaningful to speak of mental activity on the part of the U neurons because there is no way to know their status. (4) The definition of mental activity is incompatible with the idea of unconscious mental activity. There are elements of each in Searle's argument; unfortunately, they are not mutually consistent.

Two of these theories are of some interest in the anesthesiology problem. The idea that the U neurons are different is approximately the same as the idea that the C neurons have been impaired by the anesthetic and have become incompetent to support consciousness; that is, drugs have turned them into U neurons. This is an empirical claim that is not considered in Searle's paper. The second is more interesting: What if the U neurons are fully conscious but unable to communicate their situation to the only outside observer available, the collectivity that is the C neurons? How could this state be identified? It is the possibility of this state during anesthesia, easily producible with common anesthetic agents (Vickers 1987), that has motivated the search for ways to recognize consciousness. Searle does not seem to help.

What's it like to be a gutbrain?

John Limber

Department of Psychology, Duke University, Durham, NH 03824
Electronic mail: j—limbena unhh.bitnet

Searle presents a view of psychological explanation touted as radically different from the contemporary "pre-Darwinian" notions of explanation found variously in the cognitive sciences. Although I am not so confident about all this as Searle seems to be, I hardly find his ideas radical. Neither should the odd Cartesian or behaviorist who happens on this paper.

Searle argues that there is no level of mental causality that is not "in principle" accessible to consciousness – his "Connection Principle" (CP). Unless the CP holds, intentionalist accounts of behavior are incomprehensible and can be interpreted only as efforts to rationalize what the causal hardware is doing from some external perspective. Finally, he tries to convince us that this principle "radically" changes our conception of psychological explanation.

Three issues deserve further discussion. Searle has started off with a conception of consciousness that largely ignores the history of this topic. His fear of intentionality in everything from stones to infants acquiring language is absurd. Finally, I do not see this as a very radical innovation in psychological explanation. Even if it is, the vagueness of the "in principle" clause of the CP makes it of little more than therapeutic value in reminding us again of the possible errors in attributing intentionality in an explanation of complex behavior.

Consciousness in psychology. Until the eighteenth century, a standard "Cartesian" view of consciousness was widely accepted. One knew about consciousness either from firsthand experience or inferentially through the language of another – call this the Cartesian Principle (CP2). This clear conception began to fade as boundaries between species eroded both from the emerging materialism of evolution and from the behavioral perspective of modern biology and psychology. Increasingly, consciousness was linked with the acquisition of complex behavior (Limber 1978) rather than introspection or verbal report. William James (1980, pp. 5–8) reflects on these developments.

"But actions originally prompted by conscious intelligence may grow so automatic by dint of habit as to be apparently unconsciously performed. . . . The pursuance of future ends and the choice of means for their attainment are thus the mark and criterion of the presence of mentality in a phenomenon.

Moreover in an era where nearly everyone from Darwin to Freud and Piaget held some Lamarckian notion of the inheritance of acquired traits, the distinction between ontogenetic acquisition and phylogenetic behavior itself becomes very fuzzy. See my discussion of this and the "Baldwin effect" in the context of speculation on language evolution in Limber (1982). I can only remark here that Searle's distinction (section 3, Inversion) between "normative" elements internal and external to a system is, at least in phylogenetic context, likewise fuzzy. Clarifying this is especially crucial for his assault on Chomsky's LAD (language acquisition device).

It is ironic that Searle seemingly wants to return to the good old preDarwinian days of CP2 while using his own CP to argue for a "postDarwinian" conception of psychological explanation! I am not absolutely certain that Searle is advocating this, but the issues he raises about mentality in psychological explanation have their roots much more in Darwin, Peirce, Helmholtz, James, Tolman and Hull – inter alia – than just Freud or Chomsky.

Rampant intentionality? The most problematical aspect of Searle's proposal involves the claim that CP must hold for a claim of mental causality to be even comprehensible! This is of course just the point that James and generations of psychologists have been struggling with as, for the most part, an empirical matter. Look, for example, at Tolman's efforts to link the intentional terms of his theory with the facial gestures of rats making decisions. Searle must do as much. He certainly cannot just stipulate CP and its use. He must give us some direction on how to assess the presence or absence of consciousness or how to detect the manifestation of "aspectual shapes" – especially in others. We can also expect some advice on what kinds of neurophysiological stuff have the capacity to "cause consciousness." Or conversely, just – for example – what is it about "gutbrain" stuff that precludes it from anything more than "as-if intentionality?" I can empathize with Searle's not being in touch with even his own "gutbrain," but, is it remotely possible that the gutbrain's rude noises mask an untold intellect? Might not some enterprising behaviorist teach a gutbrain some signs we can understand and – to paraphrase LaMettrie – make it a perfect little gentleman? In any event, his fear of universal intentionality here is very puzzling. At some point, on his account, we will presumably have the scientific knowledge to examine gutbrain tissue and make an assessment of its "consciousness-causing capacity." Surely we have rudimentary taxonomic knowledge already. Most of us can agree that earth, wind, and fire have a low probability of consciousness, whereas various organizations of organic cells have a considerably greater probability of manifest intentionality. For all I know, even the gutbrain might manifest consciousness without behaviorist intervention if it were not so inhibited by "higher" brains. Compare the apparent intentionality of the recently liberated right-hemispheres in "split-brain" patients. Is it impossible for Searle to conceive of multiple intentional agents inhabiting one organism? Is intentionality necessarily linked with "I" rather than "one?"

Does it really matter? The real issue is how much all this matters to cognitive scientists. Without substantial details on

the use of the CP, I suspect, it will matter very little. Space precludes going into all the problems here; I have alluded to some of them already. The problem cases continue to be those where physical descriptions of the neurostuff do not lead us more or less directly to the behaviors they cause. As these behaviors become seemingly more arbitrary, symbolic, and conventional, the a priori case for intentionality or at least rule-following is strengthened. Therein lies a major difference between the VOR and LAD examples. Let me conclude by posing some questions. What, for example, does Searle have to say about the acquisition of shoelace tying? Most of us learned this following a verbal rule that faded away and tying became a "thoughtless" motor habit. Is this rule still operating every time I tie my shoes? How might this be different from language acquisition? Does Searle know much more about the mind of prelinguistic infants than about gutbrains?

Loose connections: Four problems in Searle's argument for the "Connection Principle"

Dan Lloyd
Department of Philosophy, Trinity College, Hartford, CT 06106
Electronic mail: dlloyd@trincc.bitnet

As fish are to the sea, so are representations to the received view of the mind in cognitive science: abundant, teeming, and rarely visible on the surface of consciousness. John Searle, author of this appealing metaphor, has cast his net and lo, the sea is shallow and fish are few. A "Connection Principle" construes all intentional states as accessible in principle to consciousness, thereby denying the existence of the myriad representational (and unconscious) states that populate computational theories of mind. But the arguments that defend the Connection Principle are doubtful, in my opinion. This critique outlines four main problems. If I am right about any of them, the Connection Principle is undermined. (An extensive discussion of consciousness and cognitive science appears in Lloyd, in press.)

Searle's argument for the Connection Principle is a dilemma, pitting the mental against the physical, specifically the neurophysiological. On the one hand, intentional states have "aspectual shapes," representing their objects from particular points of view and with particular features (Searle's step 2). On the other, no "third person" scientific approach can "exhaustively and completely characterize" aspectual shape (step 3). The dilemma appears through the example of a person in a dreamless sleep. The sleeper's mental lights are out: the murmurs of the brain are unconscious and purely neural. Nonetheless, the sleeper can be said to believe many things even then (step 4). How is this possible? Searle proposes that unconscious beliefs are intentional because they are potentially conscious (step 5). As he puts it, "the ontology of the unconscious consists in objective features of the brain capable of causing subjective conscious thoughts." This is the Connection Principle (step 6).

Each of the premises above is problematic. Searle, like most cognitive scientists, is a materialist: "On my view of mind-body relations, the mental simply is neurophysiological at a higher level" (Note 4). Yet he contradicts this most fundamental principle in step 3 with the conclusion that the aspectual shapes of mental states cannot be constituted by neurophysiological states, from which it follows that mental states cannot be constituted by neural facts.

Two subarguments defend the main claim of this step. The first uses examples of the indeterminacy of translation (and the general underdetermination of theory by data) to undermine any behavioral definition of mental states. Searle then extends this familiar anti behaviorist line of argument to the brain by analogy. The analogy is imperfect, however, because Searle

169

grants the basic ontological point that "neurophysiological facts are always causally sufficient for any set of mental facts." The antibehaviorist argument therefore turns out to be irrelevant.

Thus he moves to a second subargument, which turns on the observation that "the specification of the neurophysiological in neurophysiological terms is not yet a specification of the intentional." The inference from this to the ontological claim that intentional states cannot be constituted by neurophysiological states is fallacious, however, and confuses an epistemic distinction about how an entity is known or described with an ontological distinction about the constitution of an entity. If this hypostatization were valid, then (for example) automobiles would not be constituted by their mechanical parts (because one infers that something is an automobile from the observation of its parts. The *Car and Driver* version of Searle's argument would note that the specification of the mechanical in mechanical terms is not yet a specification of the automobile.) In short, the strong ontological claim apparently proposed in step 3 is unsupported. This is not surprising, because Searle's own materialism contradicts it.

Searle's description of the dreamless sleeper in step 4 is also peculiar from the point of view of his own materialism. In the sleeper "there is simply nothing there except neurophysiological states and processes." The statement is equivocal. In one sense it is true, since according to materialists it is true of all minds in all conditions that there is simply nothing there except neurophysiological states and processes. But Searle means that in the dreamless sleeper there are only "purely" neurophysiological processes, processes without "higher" intentional (mental) descriptions. How can Searle, or anyone, assert that without begging the question? In any case, Searle's sleeper is something of a straw man. For the sleeper, the only possible intentionality may be in the form of a latency or potentiality, but the hypothesis of unconscious mental processing is usually posited to explain processes in awake and active minds. Searle's example, even if correctly interpreted, leaves open the possibility of a scientific defense of the unconscious on other grounds.

At issue in both steps 3 and 4 is the elusive theme of aspectual shape, introduced as a "term of art" in step 2. Searle is wise to duck the perennial issues of qualia and subjectivity lurking behind this issue, but with the privilege of introducing technical terms comes the attendant duty of defining them. In the absence of a definition, we may consider some possible realizations of aspectual shape in neural hardware. Perhaps aspectual shape is a property of neural states themselves; perhaps it is a relation between neural states and the objects they represent; perhaps aspects are simply features of objects to which we are specifically attuned; or perhaps it is some combination of these. Regardless, it is something we can become aware of in at least some circumstances, which is to say it has some causal effects, which is to say it is in principle open to scientific exploration and explanation. Its mystery is a symptom of primitive theory and data, not grounds for the wholesale rejection of materialist theories of aspectual shape.

In any case, Searle asserts that aspectual shape is a feature of intentional states whether conscious or not. If aspectual shape is required for intentionality but missing in the dreamless sleeper, then the solution to the dilemma is to construe those unconscious neural states as potential bearers of aspectual shape – only this, and nothing more. In other words, nothing in the argument leading to step 5 logically requires a connection between neural states and consciousness. If, on the other hand, it were proposed that aspectual shape is both necessary to intentionality and necessarily conscious, then the question of unconscious mentality would be begged in step 1.

In sum, the argument for the Connection Principle is multiply flawed. Nonetheless, the Connection Principle, or something like it, may be true. (Searle's discussion of functional explanation likewise is plausible independent of the Connection

Principle.) I agree with Searle that cognitive science must attend to consciousness, and that neglect has led to mystification of this central human capacity. Any hypothesis that develops a principled link between cognitivist models of mind and states of consciousness deserves further study. Searle's proposal is simple and consequential. Perhaps there are other grounds for its support?

Does cognitive science need "real" intentionality?

Robert J. Matthews

Department of Philosophy, Rutgers University, New Brunswick, NJ 08903
Electronic mail: *matthews@cancer.rutgers.edu*

Searle claims to have discovered a fundamental mistake in cognitive science methodology, namely, the postulation of intentional processes that are in principle inaccessible to consciousness. The alleged consequences of this mistake are rather remarkable: There is, it turns out, "no rule following, no mental information processing, no unconscious inferences, no mental models, no primal sketches, no 2½D images, no three-dimensional descriptions, no language of thought, and no universal grammar." Now, what is striking about Searle's discovery is not simply the apparent import for cognitive science but also its aprioristic origins in what Searle dubs the Connection Principle: Who would have believed that one could get so much from so little!

Searle is probably right in his belief that intentional explanation cannot play the significant role in cognitive science theorizing that philosophers generally suppose. At least, it is arguable that it does not *in fact* play such a role. But Searle has provided no reason for supposing that this fact has the dramatic consequences for cognitive science that he alleges.

1. Universal grammar. Searle claims that the postulation of universal grammar is both explanatorily (and predictively) vacuous and inconsistent with his Connection Principle. He makes much of the fact that no argument has ever been presented to show that the rules of universal grammar have any different status than the rules of a "universal visual grammar" that would explain subjects' ability to see blue and red but not infrared or ultraviolet in terms of their following certain unconscious rules.

Given the striking disanalogies between the two cases, it is hardly surprising that no one has bothered to make such an argument. Rules play a crucial role in our best descriptions of linguistic competences; we don't know any better way to capture what seem to be the salient features of such competences. The same can be said of our best descriptions of both the presumably innate constraints on natural languages and of the developmental changes that language learners exhibit: We don't know any better way to formulate these constraints. Not only do rule systems offer a seemingly illuminating descriptive apparatus for linguistic (and psycholinguistic) theorizing, but we also have a number of detailed proposals as to how such rule systems might be implemented or realized computationally, thus lending support to the belief that they can also serve in an explanatory role. *None* of this is true of visual color competence. There are *no* comparable grammars for such competence, indeed, *no* comparable learning problem that the brain might possibly solve by means of a "universal visual grammar." The differences in status between the two sorts of universal grammar should be clear: For the one, there is considerable empirical evidence; for the other, there is none. Surely that's a difference that makes a difference, at least in the empirical sciences.

What Searle clearly needs here, but has not actually provided, is an argument to the effect that the rules of universal grammar cannot really be causally explanatory. His remarks about why *as-if* intentionality will not rescue the cognitive

170

science paradigm suggest the following argument: (i) The rules of universal grammar are not even in principle accessible to consciousness; hence, by the Connection Principle, (ii) these rules do not have "real" intentionality; (iii) but if they lack real intentionality, then they have only "*as-if* intentionality; but (iv) if they have only *as-if* intentionality, then "not being real, [they] have no causal powers whatever. They explain nothing." Hence, (v) the rules of universal grammar are not genuinely causally explanatory.

If this is indeed the argument that Searle has in mind, then his critique of universal grammar (and the cognitive science paradigm, more generally) requires more than his Connection Principle. It also requires the premise that *as-if* intentional states can explain nothing. But this is a dubious premise indeed. Rules, programs, data structures, and so on are the coin of computational explanation (at least for machines with classical architectures); however, by Searle's own account, they have only an *as-if* intentionality. Hence, on the premise in question, all explanations in terms of such notions are in fact pseudo explanations. Consider, for example, the word-processing program with which I am writing this commentary. It will not read files formatted for certain other word-processing programs; there are conversion programs that will reformat those files, however, so that they can be read. The usual explanations of why unconverted files cannot be read, why converted files can be read, and so on, appeal to just the sort of rules, programs, and so on that Searle seemingly wants to reject as explanatorily vacuous. If these seeming explanations are in fact pseudo explanations, then Searle owes us a (persuasive) argument to that effect.

What might really be bothering Searle is not so much the claim that there are rules of universal grammar, but rather the claim that learners might be said to *follow* such rules in the course of acquiring their native language. As Searle puts it, "the child acts *as if* he were following rules, but of course he is not really doing so." Searle is clearly convinced that whatever else may be going on, the first-language learner is not following rules in anything like the way we sometimes follow rules in the course of learning a new game or a second language. But how can Searle be so certain of this? Granted that the rules of universal grammar are presumed not to be present in consciousness. Is this obviously pertinent to genuine rule-following? We may be agreed that the paradigms of rule-following behavior are cases of *conscious* rule-following; nothing follows from this fact as regards the essentiality of consciousness, however. No one supposes that all properties of paradigm cases are essential. Once again, Searle owes us an argument, in this case for the essentiality of consciousness to genuine rule-following.

2. Doing without "real" intentionality. The notion that Searle's Connection Principle entails dramatic consequences for cognitive science clearly presumes that intentional states to which cognitive psychological explanations appeal must be really intentional, that is, in principle accessible to consciousness. But what is the argument? Why should it matter, for example, whether the rules and inferences postulated in Rock's explanation of the Ponzo illusion are accessible to consciousness? Such explanations do not postulate certain nonexistent causal mechanisms in addition to the neurophysiological mechanisms that Searle is willing to countenance; rather, they simply describe those mechanisms in a way that is intended to illuminate their function for the organism. The description of perceptual processes as inferential underscores an important function of the neurophysiological processes that subserve perception, namely, that certain information about the distal scene is preserved by these processes. The description becomes mischievous only when it leads some (e.g., Searle, sect. 3, para. 9) to wonder who is supposed to be performing these inferences.

Searle clearly assumes that the allegedly intentional states to which cognitive science explanations appeal need to have determinant "aspectual character." But why should these explana-

tions be thought to require determinant aspectual character? The underdeterminacy that Searle ascribes to the states that figure in these explanations seems on Searle's own account not to differ from the underdeterminacy of third-person intentional ascriptions in everyday life. For the purposes of rationalizing interpretations of cognitive mechanisms, the epistemic standards of everyday life presumably should suffice.

It is, of course, an open empirical question whether in specific cases these rationalizing interpretations are helpful or illuminating (e.g., whether the mechanisms of language acquisition lend themselves to an illuminating interpretation in terms of rules and representations). As Searle himself emphasizes, we cannot assume that there has to be a rationalizing interpretation of cognitive processes. Sometimes the mapping of the functional rationale onto the subserving mechanisms and processes will simply be too indirect or diffuse. But neither can we assume that there is some a priori way (e.g., via the Connection Principle) to determine when a nonvacuous rationalizing interpretation of intervening mechanisms and processes is available. Everything depends on the particular mechanisms involved. Some mechanisms (e.g., those that realize classical von Neumann architectures) lend themselves to such interpretations in a way that other mechanisms (e.g., those that realize connectionist architectures) do not.

Zombies are people, too

Drew McDermott

Yale Department of Computer Science, New Haven, CT 06520
Electronic mail: *mcdermott@cs.yale.edu*

I find this target article wrongheaded from beginning to end. But if I had to pick the pivotal mistake, it would be this paragraph near the end of section 5:

> Notice that it is not enough to rescue the cognitive science paradigm to say that we can simply decide to treat the attribution of rules and principles as as-if intentionality: because as-if intentionality states, not being real, have no causal powers whatever. They explain nothing. The problem with as-if intentionality is not merely that it is ubiquitous – which it is – but its identification does not give a causal explanation, it simply restates the problem that the attribution of real intentionality is supposed to solve.

It is crucial for Searle to try to make this point, because if he's wrong then his whole article is criticizing a straw man. And in fact he is wrong; it is precisely the case that hypotheses about rule following – and about all other forms of computation – in cognitive science are references to *as-if* intentionality. When a cognitive scientist[1] proposes that a certain phenomenon is to be explained by the existence of an unconscious algorithm, the intent is to propose the existence of a biological computer, and nothing else. Searle is right that calling such algorithms "unconscious" is misleading. Consciousness just doesn't enter into it. It's as if we called mental mechanisms "inaudible"; we would be right, but people would start imagining tiny noises. When we call them "unconscious," people start imagining tiny thoughts.

Our language for describing what computers do is notoriously fraught with *as-if* intentionality. Computers *follow* instructions; they *look for* data; they *know* a datum lies in a certain part of an array. None of this is to be taken literally. Computers don't really follow instructions (for starters, they don't have to understand them to do what they say). What they actually do is pass through several state transitions under the direction of bits from a code word. What's interesting is that it is impossible to eliminate the as-if intentionality, not just because the alternative language would be more clumsy, but because *there often is no alternative language*. Consider two different types of computer made by the same manufacturer that execute the same machine language but use entirely different finite-state

171

code interpreters. When someone explains what one of these computers did in terms of machine-language instructions, he doesn't have to know which kind of code interpreter the machine has. It may not be possible to know; the machine might still be on the drawing board, and the behavior could be hypothetical. Still, the explanation is perfectly sound and causally grounded, because everyone knows that getting a machine to interpret instructions in this sense is a straightforward problem. The details of the mechanism are irrelevant.

In other words, as-if intentional states have causal powers whenever there is a way to fill in the details of the relation between an as-if intentional state and its intended consequences, and especially when there are a variety of ways. Although computers make the point vividly, it has been obvious since the invention of the flyball governor. Consider an engineer explaining why a room was too cold in terms of the proximity of the thermostat to a hot light bulb. The engineer can explain everything just by knowing that the thermostat acts "as if" it is trying to make the room a certain temperature. It is not necessary to know how this particular thermostat works; there are too many ways to build a thermostat for it to matter.

Hence, to summarize, (a) cognitive science proposes mechanisms that are "nonconscious" because they are computational, and hence rich in as-if intentionality; (b) this level of explanation works just fine because it is underdetermined how the gaps get filled in. This whole line of argument should be familiar, having been set out at greater length by people like Dennett (e.g., 1981) and Fodor (e.g., 1975) many times.[2]

Philosophers can be like lawyers: If you don't put an obvious distinction in writing, they feel entitled to assume you don't see it. So I'd better be explicit about the difference between two kinds of as-if intentionality, *fanciful* ("The rock is trying to reach the ground") and *effective* ("The VOR is trying to keep the retinal image stable"). Obviously, only the latter can play any role in causal explanations. I know Searle will claim there's no real difference (that's the message of the word "ubiquitous" in the quote I started with), but it's crystal clear. The VOR really is "trying" to keep the image stable; the rock is not really even "trying." That is, there's a feedback loop in the one case, and not in the other.

A consequence of the reliance of cognitive science on (effective) as-if intentionality is that it is the study of people as machines – or as zombies, if you will. As Searle says in his introduction, "consciousness is . . . a 'problem,' a difficulty that functionalist or computationalist theories must somehow deal with. Now, how did we get into this mess? How can we have neglected the most important feature of the mind in those disciplines that are officially dedicated to its study?" The answer is that science tends to get into this kind of mess all the time. Take Eddington's table, an apparently solid object that turns out to be mostly empty space. It would be misguided to say the table "isn't really solid." It just turns out that the explanation of its solidity does not lie in its being made up of solid pieces. In retrospect, we should have rejected any such explanation. Similarly, the proper explanation of consciousness must start with nonconscious processes. So far, all we have is the start. Nothing to be embarrassed about, unless the enterprise fails, as, of course, it may.

It may sound incoherent to expect to explain "intrinsic" intentionality in terms of as-if intentionality, but such things have happened before. Physics supplies another example; The positions of macroscopic objects are ultimately explained in terms of the as-if positions of the elementary particles they comprise – which don't have "intrinsic" location at all, but only a quality that resembles location metaphorically. Just because *as-if X* isn't *X* doesn't mean it isn't *something*.

NOTES
1. Including a connectionist. I am baffled by the urge to see a difference between neural-net theorizers and algorithmic theorizers from a distance as remote as that of this article.

2. Although I resist Dennett's formulation of these points in terms of the "intentional stance" we take toward an information-processing system, which implies that we have other options. It's like saying that running into a brick wall involves taking a "material stance" toward it.

Somebody flew over Searle's ontological prison

Massimo Piattelli-Palmarini
Center for Cognitive Science, Massachusetts Institute of Technology, Cambridge, MA 02139
Electronic mail: *massimo@athena.mit.edu*

History ought to have taught us that it is pointless to prospect dilemmas between scientific theories and a priori conceptual analyses, because successful scientific theories invariably manage to overthrow the very canons of our reasoning. For instance, when quantum theory was discovered to be at odds with classical binary logic and with strict identity, the latter were simply set aside and a new "quantum logic" was developed. From a number of such episodes in the history of science we can also derive a second lesson: The contents and concepts of future scientific theories cannot be constrained a priori. No classical physicist could have anticipated the emergence of *fundamental* laws of physics based on exchange forces, gauge invariance, isospin quantum numbers, and the like. Therefore, Searle's somewhat apodictic claim that "A *perfect* science of the brain would be stated in neurophysiological (i.e., "hardware") vocabulary" (sect. 3, para. 10, my emphasis) is eventually destined to meet the same kind of historical refutation. Unless, of course, it amounts to the tautology that neurophysiological kinds and laws will be whatever the neurophysiologists will severally discover them to be. This is a point to which I will return later.

Another relevant lesson from the history of science is that no one can delimit a priori the kinds of evidence that can come to confirm or refute the specific ontological claims of scientific theories. Darwin could not have anticipated that the rates of substitution of third bases in the codons of introns could offer a calibration of selective forces, nor could Paul Ehrlich have imagined that monoclonal antibodies, polymerase-chain-reactions, and other refined techniques were one day to vindicate his insight into the sources of antibody repertoires. A plethora of such cases shows that it is always productive to be bloody-mindedly realist about the ontological interpretations of scientific theories. In sum, even if Searle's entire argument were flawless, I do not see how it could affect linguists and cognitive scientists. However, it is not.

For the sake of brevity, I will just concentrate on Searle's "in principle" accessibility to consciousness. There is an innermost feeling for certain truths, which we (almost routinely) develop as knowledge and expertise grow. The span, the grain, and the depth of our introspective evidence for certain truths (and falsehoods) depends on specific knowledge that we typically acquire through scientific practice and personal reflection. For instance, as a consequence of things I have learned in linguistics, I came to acquire some introspective evidence that, indeed, when in my native Italian I say something like, *"Piove, ma non importa"* (It rains, but it does not matter), there are two silent pronominals, referentially disjoint, absolutely inexpressible by any Italian audible sound, which make the sentence point-by-point mappable onto its English translation. The examples could be made more elaborate, but this simple one will suffice.

The sort of inner anecdotal evidence I have just alluded to seems genuine enough, if one adopts Searle's criteria. If he denies ontological import to this specialized self-awareness, he is contradicting his own "Connection Principle." If, on the contrary, he recognizes it, the whole argument developed in his target article collapses, since he would have to recognize as *bona*

172

fide mental, in his ontology, at least all those entities and mechanisms for which we *happen* to develop an inner feeling, by reading textbooks and articles in linguistics and in cognitive science.

Unlike Searle, I do not attribute any privileged status to introspective evidence, but if I did I would not know where, and how, to trace a principled barrier between this introspectively "vivid" intimate raw feel and a more poised intellectual awareness. I only see different shades of "in principle" accessibility to consciousness. Let's imagine that, after very long and intense training, we could become introspectively aware, instant by instant, of *how* we speak. (After all, it is claimed that three to five years of Freudian therapy can bring to consciousness fragments of the emotional unconscious, and it is a fact that certain revealing experiments à la Tversky and Kahneman (1982) succeed in introspectively exposing some of our cognitive illusions). Then, in virtue of Searle's Connection Principle, these elusive mental contents must gain a fair chance of passing the ontological test. And here comes the rub, because what shall we say of the existence of empty categories, and binding principles, and lexical projections, and $2\frac{1}{2}$ D sketches, and so on, during the eons of untutored natural thinking, seeing, and talking? Did they *already* exist before anyone had developed any such introspective awareness? Surely we do not want to claim that H_2O *became* the structure of water only when this fact was discovered!

If introspective evidence had any ontological import, then it would disprove its own relevance by generating all sorts of paradoxes. Imagine, in fact, that we lose all introspective track of these entities when we utter, or hear, or read, very complex sentences. Would we conclude that these core linguistic entities cease to exist when we process very complex sentences? What if an unenlightened 40 per cent of "mutant" linguists never develop this inner feeling? Would we conclude that these entities are absent in their minds? Linguistics and cognitive science are well advised to pay scant attention to criteria of introspective availability.

I will press this point a little further. The paradigmatic example of a true-to-God mental state is, for Searle, the thought "I want water," quite different from the thought "I want some H_2O" (II "step 2", first paragraph, and "step 3", second paragraph). Even Linus Pauling, probably the greatest chemist of this century, may well have occasionally desired water without thereby desiring H_2O. Yet, H_2O-directed thoughts typically become possible only through suitable schooling. A chemistry textbook, plus being alert, understanding what one hears or reads, and being in an H_2O-favorable mood is all we need to make such thoughts "in principle" accessible to consciousness. One fails to see the difference between Pauling's case and the case of John Doe, a consummate linguist, who can "in principle" have such conscious (and perfectly aspectual) thoughts:

(1) Who do I think saw Bill?
(2) Who do I think Bill saw?

Searle stresses the "aspectual shape" of conscious mental states as "a universal feature of intentionality" (step 2, first line of first paragraph) but fails to remark that linguists have a lot to say about the aspectual character of (1) and (2). The study of aspect is also a subfield of syntax. Just to give an idea, there are thoughts, which John Doe can also have, which stand to (1) and (2) just like the formula H_2O stands to the term water. Take (2), for which the underlying structure (Van Riemsdijk & Williams, 1986) is something like:

(2a) Who $_i$ do I think [$_t$ [$_{COMP}$ e_i that]$_{COMP}$ [$_S$ Bill saw e_i] $_S$], (the i indices mark co-reference)

John Doe is disposed to assent to the identity between overt-English (2) and its structural decomposition (2a) just as much as a chemist is disposed to assent to the identity between water and H_2O, and very much for the same reasons. We grant that (2) and (2a) do not constitute identical thoughts, and that no normal (I mean linguistically untutored) English speaker would ever *consciously* think something like (2a). One needs specialized knowhow, paper and pencil, and some tranquility to rewrite a plain sentence such as (2) in terms of its proper underlying structure 2a). But one also needs specialized know-how, proper equipment, and skill to decompose water stoicheiometrically into hydrogen and oxygen. There are some differences between these two cases, but none that is principled, and none that matters ontologically.

Nor should it matter, ontologically speaking, how *fast* we are in computing structures like (2a) (If we do not trust this sort of linguistic example, we are free to substitute our favorite formula from our favorite mental theory). It is just a contingent fact about us humans that mental events which occur below a certain time threshold, and above a certain complexity, are constitutively inaccessible to unaided introspection. They could well be, counterfactually, transparent to the consciousness of a much smarter species; and they often become accessible to us by means of some innocent device – paper and pencil, for instance.

The point is: We do not have to wait for Searle's hypothetical science-fiction cerebroscope. We already have a variety of simple, real and reliable "mind-o-scopes," such as (2a). It would be very unwise to take Searle's suggestion and happily trade all mind-o-scopes, en bloc, for the hope of a possible, perfect "brain-o-scope." Even if it existed here and now, it just would not help.

In fact, imagine that this wondrous neuronal monitor indeed materializes, and that it proves vastly more reliable than our poor, pathetic present-day mind-o-scopes. We can safely bet that, if it were sensitive enough, it would induce a lot of revisions in linguistic and cognitive theories by leading to the discovery (not the stipulation) of the neuronal correlates of trees and brackets and indexes and arrows, and of much beyond. These still would not be made accessible to unaided introspection. Nor would they, *pace* Searle, become brute "hardware mechanisms," (sect. 5, para. 13), that is, properties of our brain, not of our mind. According to Searle, they would be on a par with the myelinization of the nerve fibers and the secretion of neurotransmitters. I think this suggestion does not even begin to make sense.

The terms of a brain vocabulary, which Searle so strongly recommends, would not only miss all the crucial linguistic and cognitive generalizations, they would not even allow us to state properly the relevant linguistic and cognitive facts, not even at a proper descriptive level. It is precisely something like (2a), not (1) or (2), that is closer to what goes on in our brain. It is plain to see that the entities and processes which will one day appear in (2a)$_{perf}$ (the equivalent of (2a) in a future perfect linguistic theory), whatever they are, will be even closer to physical computations, but will not even look like candidates for entities and processes of our brain as we presently conceive them. This shotgun elimination is not the way reidentification and reduction works in the sciences, even when it does. Only suitable abstractions reduce to, or become identical with, other suitable abstractions. The abstract notion of valence in chemistry reduces to the abstract physical notion of a certain quantum state of certain electrons. But a piece of sulphur is irreducible to anything else. Similarly, only (2a)$_{perf}$ (if anything at all) will be able to meet the tribunal of identification and reduction, not (2). (2) will be forever just the linguistic equivalent of a piece of sulphur. Only theoretically transmogrified cognitive and linguistic categories may one day be tentatively mapped onto equally transmogrified nueronal analogoi. To wit, the latter (called neural networks) are here and now structures vastly more abstract than those captured by what Searle calls "hardware vocabulary." In the most sophisticated neurosciences we have already witnessed the emergence of suitably abstract entities, like cerebellar tensors, heterochrony map functions, neuronal population vectors, Darwinian algorithms, phasic attractors and so on. These are the intimations of a future cerebroscope, if anything is. Even the neurobiologists have already escaped from Searle's ontological prison.

173

Constituent causation and the reality of mind

Georges Rey

Department of Philosophy, University of Maryland, College Park, MD 20742
Electronic mail: *rey@tove.cs.umd.edu*

Searle's paper is problematic both in the structure of its argument and the nature of its claim. It would have helped had the six "steps" been deductive. As it is, there seem to be two strands of argument for the "Connection Principle," the first claiming that, without consciousness, no distinction could be drawn between literal and metaphorical ("as-if") ascriptions of intentionality; the second that, without consciousness, no distinction could be drawn between mental states that involve the same references. The trouble with both these strands is that they involve essentially "negative conceivables," which are notoriously more difficult to establish than negative existentials: Claims that "there is no *conceivable* way that (or we have no notion of) such and such unless . . ." are often the result not of sound argument, but of failed imagination. They can be refuted merely by conceiving the possibility that is being denied. I'll do this for each strand and then consider the plausibility of Searle's conclusion on its own.

1. Consider first the distinction between literal and metaphorical intentionality. I join Searle (sect. 5, para. 5) here in tying the literal to the causally efficacious. But I don't see why the following isn't one *possible* story of how this could come about: Desires are relations agents bear to causally efficacious propositional structures entokened in their brains. Mary's desire to run downhill is a genuine cause of her so running because a propositional structure, as it might be, [I'd like to run downhill], plays a crucial causal role in the production of her running. It's the sort of structure that, were it to interact with another such structure [If I'm to run downhill, I'd better get my shoes on], would cause her to go get her shoes; and, were it to interact with [If I'm to run downhill, I'd better say my prayers], would cause her to say her prayers; and so forth, for innumerable other attitudes. Moreover, it's a *propositional structure*, since it has semantically evaluable *constituents*, for example [I], [like], [run] and [downhill], that can each combine in standard syntactic ways with other constituents to form other propositional structures in Mary that produce still further behaviors: For example, she runs uphill when she wants to *run back*, walks downhill when she thinks *running unseemly*; and stays put if she thinks [*I ought to stop and watch everyone else run instead*].[1]

What makes the propositional structure intentional? What makes its constituents *about* for example *running, downhill,* and *Mary?* This is no place to review the many proposals that are currently being explored, involving actual and counterfactual, internal and external causal relations among the constituents and properties in the world.[2] Given the richness and occasional plausibility of these proposals, it is odd that Searle doesn't consider them (neither here nor, by and large, in his "Chinese Room"). If he is to sustain his negative conceivable claims about literal intentionality, he needs to show that no combination of these causal relations could *possibly* capture it. I don't see any argument here that shows *that*.[3]

Constituent causation provides one possible way to distinguish people like Mary from substances like water. People's behavior is demonstrably diverse and systematic in the ways mentioned, and they have sufficiently complicated brains in which the constituent causation could plausibly take place. There's not the slightest reason to think any of this is true of water: What, after all, is supposed to correspond in the water to the proposition [I run downhill] or to its parts? Where are the processes involving such representations that are responsible for its behavior? How do the constituents enter into the appropriate causal relations with each other and the world to provide them meaning, and recombine into other attitudes to cause water to act differently?

Precisely where constituent causation, that is, genuine propo-

sitional representation, needs to be invoked is a subtle empirical question about which one isn't entitled to many a priori intuitions. It's plausibly true for the above kinds of behavior of human beings and many animals. It's probably not true for phototropic plants or VOR movements. But there are a lot of complicated cases in between, the rules and principles of language comprehension among them. Everything depends on when a system exhibits lawlike regularities that cannot otherwise be explained.

I agree with Searle that Chomsky (1986) and Rock (1984) are not always perfectly clear on this issue, and sometimes seem to suggest that anything that obeys rules *therefore* represents them. That would be a mistake. Nevertheless, constitutive causal representation of *some* of the structure of a natural language may afford *the best explanation* of the observed regularities. There is abundant evidence, for example, that English speakers represent English sentences as possessing fairly elaborate grammatical structure, often with hidden "trace" markers indicating the position of an implicit element. This supposition helps explain, for example, why certain transformations and not others are permissible, why people are faster in processing some parts of sentences than others, why they are taken in by "garden path" sentences, why they sometimes misidentify the timing of accompanying "clicks," and so on. Perhaps Searle can explain all these regularities with his "best theory" of "association patterns that are not produced by mental representations" (sect. 5, para. 11). "Association" theories – from Hume to Skinner to connectionism – don't have a good track record, however. In any case, claims of "best theory" here will not be settled by a priori argument alone. Searle needs to do the *detailed* empirical explanatory work.

Note that it is not enough – indeed, it's utterly irrelevant – to wave, as he does (sect. 5, para. 12), toward "hardware" and "functional" levels of explanation. Of course, there are such explanations of any *individual* linguistic event – that's just standard materialist piety. The question is not whether individual events can be explained by underlying physical theories, but whether the *lawlike regularities* – in this case, the psycholinguistic regularities – into which those events enter can be so explained. For familiar functionalist reasons, there is absolutely no reason to think that those regularities will be explainable in neurophysiological or biological terms, any more than the explanation of the regularities in my word-processing program will be explained in terms of electrical engineering, or the structure of Shakespeare's plays in terms of the chemistry of ink.

2. Constituent causation also goes some way toward answering the second strand of Searle's argument concerning how *possibly* to distinguish coreferential thoughts with different "aspectual shape" ("sense"?), for example, thoughts about water from thoughts about H_2O. A child ignorant of chemistry wants "water" in a glass, where the chemist wants H_2O, since the chemist, but not the child, has thoughts about hydrogen and oxygen that are causal constituents of a representation expressing the desire. Why think this? For starters, presumably because the chemist, but not the child, can *distinguish* hydrogen and oxygen behaving in ways that are sensitive to the distinction; and we may suppose, on the present hypothesis, that the chemist does it by using the very same representations throughout. We can, as it were, spell it out: Someone who is thinking H_2O is wet is someone who is thinking a thought that has as constituents something very like [H], [2], [O], [is] and [wet]; whereas someone who only thinks *water* is, is thinking something that is spelled more simply.[4]

3. Now Searle needn't *believe* this story of constituent causation. All that I have been claiming is that it is a *possible* story of literal intentionality and coreferential "aspectual shape," one that, so far as I can see, doesn't depend in the least on consciousness, or any disposition thereto.

Actually, I'm not clear just what Searle's claim about consciousness actually comes to. That attitudes necessarily involve a disposition to become conscious could be read trivially, in a

174

way that even Chomsky wouldn't deny: Any propositional attitude can be *thought* consciously, for example if the agent is asked (this seems the most natural reading of step 5). What Freud, Chomsky, and most cognitivists are denying is of course not *that*, but rather that the possession of an attitude must be *introspectible*. But put this way, Searle's claim seems awfully implausible – for starters. introspective skills seem to vary arbitrarily in the population (see e.g. Nisbett & Wilson 1977); for another, the attempt to do psychology under this constraint, the "introspectionist" psychology of a century ago, was a ludicrous failure. Is Searle really advising a return to Wundt?

In any case, if we are to take Searle's dispositional claim seriously, we need to have some idea of the activating circumstances. The most we are told (steps 5, 6) is that they include the absence of lesions and repression since, at least in *those* cases, even Searle agrees that someone's attitudes could be unconscious. But then suppose these lesions and repression are genetically determined, and gradually spread in the population so that they become the normal condition of humankind: Wouldn't whatever explanations we now apply to these cases apply then to the whole population? And if this is possible, why shouldn't it be possible that we are in that evolutionary state right now? Indeed, one could take the work of Freud and Chomsky to show just that: that we've evolved into systems that inherently suffer from "natural" lesions and repressions that render many attitudes inaccessible to introspection. It would be scientifically most peculiar if the explanatory power of Freud's or Chomsky's theories depended upon this not being so.

Quite apart, however, from whether consciousness is *required* for mentation, a deeper problem for Searle is how it is supposed to be of any help. If there were no "objective" way to draw the above distinctions, I don't see how there could then be any "subjective" ways either. First-person stances and privileges are all very nice: I regularly enjoy them myself. But I don't see how they bring in the philosophical wash. How is *introspection* to tell me whether my introspected attitudes are literal or metaphorical, whether, for example, it is *they* that are causally responsible (recall sect. 5, para. 5) for my behavior?[5] Are we supposed to know *introspectively* that epiphenomenalism is false? *I* don't.

Moreover, if there is no *objective* fact about whether I mean *water* or *waterhood* by "water," I don't see how beating my breast and insisting on a "subjective fact" will do any good. The question is: Just what is the fact on which I'm supposed to have such a privileged grip? That by "water" I mean water and by "waterhood" waterhood? Without any objective basis, this seems no better than thinking you know how tall you are by placing your hand on your head to prove it (Wittgenstein 1953, p. 279). If there is no objective fact describable in third-person terms, there seems to me nothing for me in the first-person to be right or wrong about; and so nothing for me to know (Searle seems to have provided finally the right target for the private language argument). In any case, Searle owes us an account of precisely what these further nonobjective, essentially "subjective facts" might be, and how they don't undermine the materialism that he otherwise seems to want to believe.[6]

NOTES

1. Lest Searle reject this proposal as running afoul of his "Chinese Room," note that it is a proposal of a *(propositional) representational* theory of mind. Whether one thinks further that the causal processes involving representations are *computational* depends upon what one thinks of such "sociological phenomena" as the work of for example Turing, Simon, Fodor, and practically the whole of cognitive psychology. See Rey (1986) for a defense of these latter enterprises against the several "Room" fallacies.

2. Searle considers only very crude behavioral and neural corelate proposals. He might also consider for example Stampe (1977), Devitt (1981), Dretske (1981, 1988), Millikan (1984), Block (1986), and Fodor (1987, 1990).

3. He does suggest a kind of in-principle argument a little later, claiming that, even were there neural correlates of intentional states,

"there is still an inference, and the specification of the neurophysiological in neurophysiological terms is not yet a specification of the intentional" (step 3). Now, neural correlates may not be the best candidates for specifying intentions, but I don't see their *inferential status* as a problem. There may always be an inference between water and H_2O, but the latter is a perfectly good specification of the former.

4. This isn't to say that there aren't more difficult cases that Searle might have mentioned. For a more compelling case for Searle's claim, see Bealer's (1984) adaptation of Quine's (1960) rabbit/rabbithood argument to functionalist theories of mind. But Bealer's claims also suffer as negative conceivables.

5. Actually, it's not at all clear how introspectible facts about aspectual shape – which "cannot be constituted" by any objective facts (step 4,2) – could *themselves* be causally efficacious at all. I don't quite see how Searle thinks he's succeeded in extricating himself from his own contradiction or avoided classical epiphenomenalism.

6. I'm grateful to Michael Devitt and Ken Taylor for comments on earlier drafts of this commentary.

On being accessible to consciousness

David M. Rosenthal

ZiF, Universität Bielefeld, D4800 Bielefeld 1, West Germany and City University of New York, Graduate School, New York, NY 10036-8099
Electronic mail: *drogc@cunyvm.bitnet*

Searle believes that the idea of intentional states that are in principle inaccessible to consciousness is incoherent. "[I]t is incoherent, in the [rather special] sense that it cannot be made to cohere with what we already know to be the case" – presumably the Connection Principle and the premises that Searle believes lead to it. No explanation that posits such deep unconscious intentional states and processes (henceforth "deep explanations") can be true, he concludes, though some corresponding explanation that is not literally intentional may well be.

Searle may well be right that some deep explanations are inviting largely "because we lack hardware explanations of the auxin type." And when connections hold among states with content we probably infer too readily that the processes linking those states also have content. But Searle's examples are of questionable relevance. If, for example, the hardware circuitry that keeps eyes pointed in the same direction ran through suitable regions of the cortex, it would plainly be reasonable to suppose that mental processing plays some role.

In any case, Searle's master argument from the Connection Principle fails to show that deep explanations are never warranted. There are two main difficulties: the "in principle" clause, and the role in the argument of aspectual shape.

Searle recognizes the need for clarity about what "in principle" means. An unconscious intentional state is, Searle claims, a *"possible conscious thought or experience"* and he explains "possible" here in terms of "capable": unconscious intentional states are states *"capable of causing subjective conscious thoughts."* So an unconscious intentional state is "in principle" (he occasionally says "intrinsically") accessible to consciousness if, and only if, it is "capable of causing the conscious experience."

But what does "capable" mean? In section 6, it emerges that a state can be thus capable even if there is some hardwired obstacle to the relevant causal linkage, and that obstacle involves "nothing pathological." This is reasonable; a state may have distinctive causal powers even if, because of some hardwired, nonpathological blockage, those powers cannot be realized. But why wouldn't this degree of inaccessibility be enough for even the most hard-core proponent of deep explanations? Unconscious intentional states might well, for example, have the relevant causal powers even though some hardware blockage puts those states, as Chomsky suggests (Note 1), "beyond the reach of conscious introspection." Even granting the Connection Principle, the difficulty for deep explanation does not follow.[1]

175

The Connection Principle, however, is itself dubious at best. All intentional states, conscious or not, have aspectual shape. Searle's argument is that unconscious intentional states consist wholly in neurophysiological phenomena, but their aspectual shape "cannot be constituted by such facts." Searle also claims, however, that "it is reasonably clear how . . . conscious thoughts and experiences" have aspectual shape. So the only way for unconscious intentional states to have aspectual shape is by having the power to cause states that, by virtue of being conscious, have aspectual shape directly and in the primary sense. Thus unconscious intentional states in effect have aspectual shape only indirectly.[2]

Why does Searle think aspectual shape is unproblematic in the case of conscious intentional states? Presumably because of its tie to agents' points of view. "[A]spectual shape must matter to the agent. It is . . . from the agent's point of view that he can want water without wanting H_2O."

But what matters may not matter consciously. Conscious intentional states matter in part because of their effect on other intentional states, both conscious and unconscious; the same holds for unconscious intentional states. Thus my unconscious desire is for water but not H_2O if I believe the two to be different and my desire would be satisfied by what I believe to be water but not what I believe to be H_2O. All these beliefs may themselves be unconscious. In this case I unconsciously want something as water, and not as H_2O; similarly for other unconscious intentional states. The causal connections my unconscious intentional states have to other intentional states, which may themselves not be conscious, manifests their aspectual shape and how it matters to me. By the same token, an agent's point of view need not be wholly conscious; one's unconscious beliefs and desires partially define one's point of view. Thus the tie between aspectual shape and viewpoint does not guarantee a connection between aspectual shape and consciousness.

Searle would respond that all this makes no sense unless the relevant unconscious states can produce conscious states with the relevant aspectual shape. But why? One's conscious first-person perspective doubtless reveals the aspectual shape of one's intentional states, but that hardly shows that aspectual shape cannot exist unconsciously. Nor does consciousness in any way help explain aspectual shape. Differences in aspectual shape are differences in how something is represented; so to explain aspectual shape we must have a theory of content.

Differences in aspectual shape also emerge with speech acts; the reason speech acts can be about water but not about H_2O is because of the words they use. Searle may think this kind of explanation is unavailable for intentional states because such states have no medium corresponding to the words of speech acts.[3] But he also concedes that "the mental simply is neurophysiological at a higher level" (Note 4); as he puts it elsewhere, intentional states have physical "forms of realization" (Searle 1983, p. 15). So neurophysiological differences can help explain differences in the aspectual shape of intentional states in just the way that different words do for speech acts.

Searle insists that neurophysiological differences cannot make for differences in aspectual shape. A certain ambiguity threatens here. Searle puts his point this way: "No set of neurophysiological facts under neurophysiological descriptions constitutes aspectual facts." This is undeniably so, but only because facts are relative to how we describe things.[4] Describing things neurophysiologically is, of course, different from describing them in terms of aspectual shape.

But "the mental simply is the neurophysiological at a higher level" (Note 4).[5] So, even though we cannot describe aspectual shape in neurophysiological terms, aspectual shape is still a property of neurophysiological states. There is thus no reason why those neurophysiological states that are unconscious intentional states cannot have aspectual shape. Searle cannot invoke here the connection between aspectual shape and consciousness, since that very connection is at issue. So the neu-

rophysiological character of unconscious states does not prevent them from having aspectual shape in their own right.

Moreover, if the mental is simply the neurophysiological "at a higher level," even conscious intentional states are neurophysiological states. Since consciousness itself is a property of neurophysiological states, why can't aspectual shape be, as well? Again, it begs the question to appeal here to the alleged tie between consciousness and aspectual shape.

Searle argues that unconscious intentional states are purely neurophysiological by considering the intentional states of unconscious people. This strategy conceals an important distinction. Unconscious intentional states also occur when we are awake, and thus conscious. Despite connections between them, what it is for a state to be conscious is distinct from what it is for a creature to be conscious.

Searle assumes that no facts other than neurophysiological facts or the fact of consciousness could explain aspectual shape and neurophysiological facts plainly cannot do so. If so, we cannot understand intentionality without understanding what it is for an intentional state to be conscious. This arguably makes it more difficult, and perhaps impossible, to explain such consciousness; intentionality is plainly far harder to understand if it is essentially conscious.[6] In the absence of more compelling argument, therefore, we should reject the tie Searle hopes to forge between aspectual shape and consciousness.

NOTES
1. Nor is it wholly obvious that cognitive science requires deep explanations. In the quoted passage Chomsky himself hardly seems wedded to deep explanations: The posited states "*may be . . . even beyond the reach of conscious introspection*" (my emphasis). Shallow unconscious intentional states that are for practical purposes inaccessible to consciousness would presumably suffice for most or all theoretical purposes in cognitive science.
2. Indeed, it is unclear why Searle does not conclude that unconscious states, though accessible to consciousness, have merely as-if aspectual shape, and thus only as-if intentionality.
Searle's views about unconscious intentionality are reminiscent of his claims about intrinsic and derived intentionality. The intentionality of speech is "derived," he maintains, in that "the direction of logical analysis is to explain language in terms of [the] Intentionality [of the mental]" (Searle 1983, p. 5). Similarly, he now urges that we can understand unconscious intentionality only in terms of conscious intentionality.
3. Searle argues from the lack of a mental medium to the intrinsic intentionality of intentional states in Searle 1983, pp. 27–29). See Rosenthal 1985 for discussion of this point.
4. Similarly with phenomena, which Searle also invokes in putting the point.
5. Searle's claim that "there is no aspectual shape at the level of neurons and synapses" presumably trades on a notion of "level" such that the neurophysiological level is the level of neurophysiological *description*.
6. See (Rosenthal 1986; 1990) on this point, and also for how consciousness can be explained if intentionality is independent of consciousness.

When functions are causes

Jonathan Schull
Haverford College, Haverford PA 19041
Electronic mail: *jschull@hvrford.bitnet*

I think Searle is right in calling attention to the tacit but unacknowledged question of consciousness in cognitive science and the need for a "Darwinian" revolution in the explanation of mind-body relations. He gets the significance of Darwinism wrong, however, conflating intentionality with regularity and rule-following and intrinsic intentionality with subjective consciousness. When these errors are corrected, the implications of Searle's interesting discussion are hardly what he supposes.

176

Searle asserts that "the price of denying the distinction between intrinsic and as-if intentionality . . . is absurdity, because it makes everything in the universe mental." To pay this price, however, one must also equate intentionality with mentality, and both of these with mere regularity of behavior (or rule-following). A century ago, William James pointed out that it is a particular *kind* of behavioral regularity which invites the attribution of intentionality: "The pursuance of future ends and the choice of means for their attainment are . . . the mark and criterion of the presence of mentality in a phenomenon" (James 1896/1952, p.5). And he drew a very clear distinction between the regularities of (nonintentional) systems like bubbles in water, which act "as if" they were seeking the surface until they encounter a submerged concavity but which are then completely stymied, as opposed to the regularities of intentional systems like frogs which will change their behavior in adaptive efforts to achieve their ends.

To some extent, the intentional kind of regularity is present in systems like Searle's lawn, and this is why we feel comfortable describing it as "thirsty." For example, whereas plants normally turn toward the sun, they might turn away from the sun when they are water-deprived, to reduce evaporative water loss. There is thus a distinction to be drawn between intentional systems on the one hand, and (generally nonliving) systems which lack needs, goals, purposes, and so on, and evolved abilities to pursue them adaptively.

Although I accordingly differ from Searle in recognizing a real distinction between intentional and nonintentional systems, I agree with him that it would be a mistake to attribute intrinsic intentionality to the lawn. I also agree that the distinction is related to the logic of functional explanations; but I think Searle has this logic only half-right.

This part is right: Because the individual plant's future ends and needs (water, survival, and reproduction) do not play a causal role in the effectuation of the individual plant's behavior, it would be wrong to impute intrinsic intentionality to the plant. Searle admirably develops this criterion for intrinsic intentionality.

However, I think Searle is wrong in supposing Darwinism to imply that adaptive functions of physiological "hardware mechanisms" are just "effects" that "happen," and that functional explanation "shifts the normative component outside the mechanism to the eye of the beholder of the mechanism." In fact, Darwin's absolutely fundamental contribution was the explanation of how, in evolving species at least, *functions* (of mechanisms in individuals) *can be causes* (of the evolution of those mechanisms in a population of individuals). In slightly more detail, effects of heritable mechanisms which confer a selective advantage on their bearers (i.e., functions) can be causal mechanisms of evolutionary change in the species. The shift is not to the eye of the beholder, but to the larger system of which the mechanism is a part (Schull, in press. See also Rozin & Schull 1988 for a discussion of interrelations between functions, principles, and mechanisms).

It is precisely this which makes it inappropriate to attribute intrinsic intentionality to an individual plant: The causal loop in which functions become causes (natural selection) operates in populations of plants, not within individual organisms. (For this same reason, it may be appropriate to attribute intelligence and intentionality to evolving populations [see Schull 1990 – but that's a different question.)

Now, evolving populations are not the only systems in which functions really do play causal roles. Another such system is a conscious human being. But notice that at this point *two different* considerations have just slipped into the picture. On the one hand, we have acknowledged that the nervous system appears to be organized in a way which allows functional and normative considerations to play causal roles (analogous to what goes on in evolving populations, but mediated by physiological mechanisms and organizational properties which have yet to be identi-

fied completely). And on the other hand, we have admitted that human beings are conscious.

The connection between these two considerations is not at all clear, and until someone demonstrates why they must necessarily be yoked, we might distinguish "intrinsic" from "non-intrinsic" intentionality on the basis of either or both of these considerations. The meaning of "intrinsic intentionality" will hang in the balance.

Thus, it seems to me that Searle's invocations of subjectivity are independent of his articulation of the principle (which I would endorse) that intrinsically intentional systems are those in which goals do play a causal role, and that he has not shown why subjective consciousness must necessarily accompany all cases in which functions and purposes play genuinely causal roles. Nor has he shown why "in-principle accessibility" to subjective consciousness is a necessary condition for deciding whether or not a performance is mediated by "mental processes." (That it is sufficient, is probably true by definition.)

Once again, William James has some timely observations: One of the central themes in the *Principles of psychology*, and indeed of much of James's life's work, is the demonstration that many of the facts of subjective experience are consistent with the hypothesis that the brain creates a multilevelled, hierarchically organized arena in which natural-selection-like processes operate, and that this can account for much of the idiosyncratic adaptive selectivity which Searle calls "aspectual shape" (see Schull, in preparation). Searle wants this term to further include something like conscious subjectivity, and that is his prerogative. But while James argued convincingly that purposive selectivity of the sort just mentioned is (in humans) the very essence of subjective consciousness, he still felt it worth asking, "is the consciousness which accompanies the activity of the cortex the only consciousness which man has? or are his lower centres conscious as well?" He explicitly speculated that "the lower centres [of the brain] themselves may conceivably . . . have a split off consciousness of their own . . . which does not mix with that which accompanies the cortical activities, and which has nothing to do with our personal Self." (James 1890/1952, p.43) While James felt the question was a genuine one (after all, we have subcortical centers in our brains which are as large and complex as that of birds and rodents who may well have some kinds of consciousness), he concluded that it is unanswerable: "If there be any consciousness pertaining to the lower centres, it is a consciousness of which the self knows nothing" (James 1896/1952, p.44).

I would like to understand how, or whether, Searle would deny this possibility. Lest he argue that such consciousnesses would be accessible, "in principle" if not in fact, I will point out that, if they exist, these "subconsciousnesses" could be so alien and distinct from our own consciousness as to preclude the possibility of access. The "problem of other minds" could be a domestic one, and not just a matter of "foreign relations."

Searle seems to take up this issue in his discussion of the vestibular ocular reflex (VOR), which not only keeps our eyes on the road, but is capable of some adaptive learning. However, his exposition of why it would be a mistake to impute intentionality to the VOR ("What actually happens is that fluid movements in the semicircular canals") only seems persuasive because he does not give an equally compelling mechanistic account of an act of noncontroversial intentionality. It is not hard to imagine what such an account might be like. What actually happens (I might speculate) is that mismatches between patterns of activity in the frontal cortex (which specify the appropriate object of attention) and the visual cortex (which provide the content of visual attention) trigger circuits which potentiate road-relevant feature analyzers, thus bringing the driver's attention (and possibly the car) back to the road.

The question remains why the activity of this (realizable but not-scientifically-established) orienting process is conscious, while that of the other (realizable and established) orienting

177

process is not. But the point is that is *does* remain, Searle's discussion notwithstanding.

In conclusion, I share Searle's inclination to impute "intrinsic intentionality" to systems in which functions, goals, intentions, and so on really do play causal roles (but I do worry – and Searle should worry – about the implications of this for biological species). Unlike Searle, I think the difference between conscious and unconscious brain states is a different question. Once one distinguishes intrinsic from nonintrinsic intentionality, and both of these from the nonintentional, and all of these from subjective consciousness, the connection principle reduces to the interesting but less significant claim that all of our mental states must, in principle, be accessible to our own subjective experience. Otherwise, they are either not mental, or not ours! (If they are not ours, they might still be mental states of brain centers to which we do not have conscious access.)

ACKNOWLEDGMENT
I thank Brian Knatz for his advice, argument, insight, and acumen.

Unconscious mental states do have an aspectual shape

Howard Shevrin

University of Michigan Hospitals, Department of Psychiatry, Adult Outpatient Psychiatry, Ann Arbor, MI 48109
Electronic mail: *Howard.Shevrin@UB.CC.UMICH.EDU*

My argument is written from the vantage point of someone who has been deeply engaged in the study of unconscious mental states, both as an investigator and as a psychoanalyst. I have in the past learned much from Searle's incisive critique of strong AI and see his present argument as a logical extension of his position. He has been calling insistently for a closer look at exactly how the mind and the brain *do* work. Computers provide no shortcuts, although they can help in modeling ways we might *imagine* the mind to function. I am puzzled, however, by several aspects of Searle's current effort to encourage cognitive psychologists to address the nature of consciousness. I again find myself very much endorsing the goal, but this time I have some qualms about the conceptual tools he provides.

My main problem is with Searle's argument in support of his basic contention that "shallow" unconscious mental states are definable as mental insofar as they have access *in principal* to consciousness or indeed cause consciousness, whereas the "deeper" unconscious states are not mental at all but essentially physiological and mechanical in nature. Searle further claims that "shallow" unconscious states, although mental and intentional and thus similar to conscious states, are nevertheless qualitatively different in one important respect; whereas conscious mental processes have aspectual shape, unconscious mental states do not. I understand Searle to mean by aspectual shape that whenever one is conscious of any mental content it is always in some highly specific context determining the particular way that object is experienced in consciousness (the water versus H$_2$O example).

It does not seem to me that Searle has convincingly established that unconscious mental states do not have aspectual shape, however. I do not think his argument is internally consistent; I will also cite research demonstrating that unconscious mental states do have an aspectual shape. Thus, even if I am mistaken about the logic of Searle's argument, I still believe it is wrong in the light of empirical evidence.

With respect to the argument itself, Searle states that he is a monist and that conscious mental states must therefore also be instantiated by brain events. He then identifies a potential contradiction in his position because he points out that insofar as

"shallow" unconscious states are both mental and intentional they should also possess an aspectual shape. But, he asserts, since unconscious mental states are entirely physiological in nature, and since aspectual shape cannot be attributed to neurons and synapses, unconscious mental states cannot have aspectual shape. This contradiction can be resolved, he argues, by attributing to unconscious mental states in their purely neuronal and synaptic state, the capacity to *cause* an aspectual shape to occur consciously.

Here is my problem with this resolution: Since conscious mental states are themselves instantiated by neuronal and synaptic events, it would appear that in conscious mental states neurons and synapses can in fact instantiate an aspectual shape. If they can do so consciously, why can't they do so unconsciously? It seems to me that at the very least, Searle would need to assert that the neurons and synapses associated with unconscious mental states lack the capacity to instantiate an aspectual shape but *other* neurons and synapses, those associated with consciousness, do have this capacity. But this does not appear to be his position. Rather, he wants to claim that *no* neurons and synapses have the capacity to instantiate aspectual shape. The other alternative is to conclude that neurons and synapses at any level are capable of instantiating aspectual shape and this is as true of unconscious mental states as of conscious mental ones. I concede that once this position is reached no clear difference then exists between Searle's "shallow" unconscious mental states and conscious ones other than consciousness itself. This would be the position Searle attributes to Freud.

I would next like to cite research supporting my hypothesis that unconscious mental states are characterized by aspectual shape much as conscious states are. We have found that event-related potentials obtained in response to words presented subliminally, and thus out of awareness and in an unconscious state, contain information related to the individual, autobiographical, private significance of these words (Shevrin 1988; Shevrin et al. 1988). The words would not have the same meaning or emotional significance to anyone else; experienced psychoanalytically oriented clinicians select them from verbatim transcripts of in-depth interviews and testing sessions conducted with patients suffering from social phobias, a common psychiatric symptom. One category is composed of words selected to capture the patient's conscious experience of the symptom and a second category is selected to capture the hypothesized dynamic unconscious state (in the Freudian sense) causing the symptom. When these words are presented subliminally and supraliminally, event-related potentials are obtained and then, using an information-theoretic approach, analyzed for their information content in relation to their word category (Kushwaha et al. 1989; Shevrin 1988; Williams et al. 1987) as well as a time-frequency feature-detection procedure (Shevrin et al. 1988; Williams & Jeong 1989). With these measures we have found that different event-related potential patterns discriminate the two word categories when they are presented subliminally. The unique meanings distinguishing these words would appear to be what Searle refers to as aspectual shape. If so, a neurophysiological event associated with an unconscious mental state can instantiate aspectual shape.

These results fit well with the clinical psychoanalytic concept of the dynamic unconscious. The root cause of neurotic difficulties is hypothesized to be highly specific unconscious fantasies that actively influence the direction and content of conscious states, albeit in a heavily disguised form. These unconscious fantasies are not simply dispositional but are active largely because of repressed unconscious motives and should therefore be detectable with subliminal probes quite independently of what the fantasies might be causing consciously. Ordinarily we only learn about these fantasies through their conscious manifestations, but our research demonstrates that they may be probed and measured in an entirely unconscious state through subliminal and neurophysiological methods. These unconscious fantasies appear to have aspectual shape.

The neurophysiology of consciousness and the unconscious

Christine A. Skarda

Molecular & Cell Biology, Life Sciences Addition, University of California, Berkeley, CA 94720
Electronic mail: *wfreeman@garnet.berkeley.edu*

Searle adopts the physiological point of view that there is no basis for the claim that behavior is driven by programs like those used by digital computers. But how well does he characterize the neural mechanisms of such conscious states as perceptual recognition, and such unconscious ones as memories?

Laboratory data (Freeman & Schneider 1982; Freeman & Skarda 1985; Skarda & Freeman 1987) suggest that each conscious mental state begins *within* the brain as an internally generated, self-organized process that is projected through the brain both as a motor command that orients exteroceptors optimally in the light of past sensations, and as reafferent messages to all sensory systems that prepare them for the consequences of expected action on the basis of past experience. Conscious mental states also require chemically mediated synaptic changes during learning that lead to the formation of nerve cell assemblies (NCAs) that sensitize sensory cortices to particular stimulus configurations, but these states should not be identified with NCAs (Skarda & Freeman 1988). In the olfactory system, a perceptual process is initiated by an inhalation that causes a volley of excitatory receptor input. When a critical threshold of excitation is reached in the bulb, a state change occurs in which the entire bulb abruptly changes to a globally distributed, stereotypical pattern of activity. The role of the NCA is to mediate the selection of this pattern, but it is the globally distributed activity pattern that constitutes perceptual recognition and that is transmitted to the limbic system, comprising neocortical and subcortical structures of the forebrain in an interactive hierarchy. Its self-organized, global activity patterns are the best candidates we have to identify with consciousness. To summarize: Conscious mental states are inaugurated from within, involve globally distributed dynamics, and are self-organized, hierarchical, interactive states of dynamic patterned activity.

Contrast this with such unconscious, nonmental states as reflex behaviors (Sherrington 1906). A reflex is stimulus dependent; it is initiated from *without* by the stimulus rather than by the brain. It involves a series of passive, feedforward transformations performed on the input pattern by effecting each link in the neuronal chain in turn like a string of dominoes (Thach 1978). Most reflexes do not involve the cerebral cortex; those that do involve limited portions of it, not the whole. This neurophysiological difference is to be expected: For a state to be conscious and mental it has to have the right sort of neurophysiology, something reflexes and other kinds of nonmental, unconscious phenomena do not have.

Searle's target article makes a distinction between a neurophysiology that has the "capacity" for consciousness and that which does not, and in this respect is consistent with what we know about brain functioning. But what of his claim that unconscious, mental processes are "going on" in my brain and that they are mental precisely "because they are capable of causing conscious states" (step 6, last paragraph)? Is Searle's characterization of unconscious mental phenomena consistent with what we know from neurophysiological research? I think not.

What leads Searle astray in his attempt to characterize unconscious mental phenomena is his causal account of the

mind/brain. Neurophysiological processes do not *cause* mental states, they *are* mental states at the neurophysiological level of description. Phenomena described at different levels of description are not causally related (Rose 1987). To assume otherwise is both bad science and bad philosophy. Searle is aware of this even if he waffles on the issue, but his causal account still gets him into trouble.

Causal explanations are inappropriate not only when applied between explanatory levels, but also for phenomena *within* a level of description. Nonlinear dynamical systems theory has shown that brain dynamics preclude causal explanations for two reasons. Recent connectionist models of memory have focused on the mechanisms of neural network formation (Hinton 1985; Hopfield & Tank 1986; Kohonen 1984), with the result that many today believe that memory consists entirely of synaptic changes within an NCA. Searle may have had this in mind when formulating his causal account of mental states. Yet while activity in the NCA temporally precedes the globally distributed patterned activity we identify with conscious, mental phenomena, these microphenomena cannot explain – are not the "causes" of – the globally distributed activity patterns of the conscious state that follows (Skarda 1986). Brain dynamics are self-organized. Self-organized phenomena cannot be explained in reductionistic terms, as Searle claims (sect. 6, para. 3). Explanations of such phenomena can be given only in terms of the qualitative forms of behavior of the system as a whole, and not in terms of properties of its parts, whether these be neural networks or individual neurons. Second, the observation that neural dynamics are "chaotic" (Skarda & Freeman 1987) further undermines Searle's causal account of the unconscious (Skarda & Freeman 1988). Chaotic phenomena are inherently unpredictable because small uncertainties are amplified over time by the nonlinear interaction of a few elements (Crutchfield et al. 1987). The upshot for neurophysiology is that we cannot make strict causal inferences from the level of neural networks to that of neural mass action. The impact of this explanatory revolution in neuroscience does not seem to have reached Searle. A causal account of conscious or unconscious phenomena is doomed to failure.

What sense can we give to the notion of unconscious mental phenomena if we reject the causal account, then? How do unconscious memories differ from reflexes? The globally distributed, self-organized activity patterns required for conscious recall do not persist in the brain when the state is unconscious. Memories arise new each time by self-organized processes and are not retrieved from a "memory store" as in a digital or analog computer memory. What remains when memories are unconscious is a "space of possibilities"; in mathematical terms, a structured, interdependent global system of "attractors." What we refer to as unconscious memories are "tendencies" or predispositions to engage in particular forms of patterned activity that are made available to the system at the point when it is destabilized by its interaction with the environment (Skarda & Freeman 1988). Unconscious mental phenomena are not "going on" in the brain and they do not "cause" consciousness, but they are available forms of dynamic activity that define what the system can do and that are constrained by each new conscious experience that further shapes this space of possibilities. And because these forms of patterned activity are merely available to the system whenever it is destabilized, they preserve our notion of the unconscious in that they may never be actualized.

The key point is this: The global structuring process in which past experience changes current neural dynamics does not occur with such unconscious, nonmental phenomena as reflex behaviors. The "structured space of possibilities" is how unconscious mental phenomena operate in brain dynamics. This process is unique to unconscious mental phenomena, distinguishing the neural dynamics of the unconscious. So there is neurophysiological reality to unconscious mental phenomena that

179

sets them apart from unconscious nonmental ones, but it is not the one Searle suggests. The unconscious cannot be understood in terms of causes; it must be understood in terms of future possibilities for the system to engage in self-organized patterned activity that is created anew each time. If Searle wants an account that is physiologically sound, he would do well to reject not only the digital computer paradigm, but also the notion of causality that philosophers still too often mistake as the hallmark of "scientific" explanation.

The possibility of irreducible intentionality

Charles Taylor

Department of Political Science, McGill University, Montreal, Quebec H3A 1G5 Canada

I agreed with much in John Searle's target article, in particular, with his insistence on the distinction between intrinsic and as-if intentionality. I think I can accept the Connection Principle under some formulation. But I'm still worried about whether there is not some a priori neurophysiology lurking in Searle's account.

I can get at this worry either by going at the issue of what it is that, inaccessible to consciousness, underlies our conscious thoughts and actions; or by questioning Searle's invocation of Darwinian explanation. Let me try both of these routes, in the order introduced.

I'll take an example that I don't really accept, but that could be true in some form: a Durkheimian explanation of religion. The sense of religious awe and allegiance "really" reflects the sense of dependence on society, not just for survival but for our being fully human agents operating on a moral level. Now Durkheim may have thought that we moderns come to recognize this consciously. But according to one interpretation, even the enlightened, lay citizens of the Third Republic didn't have first-person insight into what made them such fervent supporters of the republican tradition. Then why believe the theory? Because it *could* make sense of the whole history of religious development, including the rise of lay ideologies, better than any rival view.

In other words, the theory points to some factor, here an attachment to something, which is posited as what is really moving people, even though these people can't call it up to first-person consciousness. What I mean by this last phrase is that they couldn't see the desirability of the objects of their religious or moral fervor under the aspect, "the social bond that makes us human"; although clearly they could come to accept the *theory* as a third-person account of their and others' behavior. It is clear that a Freudian-type theory could also be constructed with this feature, as could any theory supposing an "unthought."

Now this factor has a very definite aspectual character, in Searle's sense. The theory purports to identify a desirability-characterization. We don't have the kinds of cases Searle considers, where some being seeks water, and we rightly argue that we have no way of determining whether he seeks it under the aspect "H_2O." Here we have triangulated, as it were, to a determinate aspect from out of conscious, full-blown, intentional behavior.

If Searle could be got to accept this social bond factor as a genuine example of the in-principle unconscious, then his answer would be to ask what facts correspond to the claim that it explains religion. Since by hypothesis we don't have a conscious thought, we must have a brain state. And so why can't we settle for some brain condition which always produces social-bond-cathecting behavior, along with conscious thoughts identifying a certain range of things (including God, Republican principles, but not the social bond) as objects of devotion? This would be the

underlying mechanism, analogous to the VOR in the vision case, and we could dispense with unconscious desire.

Maybe we should settle for this, thus espousing the Connection Principle. But this brings us to the second line of approach, Searle's appropriation of the Darwinian analogy. What does it mean to make a hard and fast distinction between hardware explanations and functional explanations? Searle wants to argue that the functional consideration tells us nothing about the causation of the phenomenon. The plant turns because of the secretions of auxin, not because the turn maximizes sun exposure. Of course, over time the survival effects of sun exposure ensure that this kind of plant proliferates, but Searle's point concerns the explanation of the individual plant's behavior.)

But what do we mean by a hardware account? One plausible interpretation is that we mean accounts that have recourse only to factors recognized by the disciplines of neurophysiology or organic chemistry as they now exist. The firings of different neurons will be explained by local chemical changes, and the larger patterns of firing will be accounted for by the concatenation and mutual interactions of such local effects. Or we might admit larger field effects into our causal story, but these fields would be defined anatomically. Imagine, however, that a further step is necessary: that we need to invoke field effects where the "fields" in question cohere just in being embodiments of some state defined in intentional terms – a thought of something, or a desire for something.

Now the latter step would take us well beyond the bounds of mainstream neurophysiology, by incorporating intentionalist concepts into the science. Is the notion of a hardware account meant to exclude this possibility? If so, then it may be excluding a priori a promising line of advance. What is more, if all neurophysiological function is explained in hardware accounts so defined, and if all conscious states are realized in neurophysiological states (and the latter premise is pretty universally shared), then it is hard to see how accounts of conscious action can avoid becoming derivative of these hardware explanations. Epiphenomenalism looms.

If, on the other hand, "hardware accounts" can be enlarged to include any such useful intentionalistically defined factors, then it is hard to see how the rigid separation between hardware and the functional can be maintained. To play my sub-Durkheimian fantasy out to the end, what if the underlying neurophysiological account of our cathexis of the social bond had recourse to field effects of neuronal clusters, where these were identified at least in part inescapably in terms of what they related to – for example, the social bond? Would it be so clear that we shouldn't speak of in-principle unconscious states? Perhaps indeed, it would, and we would be wise to follow Searle's warnings about the confusions that can arise here. But the grounds for this exclusion would no longer be that the intentional, the aspectual, the functional cannot extend beyond what can be called to consciousness. To lay that principle down at this stage is to slide into a priori theorizing about the future development of science.

The causal capacities of linguistic rules

Alice ter Meulen

Department of Philosophy, Indiana University, Bloomington, IN 47405
Electronic mail: *atm@ucs.indiana.edu*

1. Degrees of unconsciousness and the nature of linguistic rules. In conversation we are usually aware of the meaning of the sentences uttered, but not of the syntactic, phonological, phonetic, and semantic principles and rules we use in computing it. We have conscious access only to products of linguistic rules and principles, not to the rules themselves nor to the computational

180

processes they cause. We are conscious of these products because we use them in making inferences, drawing conclusions and communicating information to others. But they need not be manifested in observable behavior, for we may keep them to ourselves. Hence products of linguistic rules cannot be so simply equated with symptoms of intrinsic intentions, which, Searle advocates, are reducible to hardware level explanation. So we do attribute to these linguistic rules and principles a causal capacity to produce consciousness. Attributing a causal capacity to something does not necessarily mean it is accessible to consciousness, nor the other way around. But even for Searle's "relatively unproblematic cases of shallow unconscious mental phenomena" – his third grade unconsciousness, in principle accessible and causing conscious states – the picture sketched is much too simplistic. In what sense are my nonoccurrent beliefs, desires, and memories unconscious? As Searle argues that they already have "aspectual shape" (seemingly just another word for intentionality or opacity) as unconscious states, the processes which make them conscious must preserve this. Consequently these shallow states must be individuated in their neurophysiological base just as they are when conscious or overtly asserted. If I fail to draw a conclusion from certain information however, either while waking or in a dormant state, do my neurological nets get altered in such a way that I can still retrieve the same information whenever I desire, use it to draw my conclusion and in doing so make it conscious? This would suggest that given a certain conscious belief state of a system, all logically possible, compatible belief states are unconscious, floating somewhere deep down in its neurophysiological nets. Here lurks the danger of metaphor! There is no such thing as a shallow pool of all logically possible beliefs in which we may dip freely to bring a new belief to the surface while preserving its aspectual shape and subsequently alter our current conscious belief-state by paying attention to it. Searle has not advanced any convincing argument that his third-degree unconscious states can have aspectual shape even as neurophysiological phenomena or that the processes which cause the conscious belief must preserve aspectual shape. The fourth degree of unconscious states is characterized by their essential inaccessibility to consciousness, and includes the (according to Searle nonexisting) linguistic rules and principles. But why should linguistic rules and principles have to be either "in the mind" or in the external phenomenal world? They are of a more abstract nature; they are not the sorts of things that need to have any location at all. They form an abstract, structured system of relations between linguistic expressions, parts of the world, common knowledge we have of the world and our own individual background of experiences, formed prejudices and beliefs. As such they constrain our cognitive processes that come into play when interpreting linguistic expressions in their context of use. In any given situation a linguistic rule is used when triggered by conditions, like a desire to reply to someone, and perhaps various additional conditions in the phenomenal world which in combination cause the product. Furthermore, linguistic rules can be the subject of conscious thought – any linguist presents abundant evidence of conscious reflection on rules and principles. It is only in their individual application in cognitive processes of language comprehension and production that users need not have access to the rules.

2. Explanatory inversion and linguistic rules. Linguistic rules are constitutive of our linguistic competence in the sense that any change in the rules and principles necessarily constitutes a change in our competence. Contrasting an ability to a skill may help clarify the underlying issues. What distinguishes the relation between linguistic abilities and linguistic rules from the relation between driving skills and the rules of traffic? Driving is a skill we learn to automatize gradually. As a mere skill, it can be hampered by certain neurophysiological conditions like being intoxicated, being drowsy or being blinded. We can be out of

practice after a period in which we did not exercise the skill, and we can always improve on our skill by more practice. Our linguistic ability cannot be so directly affected by temporary neurophysiological conditions. For we do not say of a raving drunk that he can't speak English, but rather that his tongue is thick, his mind is blurry or that he does not know what he says, or simply does not make sense. Linguistic ability, once acquired, remains constant under our continuously changing neurophysiological conditions; further training and practice do not affect it significantly. Furthermore, our brain can compensate for a loss of linguistic performance caused by localized neurophysiological damage by accommodating certain linguistic tasks elsewhere in the cortex.

Now, in explaining car-driving behavior, the Darwinian conversion Searle proposes is quite feasible and may even offer a satisfactory account. Adhering to the rules of traffic while driving certainly has survival value. But traffic rules may be changed by fiat without affecting our driving skills. Even if you find yourself driving in an environment with quite different rules, it just takes a bit of practice to get adjusted. Traffic rules are obviously not constitutive of driving skill. They can certainly be made accessible to consciousness, as required in written driving tests, and we use them all the time to make sense of what other drivers do and why. Traffic rules serve to coordinate traffic, which is a practical matter. Coordination of meaning is not such a practical matter. We use linguistic rules to make sense of what others say, but they cannot serve in explaining why they say it. Although there may be evolutionary advantages in adhering to the linguistic rules of one's environment, such conformity to linguistic practice cannot be the goal of using one's language. Language is used because it is an efficient, readily accessible, and widely used medium of communication.

Unintended thought and nonconscious inferences exist

James S. Uleman[a] and Jennifer K. Uleman[b]
[a]*Department of Psychology, New York University, New York, NY 10003* and [b]*Department of Philosophy, University of Pennsylvania, Philadelphia, PA 19104*
Electronic mail: *uleman@nyuacf.brtnet*

What distinguishes "mental" events from other phenomena, and what does this tell us about cognitive science? These are important questions. Useful answers will have to include a wider range of putatively cognitive phenomena, however, and use a more differentiated set of analytic categories than that provided by Searle.

We agree with one of Searle's conclusions: that the idea of intentional mental states which are "in principle" inaccessible to consciousness is incoherent. The distinction between "intrinsic" and "as-if" intentionality is important. One of us has written as much (Uleman 1989, especially pp. 430–34). But we are dissatisfied with the way he reaches this conclusion, the implications he draws from it, and the limited categories he provides.

Searle argues that neither computational machinery nor "brute, blind" neurophysiology can be intrinsically intentional, precisely because they are not *conscious*. The claim that plant leaves and the VOR (vestibular ocular reflex) lack intrinsic intentionality is the same, for him, as the claim that when they "act" they have nothing in mind, they follow no rule, they are not conscious.

To develop this argument, Searle introduces aspectual shapes. These are (a) uniquely characteristic of consciousness, (b) impossible to account for exhaustively with "objective" evidence ("third person, behavioral, or even neurophysiological"), and (c) not even detectible from such evidence (cf. sect. 6, para. 5 on Quine and zombies). This prevents equating intrinsic and as-if intentionality. No amount of looking or acting intentional warrants our calling a system intentional. We must (somehow) know that it performs consciously. (One might wonder whether aspectual shapes aren't just the sort of ethereal apparition that commit Searle to a dualism he wants to avoid; see Note 4.) This tack has the unfortunate effect of obscuring how or why we attribute intentionality to each other, because surely we have only "objective" information about each other.

Searle identifies unconscious intentional states with neurophysiology, given a puzzle – the sleeping man who knows the capital of Colorado – that might equally well be solved by dropping the assumption that intentional content (aspectual shapes) must be located in a head, and cannot reside in shared linguistic practice or other recurrent features of the environment. Dropping this assumption is tantamount to abandoning the view that intentionality must refer to some incorrigibly subjective feature of the agent. But the "subjectivist" tendencies of Searle's analysis cannot be pursued further here.

Is it correct, however, to insist that a given process, to be *mental* at all, need be in principle accessible to consciousness? Are there phenomena (of social cognition and cognitive psychology) to which Searle's categories cannot do justice?

First, there is Searle's assumption that "it is reasonably clear how [an intention] works for conscious thoughts and experiences." But Nisbett and Ross (1980) have shown that it is not at all clear how intentions work. People's best attempts at describing how they make intentional judgments (what rules they followed, or which factors they considered) are often wrong. They may offer plausible but demonstrably inaccurate theories, as when they "intentionally" judge how influential someone is in a two-person discussion, but are in fact responding only to that person's salience in their own perceptual field (Taylor & Fiske 1975). Or they may intend to perform some cognitive operation, being incapable of doing so but not knowing this, and so believing in the end that they have actually succeeded "intentionally." For example, beliefs may persist even after the basis for them has been discredited and the believer thinks he has taken the discrediting into account (Ross et al. 1975).

Are these cases merely neurophysiological, or are they accessible "in principle" to consciousness? They are neither, but are clearly mental events. Searle offers only a dichotomy, however: "There are brute, blind neurophysiological processes and there is consciousness; but there is nothing else" (sect. 3, para. 1).

Consider a second set of cases, illustrated by the Stroop effect in which people must quickly name the color of the ink while ignoring the printed words' meanings, and are unavoidably slowed down by the unintended activation of those meanings. Or consider priming effects, in which supra- or subliminal exposure to relevant concepts biases subsequent intentional interpretations of ambiguous behavior. (See Bargh, 1989, for a review and conceptual suggestions.) Is concept activation merely a blind neurophysiological process that influences conscious intentional processes? And does such a classification enrich or diminish our understanding of the phenomenon? Again, what can we do with skilled typists' or musicians' performances, in which behavior that is initially guided intentionally is replaced by relatively automatic routines that are *disrupted* when the performer tries to make them conscious?

Third, this scheme denies the possibility of "unconscious inferences." Yet when people read about actions that imply personality traits, they infer those traits without any intention or awareness of doing so; they do so although these inferences are not called for, and may even interfere with another primary concurrent task. (See Uleman, 1987, for a review of cued-recall results; we also have unpublished evidence from increased recognition reaction times, using McKoon & Ratcliff's [1986] procedure.) Are these somehow not inferences, but mere neurophysiological processes? Or will the gray area to which Searle

assigns our attributing intentions to grasshoppers also include such cases? How much expansion can this marginal zone withstand before it alters the distinction between intrinsic and as-if intentionality?

Fourth, Searle apparently assumes that intentional processes operate in the same way whether or not they are actually conscious. But we (Newman & Uleman, in press) and others have shown that primed concepts have different effects on subsequent judgments, depending on whether they are conscious when the judgment occurs. Furthermore, the extensive literature on implicit memory (e.g., Schacter 1987) contradicts this assumption.

Only space limitations preclude extending this list of phenomena that are mental events but are not intrinsically intentional. In throwing out the bath water of as-if intentionality, Searle has also dumped a tubful of babies that cry out for recognition and coherent organization.

Conscious and unconscious representation of aspectual shape in cognitive science

Geoffrey Underwood

Department of Psychology, University of Nottingham, Nottingham NG7 2RD, England

Electronic mail: *gju@psychology.nottingham.acouk*

Little serious attention has been paid to the subject of consciousness, in this we are agreed, but if Searle had considered some of the recent studies of conscious and preconscious processing which are available, he would have drawn different conclusions. Recent recognition of the neglect of consciousness was what prompted a number of psychologists to attempt descriptions of functions and processes (see, for example: Baars 1988; Cheesman & Merikle 1985; Dixon 1981; Klatzky 1984; Marcel 1983; Pope & Singer 1978; Shallice 1972; Underwood 1978; 1982; and many others). The following comments attempt to interpret Searle's position in the light of the current evidence from cognitive psychology. The central question is whether we have any mental processes that are inaccessible to consciousness: Searle's position is that we do not, and that any unconscious process which can be regarded as a mental process must be, in principle, capable of representation as a conscious process. It is necessary for us to look for empirical evidence of unconscious states for which we have no conscious equivalent.

Suppose we found evidence for some process, in its effects upon some other behaviour perhaps, and no evidence of a conscious representation of that process. This may not be satisfactory because we would not have demonstrated that the process could *never* be accessible to consciousness. The notion of being "in principle" accessible to consciousness provides us with a problem, and the criteria for admission of evidence need clarification. Otherwise we can find ourselves repeatedly providing evidence of inaccessibility, only to be told that we are not trying hard enough. What would constitute an acceptable test of accessibility, in which we found a null result? Does such a test exist? Given the unclear criteria for acceptability of this test, we may start by considering a number of cases which appear to provide evidence of processing without conscious accessibility.

Examples to be accounted for include experiments that demonstrate: implicit learning of an artificial grammar (Reber et al. 1980), or of configural relations in a dynamic control task (Hayes & Broadbent 1988), or of covariations in personal traits (Lewicki 1986); influences of masked words on the processing of target words (Marcel 1983; Underwood 1981); and the performance of such highly skilled actions or "learned instincts" as tying shoelaces (Underwood 1982). In these cases there is evidence of processing that is not available to conscious report, but does this mean that the processes *could not* be available? How about those processes that do not lead to a conscious representation under one set of experimental conditions but do lead to such a representation when, for example, the exposure duration is increased? These demonstrations would presumably be evidence of "shallow unconscious" processing (sect. 2), except that in some of them the experimental conditions were set so that the effective stimuli were unreportable, and in the case of skilled actions performance can be seen to deteriorate when we become aware of processing. In these cases performance suffers as a result of awareness of performance.

These demonstrations also provide a methodological comment on Searle's views of the investigation of mental states by observing behaviour (step 3). His suggestion is that we can use neurophysiological evidence, or we can use introspective techniques, and an individual's behaviour cannot tell us about the "aspectual shape" representations. (By "aspectual shape" [step 2] I take it that Searle means the necessarily selective nature of conscious processing, a property well described by William James in 1890). Is there really nothing between the hypothetical, crude brain-o-scope and the vagaries of the interview? The mistake here is to neglect a family of techniques which can provide reports of representations by indirect observation. Priming the presentation and response to one stimulus by the prior or simultaneous presentation of an unreported stimulus can reveal the individual's representation of the relationship between the two. When the prior presentation of the priming word *bird* aids a response to the target word *robin* we have learned that for this individual there is a relationship – we can say that the individual *knows* that the concepts are associated or, in this particular example, we may be able to conclude that he *believes* that one is a subordinate member of the category represented by the other. By finding the conditions of effective priming we can find those attributes which are represented. Indirect techniques can be used to provide reports of aspectual shape without individuals having to volunteer their opinions. These representations may result from conscious processing, during which the salient aspects of relationships between concepts come to be selected and remembered and available as influences on our behaviour without our needing to be aware of them. Studies of implicit learning, however, suggest that conscious representations may not be necessary for the selection and storage of these influential aspects.

The distinction between "shallow unconscious" and "deep unconscious" processes seems uncontroversial, with the deep processes describable *only* in neurophysiological terms. Shallow unconscious states are mental states which, for some reason, are not conscious at that moment, and the question is whether there exist shallow states that are completely inaccessible. There is a problem here, however; the introduction to section 3 asserts that "there are brute, blind neurophysiological processes and there is consciousness; but there is nothing else," and this appears to exclude shallow unconscious processes altogether.

On this basis we are directed to reject notions such as mental models, because although they appear to guide behaviour and result in the appearance of rule-governed performance, these appearances are the result of an anthropomorphic fantasy that mistakenly assumes that purely neurophysiological events can have intrinsic intentionality. Mental models may exist, but if so they must be accessible to consciousness. Can neurophysiological states have the aspectual shape necessary for intentionality? They can certainly represent an intentional state, because all mental states are represented by neurophysiological states (Searle disclaims dualism, and appears to be as much of an emergent materialist as the rest of us). Even intentional states that are no longer conscious states, such as those things that we know but are not thinking about right now, are represented in the hardware of the brain. In this sense the brain holds its user's mental models in a neurophysiological code, and in this sense the code represents intentional states, even when we are in

dreamless sleep, or thinking of something else, or transformed into one of Ned Block's (1978; 1980) zombies, or, we have to say, when we have our aspectual shapes represented in artificial minds. Representations that are available as influences on our behaviour, and which are currently unconscious, still have aspectual shape. It would be nonsense to say that these representations have aspectual shape when imposed upon the electrochemical code of our neurophysiology, but no such shape when imposed upon the electromagnetic medium of an artificial mind.

Is the mind conscious, functional, or both?

Max Velmans

Department of Psychology, Goldsmiths College, University of London, London SE14 6NW, England
Electronic mail: *bnanm@gold.lon.ac.uk*

What, in essence, characterizes the mind? According to Searle, the potential to be conscious provides the only definitive criterion. Thus, conscious states are unquestionably "mental"; "shallow unconscious" states are also "mental" by virtue of their capacity to be conscious (at least in principle); but there are no "deep unconscious mental states." Those rules and procedures inferred by cognitive science to characterize the operations of the unconscious mind, are not mental at all – indeed, according to Searle, they have no ontological status; they are simply ways of describing some interesting facets of purely physiological phenomena.

Given the thrust of this argument, one might be forgiven for believing that for Searle, conscious states and shallow unconscious states are *not* purely physiological phenomena. But one would be wrong. Searle is a physicalist. Deep unconscious states are purely physiological, shallow unconscious states are purely physiological (but with the capacity to be conscious) and conscious states are *also* physiological, although with higher order emergent properties (see Note 4). In short, when Searle, contrasts conscious mental states and unconscious mental states with purely physiological states, he means to contrast the physical and the physical with the physical.

So what is all the fuss about? What is crucial, according to Searle, is whether a state has *intentionality*, and in the target article this is determined by whether or not the state has *aspectual shape* (sect. 2, step 2). What characterizes the "mental" is that "whenever we perceive anything or think about anything, it is always under some aspects and not others that we perceive or think about that thing." When mental states are conscious it is easy to see how this works out in practice. A conscious desire for water, for example, is not the same as a conscious desire for H_2O, although the referent of the desire may be the same in both cases. But how can an unconscious state have aspectual shape? Only in so far as it has the potential to be conscious, claims Searle, for aspectual shape "cannot be exhaustively or completely characterized solely in terms of third person, behavioral, or even neurophysiological predicates" (sect. 2, step 3). Without reference to consciousness, he argues, there would be no way of distinguishing a desire for water from a desire for H_2O. Shallow unconscious states, therefore, are rendered "mental" solely by virtue of their connection to consciousness (the Connection Principle); deep unconscious states, lacking this connection, are not mental.

The committed physicalist reader might well feel uneasy at this point. If shallow unconscious states cannot be characterized entirely in terms of physiological predicates, how can they be entirely physiological, as Searle claims? Even worse, what is so special about being conscious that lies forever beyond the domain of any conceivable third-person description? But Searle sails on. The problem, he says, has nothing to do with *ontology*, but with *epistemology*. Shallow unconscious states and conscious states *are* just physiological states, but their aspectual shape cannot be *known* without subjective access to consciousness.

In one sense, of course, Searle is right. Insofar as aspectual shape is manifest in conscious experience, it cannot be fully known from the outside, as we do not have full knowledge of other people's conscious experience. Searle's argument that aspectual shape can *only* be known via subjective conscious experience, however, is essentially circular. Although the fact that, "Whenever we perceive anything or think about anything, it is always under some aspects and not others that we perceive or think about that thing" might, *in principle*, apply to both conscious and unconscious perceptions and thoughts, unconscious perceptions and thoughts are just physiological states made "mental" by their potential connection to consciousness, according to Searle. His claim that apart from the way they are known in consciousness, such unconscious states *have* no aspectual shape then establishes his case by definition. It is possible, however, to disagree.

One cannot *fully* know the shape of another's conscious thoughts, but this does not preclude knowing *something*, from observations of behaviour, about whether their thoughts are directed to one aspect of a thing rather than another. Perhaps in some future neuroscience such behavioral observations could be combined with direct observations of the brain. But we don't have to wait for the arrival of Searle's "brain-o-scope" to determine, say, whether a chemist wants water or H_2O. If he drinks it we assume it's water, and if he sticks it in a test tube, it's H_2O!

Nor is it self-evident that the *only* sense in which physiological states have an aspectual shape is in so far as these are manifest in conscious experience. It is likely, for example, that *all* neural representations of internal or external events code those events under some aspects and not others. Indeed, it is difficult to envisage how any representational system could be constructed differently. And if this is so, it is difficult to argue on these grounds that shallow unconscious representational states are "mental" whereas deep unconscious representational states are not.

Moreover, Searle's intuition that only that which is potentially conscious is truly "mental," needs to be set against ancient, competing intuitions. To have a "mind" is also to have certain modes of functioning and capacities, an intuition dating back to Aristotle, which recurs in Descartes' attempts to demonstrate that man cannot be just a machine, on the grounds that no machine could ever use language or respond appropriately to continually changing circumstances in the ways that humans do. It is hardly surprising, therefore, that modern cognitive science has attempted to uncover the mental processes that enable human adaptive functioning (Whether these be conscious or not). At what point mental "software" is better thought of as neurological "hardware" remains an open question, but this does not make the attempt to *specify* the software any less legitimate.

In focusing on consciousness, Searle usefully draws our attention to a central facet of the mind that is largely missing in current functionalist cognitive science. In dismissing cognitive capacities or modes of functioning as further criteria of the "mental," however, he plumps for a definition that is equally incomplete. Mental processes do not *either* produce consciousness *or* permit us to function in certain ways, but under certain circumstances achieve *both*. A science of the mind, therefore, could never be complete without addressing this duality.

Consciousness, historical inversion, and cognitive science

Andrew W. Young
Department of Psychology, University of Durham, Durham, DH1 3LE, England
Electronic mail: *a.w.young@mts.durham.ac.uk*

The thrust of what Searle has to say in his target article is right. We do need to think much more carefully about what we mean by claims of unconscious mental phenomena. Serious conceptual errors have undoubtedly been made. I don't see why these are blamed on cognitive science, however, and I'm not sure that Searle hasn't perpetuated some of the misconceptions in his attempt to clear up others.

With its emphasis on the computational metaphor, cognitive science cannot be considered to have its origins much further back than the 1940s and, in practice, its pervasive influence on thinking about the mind has only been achieved in even more recent times. Yet the conception of the unconscious that Searle attacks has its roots much deeper in Western psychology and philosophy, and was prevalent long before cognitive science was a twinkle in anybody's eye (e.g., Ellenberger 1970; Reeves 1965). As Reeves (1965) and Ellenberger (1970) show, ideas about the unconscious mind achieved widespread prominence in scientific thought in the nineteenth century, and were themselves extensively drawn on and adapted by Sigmund Freud.

It is thus misleading to state that this conception reflects a shift in thinking between Freud and Chomsky (target article, sect. 1, para. 5), if the implication is meant to be that something fundamentally new happened; in many ways it is more properly seen as the continuation of an altogether older tradition. The steps from "the belief that we are dealing with unconscious mental processes to the belief that these processes are computational, and to the belief that the mind is a computer program operating in the hardware or wetware of the brain" (target article, sect. 1, para. 5) have been much larger and taken far longer than the target article claims.

Consider, for instance, the example of the Ponzo illusion (sect. 3). Searle rejects the explanation that the converging lines create an impression of depth, leading to the unconscious inference that if the top line is further away then it must be larger (using inappropriate constancy scaling; see, for example, Gregory 1963; 1966). He regards this type of explanation as inconsistent with the Connection Principle. It may well be so; it is hardly a typical example of an explanation in cognitive science terms, however, and neither was it dreamt up by cognitive scientists. Instead, this type of explanation is widely recognised as deriving from Helmholtz's *Treatise on Physiological Optics*, first published between 1856 and 1867.

Helmholtz anticipated some of Searle's concerns. He carefully pointed out that we should not simply regard the unconscious inferences he thought necessary for visual perception as the exact equivalent of conscious conclusions. But Helmholtz did emphasise that precisely because they are not open to conscious introspection, the effects of unconscious perceptual inferences are irresistible, which Searle might not like so much. All this in the mid-nineteenth century.

In addition to concern about this historical inversion, which seems to visit the alleged sins of its distant intellectual parents on the infant cognitive science, I am worried that Searle's analysis may be creating as many problems as it solves.

What, for example, does it mean to say that in our skulls there is "just the brain with all of its intricacy, and consciousness with all its color and variety," or that there are "brute, blind neurophysiological processes and there is consciousness; but there is nothing else" (sect. 3, para. 1)? Just the brain *and* consciousness? Would one want to say that in a clock there is just machinery and time? Hardly. Would you trust a mechanic who told you that your car was only metal, rubber, petrol, and travel?

184

I doubt it. We recognise that indicating the passage of time and travelling from A to B are functions of clocks and cars, not components. It may be equally misleading to reify consciousness.

We also recognise that describing the mechanism of a clock as "just intricate machinery" contributes little to understanding how it works. It is correct, but largely empty. For a real sense of understanding, we require a quite different level of description, and the appropriate level will itself depend on whether one has only a passing interest in clocks or intends to try to repair one.

I think this is important, because it undermines step 4 in Searle's argument (sect. 2). This holds that "the ontology of unconscious mental states, at the time they are unconscious, consists entirely in the existence of purely neurophysiological phenomena." Don't we expect that when we understand more about conscious states we will arrive at the same view, that their ontology, at the time *they* are conscious, consists entirely in the existence of neurophysiological phenomena? The neurophysiological level may well prove a hopelessly clumsy and inappropriate level of description, just as would trying to describe a word processor in terms of the states in the computer's electronic components, but it could be done.

Ought we in any case to make such a simple dichotomy between conscious and nonconscious states? The target article recognises that there have been a number of different uses of "unconscious," but the notion of consciousness itself seems to be taken as unproblematic, apart from the reference to its "color and variety." Is it really so easy? Many don't think so (e.g., Allport 1988).

To take an obvious example, what about dreaming? Are we conscious or unconscious when we dream? Don't dreams have a kind of aspectual shape? And so on.

I don't pretend to know the answers to these difficult questions and, probably, neither does Searle. But I do think that until we have more clear answers about consciousness itself we will need to be cautious about proposing that certain views of nonconscious or unconscious mechanisms are untenable in principle. Searle has performed a service in charting some of the relevant territory, but he was naughty to set fire to all the trees he encountered.

Ontogeny and intentionality

Philip David Zelazo and J. Steven Reznick
Department of Psychology, Yale University, New Haven, CT, 06520
Electronic mail: *reznick@yalevm.bitnet*

The need for explanatory inversion is all too apparent; just consider the proliferation of intentional "explanations" by students of cognitive development. In this commentary, we examine ways in which Searle's argument can inform, and be informed by, an ontogenetic perspective on intentionality.

Our first point concerns the implications of the Connection Principle for attributions of intentionality to infants. Searle (1983) is not alone in his reluctance to "deny that small babies can literally be said to want milk." However, according to the Connection Principle, only states that are *in principle* accessible to consciousness are candidates for intentionality. At present, we not only lack direct evidence of infant consciousness, but we also have no clear idea of what that evidence would look like. Therefore we cannot decide which states, if any, are in principle accessible, and so have no basis for or against ascribing intentionality.

One way to sidestep this problem is to assert that infants' neurophysiological states are potentially conscious because the same states are consciously experienced by adults. This assertion is weak, however. There is no guarantee that neurophysiological states survive ontogeny with their integrity intact. In fact, considerable evidence suggests that infant brains are qualitatively different from adult brains (e.g., Goldman-Rakic 1987; Huttenlocher 1979). The functional (and, a fortiori, phenomenological) significance of these differences is largely unknown. But if infants are to adults as caterpillars are to butterflies, then introspective evidence from adults about the accessibility of neurophysiological states to the issue of whether infants' states are potentially conscious; to claim that infants are potentially conscious of these states might be analogous to claiming that caterpillars can potentially fly. It remains an open question whether consciousness and intentionality *per se* develop or change. In addition, although it is commonly assumed that both consciousness and intentionality are indivisible – that is, they are either things one is or is not capable of – it is conceivable that they are acquired gradually, or even differently across different domains. Moreover, some intentional states may arise earlier in development than others (e.g., belief versus doubt). This indeterminacy suggests that it is important to investigate topics (such as the infant's self concept) that point to the subjective quality of infant experience and topics (such as future-oriented behavior) that seem to demand intentionalistic explanations.

Commensurability across age-related changes can certainly be assumed to hold beyond some (unspecified) age. Psychological change based on this assumption motivates our second point, which concerns the relation between consciousness and intentionality. Infants and young children perform a wide variety of apparently goal directed "actions" that resemble actions later undertaken consciously: They reach for objects, anticipate future locations of objects, form and use concepts or categories, use means to achieve ends, treat sounds and gestures as referents for things, and constrain their hypotheses about the referents of words. It is unclear whether these types of behavior result from conscious processes. It seems prudent to assume, however, that at least some do not, at least at some ages. As a *reductio*, consider a fetus kicking inside the womb with the result that it attains a more comfortable position. The relation between these types of behaviors and their subsequent conscious manifestations in older children and adults seems to be analogous to the relation between an unconscious (or preconscious) idea of an instinct and its conscious cognate – with time and experience, the prior forms become conscious. Because both are in principle accessible to consciousness, both are (Searle would claim) subject to intentionalistic interpretation.

We find this claim problematic. Intentionality consists in standing in the right sort of relation to an intentional object or content or representation. It may be thought of as having a perspective on an object. The fact that the perspective is necessarily limited gives the object its aspectual shape. But in the case of putative unconscious intentional states, who (or what) stands in relation to the object? Barring the existence of a homunculus, the object is not represented *by* anybody and thus there is no intentionality, no content, no subjective, mental experience to figure in explanation.

We agree with Searle that in the absence of consciousness "there is simply nothing there except neurophysiological states and processes" (sect. 2, para. 20). We further agree that "there is no aspectual shape at the level of neurons and synapses" (sect. 2, para. 21). Unlike Searle, however, we do not see how it is possible to talk about the intentionality of shallow unconscious states. Potentially conscious states are not intentional, merely potentially intentional.

Unconscious processes (and, a fortiori, processes that have never been conscious) are just neurophysiological states unaccompanied by understanding. They are analogous to the deep unconscious processes producing, for example, certain linguistic intuitions or visual phenomena. They may be "run through" just as a computer (or Searle in a room) may "run through" a program for understanding Chinese. Any intentionality (content) that there is must reside in the conscious interpretation of these otherwise uninterpreted and unintentional processes. This claim may be counterintuitive in the case of a sleeping man who seems to continue to believe that Denver is the capital of Colorado even while he sleeps. The claim accords better with our common sense account of those cases in development, however, in which we come to do deliberately something previously done automatically (perhaps because it was learned implicitly and was not under conscious control).

If correct, our claim supplements Searle's arguments for the importance of consciousness and against the notion of unconscious intentionality. We wholeheartedly endorse Searle's position, but add that intentionality implies concurrent consciousness. Intentional states consist in the actual representation of something from a perspective, and the only perspective coherently available to us is that of our conscious mind. Explanatory inversion should be generalized to *all* unconscious states. Someone acting because of a shallow unconscious state (e.g., what is metaphorically referred to as an "unconscious belief") does not act *for a reason*, in the usual, intentional sense. Only hardware and functional accounts, which capture important aspects of the organism's phylogenetic or ontogenetic history, are relevant to explanation in this case. As it stands, Searle's argument leaves an entire class of mentalistic metaphors in need of materialization; even Searle may be accused of confusing shadow and substance.

Editorial Commentary

Some of the things we can do and say, we know (or think we know) how and why we do and say. Others we don't. But, in a sense, we don't really know how or why we do any of it: We're waiting for science to tell us, the same way it tells us how and why a molecule, a planet, a machine or a heart does what it does; we're waiting for a causal explanation. Let's call what we can or could do knowingly "mental" and what we can only do unknowingly "schmental." The domain of the mental may well turn out to be small in comparison with the domain of the schmental. Searle's strictures, which are only on the mental, would accordingly not be binding on most of cognitive science (which is surely the science of both the mental and the schmental). In particular, whereas the Chinese Room Argument may have shown that the mental cannot consist of symbol manipulation *alone*, the Connection Principle does not seem sufficient to show that the schmental – or, for that matter, the mental – cannot consist of symbol manipulation *at all*.

Author's Response

Who is computing with the brain?

John R. Searle
Department of Philosophy, University of California at Berkeley, Berkeley, CA 94720
Electronic mail: *searle@cogsci.berkeley.edu*

I. Seven objections

I presented the target article, at the invitation of the editor, to mark the tenth anniversary of the publication of the Chinese Room Argument (Searle 1980). My objective

was, in part, to continue the general spirit of my earlier investigation but with new arguments and new issues. I hope I have succeeded to that extent, but it seems clear from the commentaries that this article is more difficult to understand than its predecessor and that I will have to devote some effort to clarification. In what follows, I cannot hope to answer every point made by every author, but I would like to answer all the major objections and remove every major misunderstanding. Some really obvious misunderstandings I can remove right at the beginning. I am not denying the existence and causal efficacy of unconscious mental states, including unconscious knowledge (**Carlson, Glymour,** and **Higginbotham**), nor am I denying the existence of unconscious inferences (**Uleman & Uleman**), nor do I deny that aspectual shape can be instantiated in brain states (**Shevrin**); and my investigation is not concerned with epistemic questions (**Kulli**). On the contrary, I think most of our mental states are unconscious most of the time, that they can be causally efficacious even when unconscious, that most of our inferences are unconscious, that all our mental life including aspectual shape is instantiated neurophysiologically and that such epistemic issues as, "How do we know of the intentional states of others?" are largely irrelevant to the ontological problems that concern me in this article.

I will begin by listing the most common objections and misunderstandings together with the names of their authors.

Objection 1. The Connection Principle was not conclusively established, and the relations between aspectual shape and neurophysiology are still unclear. These points were made in different forms by **Bridgeman, Carlson, Lloyd, Rey, Rosenthal, Shevrin** and **Velmans.**

Objection 2. Aspectual shape can function as aspectual shape even when unconscious, and this seems to be inconsistent with the "dispositional" account of the unconscious given in the target article. **Dreyfus, Shevrin, Underwood,** and **Carlson** all made this point.

Objection 3. The Connection Principle is vacuous because my account has the consequence that anything could be conscious. We could imagine a superconsciousness that was conscious of its own unconscious operations (**Block**); and in any case, the theorist who formulates a theory about the deep unconscious has already become conscious of the principles he has stated. If the theorist can become conscious of the deep unconscious principles then anybody can become conscious of them. We could even imagine a cactus plant with a hardware attachment that enabled it to become conscious of its own growth (**Clark**). In addition to Block and Clark; **Higginbotham, Hodgkin & Houston,** and **ter Meulen** all made this objection.

Objection 4. The notion of consciousness in the target article is unclear (this point was made by **Block, Hodgkin & Houston, Kulli, Schull, Uleman & Uleman** and **Young**), and furthermore the sense of consciousness in question relies on some doctrine of introspection. It presupposes, in Chomsky's words, an "inner eye" of consciousness, (this was charged by **Chomsky, Limber, Piattelli-Palmarini,** and **Rey.**)

186

Objection 5. Cognitive science explanations can – or already do – dispense with intentionality and intentionalistic explanations altogether. They can use computational (program, symbol processing, algorithmic) explanations instead. These sorts of explanations are not subject to the objections made in the target article. This was the most common objection. It was made in different forms by **Chomsky, Clark, Dresher & Hornstein, Glymour, Hobbs, Higginbotham, Matthews, McDermott,** and the **EDITORIAL COMMENTARY.** According to Chomsky, Dresher & Hornstein, Velmans and the **EDITORIAL COMMENTARY,** the dispute to some extent may be just a terminological dispute about the use of the word "mental."

Objection 6. We have a lot of evidence that there are deep unconscious rules of linguistic grammar. Such rules are not like my ironically imagined rules of visual grammar, because we have very good scientific evidence for the linguistic case and none whatever for the visual case. This point was made explicitly by **Chomsky** and **Matthews,** and to some extent it was also made by **Block, Freidin, Higginbotham,** and **Piattelli-Palmarini,** all of whom gave examples of rules that they do not think I can account for. Freidin gives the example of the rule against split infinitives, Block gives the linger-singer rule that I also used in *Speech acts* (Searle 1969, p.41–42), Higginbotham gives the example of the rule of agreement in number between subject and predicate, and Piattelli-Palmarini gives examples from Italian pronunciation.

Objection 7. The target article exhibits a misunderstanding of the nature of scientific explanation and indeed may even be guilty of trying to legislate the future of science. This objection, in different forms, was made by **Piattelli-Palmarini, Freeman,** and **Taylor.**

II. The Connection Principle restated: Reply to Objections 1–3

The two central claims in the target article I call the Connection Principle and Explanatory Inversion, respectively. In the commentaries, there was little challenge to Inversion but much to Connection, so in this section I will restate the Principle in a way that will relate it directly to the objections, starting with Objection 1.

The brain (together with the rest of the CNS) is a physical system, and like any complex physical system it has many different levels of description. These range from the level of quarks and muons to such gross molar suborgans as the cerebellum and the hippocampus, all the way to the entire central nervous system. As in other physical systems, different levels within the system can function causally. The brain has causal properties at some levels, for example, long term potentiation, which cannot be ascribed to individual quarks and muons, even though all the phenomena at the higher level are realized in phenomena at the lower levels. I hope everyone agrees with these points, and I emphasize them here because **Hobbs, Young,** and perhaps some others mistakenly think that I am denying the possibility of numerous levels of description.

The brain differs from all other known physical systems, however, in that it also has a mental level of description that is likewise causally real. At this level people have beliefs and desires, hopes and fears, and these actually cause their behavior. This level of causation, though realized in the physical system and, in that sense, physical itself, is nonetheless different from other forms of physical causation. For example, the actual intentional or semantic content of the mental state can function causally in the production of the very state of affairs that it represents; it can cause its own conditions of satisfaction. Thus, for example, my desire to drink can cause my drinking. Furthermore, intentional causation admits of assessment in terms of rationality, and this is a form of assessment that ordinary physical causation, billiard ball causation, does not admit of.

Characteristically, cognitive science, correctly in my view, trades heavily on intentional causation. For example, cognitive science explanations often cite rules in the explanatory models of human behavior. Rule governed behavior as a form of intentional causation, however, has the special features common to other forms of intentional causation; specifically, the content of the rule must play a causal role in the production of the very behavior that constitutes following the rule. All this apparatus – rule following, intentional causation, conditions of satisfaction, and so on – presupposes aspectual shape. Any intentional state represents its conditions of satisfaction under certain aspectual shapes.

Now here is the puzzle: When my mental states are conscious, the nature and role of aspectual shape is reasonably clear. (I do not say that there are no problems connected with conscious aspectual shape, but it is not difficult to figure out what we have in mind when we say, for example, that I am consciously wanting water but not consciously wanting H_2O.) Most of my intentional states are unconscious most of the time, however. They continue to exist even when unconscious, and furthermore, they sometimes even cause my behavior when they are unconscious, as **Dreyfus, Carlson, Shevrin,** and **Underwood** point out in Objection 2. But though aspectual shape is always instantiated in neuronal structures, just as neuronal structures are instantiated in molecular structures, the specification of the neuronal features is not yet a specification of the aspectual shape. Aspectual properties are just different from neuronal properties. When I say of someone, "He is thinking about water," I am just describing a different sort of phenomenon from, "His synaptic receptors are reabsorbing seretonin."

All aspectual phenomena are realized in neurobiological properties; but there is no aspectual shape at the level of neuronal properties; and when a state is completely unconscious, there is no aspectual shape that is manifest, so to speak, then and there. **Velmans** and **Rosenthal** are both in agreement with me that aspectual shape is encoded in neuronal structures, but the question they do not face and that I am trying to answer is: What sort of fact about this cluster of neurons could make it the case that it encodes, for example, the desire for water but not the desire for H_2O? And **Bridgeman,** with his computer analogy, provides a way of stating my answer: The computer disk has no margins, lines or paragraphs; but given the whole apparatus of the formatting program, and so on, the system can produce something that has margins, lines, and paragraphs. He concludes, "We need not expect aspectual shape at the level of neurons." He presents this as an argument against me, but it is my point exactly.

Now let us pause for a moment and reflect on these points. When a state is totally unconscious, the only phenomena occurring right then and there are neurophysiological events in neurophysiological architectures. On my reading of Freud, he denies this very point. He apparently thought that subjective but unconscious mental events were going on in addition to any neurophysiological occurrences that might also be going on. I think this amounts to a form of dualism, and I believe it represents an abandonment of his earlier project of a scientific psychology. But whether or not I am right about the interpretation of Freud, it seems clear to me that within our current nondualistic conception of reality no sense can be attached to the notion that aspectual shape can be both manifest as aspectual shape and yet totally unconscious. But since unconscious intentional states with aspectual shape exist when unconscious and cause behavior when unconscious, what sense are we attaching to the notion of the unconscious in such cases? I have argued that we can attach to it the following perfectly adequate sense: The attribution of unconscious intentionality to the neurophysiology is the attribution of a capacity to cause that state in a conscious form. This point holds whether or not the unconscious intentionality causes an unconscious action without causing a conscious mental event.

My reply to Objection 2 is thus that the unconscious mental states can burst out in the form of unconsciously motivated behavior even though there is no aspectual shape manifested then and there. There is instead just the capacity to cause the behavior or to cause the conscious mental state, either of which may occur without the other. Notice that in the sorts of cases described by **Dreyfus** (or Freud for that matter) we require an *interpretation*. The behavior, unlike the conscious intentional state, does not wear its aspectual shape on its sleeve; we have to interpret it by postulating an unconscious state to explain it. But that postulation must meet certain conditions of intelligibility that I specify in the Connection Principle. That is, if we try to state clearly what facts in the brain are supposed to correspond to the ascription of unconscious intentional states with determinate aspectual shapes, we come up with the result that the only things that could correspond to those claims are neurophysiological structures capable of causing the states in conscious forms; and in that sense, I said, such states must be in principle accessible to consciousness.

Various commentators point out (Objection 1) that the argument does not demonstrate that the Connection Principle with absolute certainty. They are right about that. The argument is explanatory though not demonstrative, but in the absence of any telling objections to the argument or any better explanations, I really do not have any doubts about it.

I now turn to Objection 3 (an ingenious one!). The objection is that, as I have stated it, the Connection Principle is vacuous because any unconscious state, deep or shallow, can be brought to consciousness. **Block** imagines a case in which we have a superconsciousness, and other writers point out that, because cognitive scientists are able consciously to state deep unconscious principles,

it follows that these principles are accessible to consciousness.

This objection rests on an important misunderstanding of the relation of consciousness to unconsciousness, which I will now try to clarify. There is a systematic ambiguity in such expressions as "conscious of" or "accessible to consciousness." I can be "conscious of" my pain and "conscious of" the cup on the table, similarly my pain is "accessible to consciousness" and the cup is "accessible to consciousness." But there is an important difference. The pain is a mental state, but the cup is not the sort of thing that could be a mental state, though it can be the *object of a mental state*. In the one case the noun names something identical with a mental state, in the other case it names the intentional object of a mental state. Applying these distinctions to deep unconscious states we get this result: they cannot be *identical* with any conscious states, by definition, but they can be the intentional *objects* of conscious states, as can anything in the universe. But simply because they can be the objects of intentional states, in the sense in which cups can be, it does not follow that they are themselves mental states. Analogous to the distinction between pain and cup is the distinction between my unconscious belief that Bush is President and the nonconscious myelination of my axons. My belief is a mental state, and it can become conscious. My myelination is not a mental state, and it cannot become conscious even in principle. It is no objection to this, indeed it is irrelevant to point out, that we could imagine a being that could become aware of its myelin sheathes in the sense in which it could become aware of a cup on the table. The myelin sheath still fails to qualify as a mental state.

The distinction is obvious in the cases of the contrasts between cups and pains, between myelination and beliefs. It is less obvious when we try to postulate deep unconscious entities that are supposed to be mental in the sense of beliefs but that are inaccessible to consciousness in the way in which myelination is. I have argued that there cannot be any entities satisfying both these conditions. The postulation of such entities ends up as the postulation of such phenomena as myelination or cactus growth. There is nothing mental about them.

According to **Chomsky, Velmans, Dresher & Hornstein** and the **EDITORIAL COMMENTARY**, the issue here is in part a verbal dispute about the use of the word "mental." It is not. I could say the whole thing without using the word "mental." The point is this. There is a certain apparatus of intentionalistic causal explanation. Now if you give up on that apparatus, the specification of the "rules" no longer specifies a cause that functions in the production of cognition. It just specifies an association pattern. The association pattern may be functional, as I suggest in my discussion of explanatory inversion, but it is a confusion to suppose that a pattern of events that is functionally useful in the life of the organism is thereby the result of intentional causation.

III. Consciousness without Introspection: Reply to Objection 4

Several commentators, specifically **Block, Hodgkin & Houston, Schull, Uleman & Uleman, Velmans,** and

Young in various ways question my notion of consciousness; and several others, specifically **Chomsky, Limber, Piattelli-Palmarini** and **Rey**, think I am relying in some way on the notion of introspection.

By consciousness I simply mean those subjective states of awareness or sentience that begin when one wakes in the morning and continue throughout the period that one is awake until one falls into a dreamless sleep, into a coma, or dies, or is otherwise, as they say, unconscious. On my account, dreams are a form of consciousness (this answers the queries of **Young** and **Hodgkin & Houston**), though they are of less intensity than full blown waking alertness. Consciousness is an on/off switch: You are either conscious or not. Though once conscious, the system functions like a rheostat, and there can be an indefinite range of different degrees of consciousness, ranging from the drowsiness just before one falls asleep to the full blown complete alertness of the obsessive. There are lots of different degrees of consciousness, but door knobs, bits of chalk, and shingles are not conscious at all. (And you will have made a very deep mistake if you think it is an interesting question to ask at this point, "How do you know that door knobs, etc., are not conscious?") These points, it seems to me, are misunderstood by **Block**. He refers to what he calls an "access sense of consciousness." On my account there is no such sense. I believe that he, as well as **Uleman & Uleman**, confuse what I would call peripheral consciousness or inattentiveness with total unconsciousness. It is true, for example, that when I am driving my car "on automatic pilot" I am not paying much attention to the details of the road and the traffic. But it is simply not true that I am totally unconscious of these phenomena. If I were, there would be a car crash. I need therefore to make a distinction between the center of my attention, the focus of my consciousness on the one hand, and the periphery, on the other. William James and others often use the notion of consciousness to mean what I am referring to as the center of conscious attention. This usage is different from mine. There are lots of phenomena right now of which I am peripherally conscious, for example, the feel of the shirt on my neck, the touch of the computer keys at my finger tips, and so on. But as I use the notion, none of these is unconscious in the sense in which the secretion of enzymes in my stomach is unconscious.

A remarkably large number of commentators think that introspection plays some role in my account. They are mistaken. I make no use of the notion of introspection at all. **Chomsky** in particular thinks that I assign some special epistemic priority to introspection, that I am committed to the view that we have some special knowledge of our own mental states by what Chomsky calls an "inner eye." That is not my view at all. Except when quoting Chomsky, I was very careful never to use the word "introspection" in the course of my article, because, strictly speaking, I do not believe there is any such thing. The idea that we know our conscious mental states by introspection implies that we spect intro, that is, that we know them by some inner perception. But the model of perception requires a distinction between the act of perceiving and the object perceived, and that distinction cannot in general be made for our own conscious states. This point was already implicit in our discussion of Objection 3.

I assign no epistemic privilege to our knowledge of our own conscious states. By and large, I think our knowledge of our own conscious states is rather imperfect, as the much cited work of Nisbett and Wilson (1977) shows. My points about consciousness, to repeat, had little or nothing to do with epistemology. They were about the ontology of consciousness and its relation to the ontology of the mental.

Chomsky, by the way, is mistaken in thinking that the discussion in this article is the same as our dispute of several years ago. That issue was genuinely epistemic. This one is not.

IV. Program explanations: Reply to Objection 5

One of the most fascinating things in this discussion is the extent of the disagreement among the objectors about the nature of cognitive science explanations. Most of the commentators accept my claim that cognitive science typically postulates deep unconscious rules and that *as if* intentionality explains nothing. But several have a different conception of cognitive science explanation. They see the computational paradigm not as an implementation of intrinsic intentionality but as an alternative to it or even a rejection of it. **Matthews, Higginbotham, Glymour** and – to some extent – the **EDITORIAL COMMENTARY** take it that computational forms of explanation might be a substitute for intentionalistic explanations, and **Hobbs** and **McDermott** think we have already superseded intentionalistic explanations, that the ascriptions of intentionality in cognitive science are entirely *as if*, but that this does not matter because the causal account is given by the program explanation. We substitute an algorithmic explanation in terms of formal symbol manipulation for the intentionalistic explanation.

The logical situation we are in is this: We are assuming that the Chinese Room Argument shows that the program level is not sufficient by itself for intentionality and that the Connection Principle argument shows that there is no deep unconscious intentionality. Well, all the same, the question remains open whether or not the brain processes might still be in part computational.

And the underlying intuition is this: As a matter of plain fact, there are a lot of physical systems out there in the world that are digital computers. Now maybe, as a matter of plain fact, each brain is one of those. And if so, we could causally explain the behavior of the brain by specifying its programs the same way we can causally explain the behavior of this machine by specifying its programs.

I can, for example, causally explain the behavior of the machine on which I am writing this by saying that it is running the vi program and not the emacs program. **Hobbs** and **McDermott**, more strongly than the others, concede my points about the nonexistence of the deep unconscious and think that cognitive scientists in general should also, but they think the Darwinian Inversion could be avoided because they think we might just discover that the brain is a biological computer.

Notice that this point is logically independent of the main argument in the article. I could, in principle, just concede it as irrelevant to the present issue but I want to discuss it, at least briefly, because I think the hypothesis of computationalism as a causal explanatory model of cognition has some real difficulties.

The idea is that just as we discovered that the heart is a pump so we might discover that brain processes are computational. But if you look at the standard definitions of computation in terms of syntax or symbol manipulation, there is no way we could make any such discovery because syntax and symbol manipulation, unlike pumping, are not defined in terms of physics. Any physical object can be used as a symbol, but for the same reason, no object is a symbol just by virtue of its physics. Syntax and symbols are matters of the *interpretations* that we assign to the physics; we do not discover, for example, 0's and 1's in nature the way we might discover circles and lines, because syntax is not intrinsic to physics. Some physical systems facilitate a syntactical interpretation better than others, just as some physical substances are easier to write on than others, and some are easier to sit on than others; but this does not make syntax, writing, symbols, chairs, or computation into natural physical kinds.

Trying to discover whether or not some object in nature is intrinsically a symbol manipulating device is, therefore, misconceived. We think of it as similar to trying to discover whether some object in nature is a pump or is performing photosynthesis, but it is more like trying to discover whether some object is intrinsically a chair or a book. It is indeed a matter of plain fact that (a) the object I am sitting on is a chair, (b) the object on the table in front of me is a book, and (c) the machine I am writing this on is running vi. But all those facts make implicit reference to a homunculus. If you tried to discover such facts in nature independently of any intentionality, the most you could find would be systems that we could use or interpret or that could function as chairs, books, and symbol manipulators. In the case of computation, the most you could discover in nature would be a set of patterns to which you could *assign*, not *discover*, a computational interpretation.

Now there is certainly no a priori limit on the number and complexity of patterns you could discover in physical nature. Given a free hand in selecting elements of the pattern, you could, no doubt, find a pattern of events in the brain that was exactly isomorphic with the pattern of events in this computer when it is running vi. But such patterns are everywhere. The same free hand will give you a pattern of molecule movements in the wall behind me or in the Milky Way that is also isomorphic with vi. This does not show that you have discovered a lot of computer programs running in nature; it shows that a homunculus is necessary to assign a computational interpretation to a set of physical events, before those events can be described as computational.

In short, syntax and symbol manipulation are in the eye of the beholder. They are observer relative, and among observer relative concepts some are, so to speak, essentially homuncular. The application of such concepts as chair or computer to physical systems differs from such concepts as shape or voltage level, because the former but not the latter make implicit reference to an actual or possible user or interpreter. Just as the Chinese Room Argument shows that semantics is not intrinsic to syntax, so this extension of it shows that syntax is not intrinsic to physics.

This point has to be understood precisely. Every symbol is indeed a physical entity and every symbolic process, such as the implementation of a program, is a sequence of physical events. This physical shape and this voltage level are both symbols, and both are physical. But if you ask, What physical fact about this entity makes it a symbol and makes that entity not a symbol? and What physical fact do this shape and that voltage level have in common that makes them both tokens of the same symbol? there is no purely physical answer to those questions; no features stated in terms of mass, charge, shape, and so on, constitute symbolhood. Syntax consists of those features of the physics that you have *decided to treat as syntax*, and the reference to a homunculus in such a characterization should be obvious.

Glymour's analogy with a Mount Rushmore formed by wind erosion is a good one. The sheer physics of the situation might facilitate an observer relative attribution of representational properties to the natural phenomena, but the brute physical features of the situation are not themselves intrinsically representational. You can't discover that a natural phenomenon is intrinsically a picture, though you could decide to use it as one.

Similarly with the syntax of the program explanation: Because the syntax is always in the eye of the beholder, you can't discover that computations are going on in nature, though you could decide to treat certain natural processes computationally.

The point of this discussion can now be stated: Because computation does not name a process of physics, like pumping or photosynthesis, there cannot be any *causal explanation* provided by the specification of the program level in addition to the level of the physical implementation and the level of the intentionality of the observer or user of the system. The level of the physical implementation is indeed a causal level, but program explanations of the sort I cited above are always relative to the fact that we program and use certain systems to perform in certain ways. We identify the program as an equivalence class defined formally or syntactically, independently of its physical realization. But if you take away our intentionality, if you remove the homunculus from the system, then there is no way that you could discover that the brain intrinsically has a syntax. The most you could find would be a pattern of events that you might decide to interpret symbolically. This is not to say that brains are not digital computers but rather to say that the question is empty. Because you can assign a syntax to anything, you can certainly assign a syntax to the brain. But for the same reason that anything can be described as a digital computer, nothing can be intrinsically a digital computer.

In these respects, the program level is quite different from the mental level in human beings. Mental causation is indeed in some sense "emergent" from the underlying physical features of the system, but it is a separate causal level, intrinsic to the system.

McDermott's analogy with the thermostat illustrates these points precisely, though not in the way he intends. We can explain the temperature in the room by referring to the fact that it is controlled by a thermostat, and we can do this without knowing the physical realization of the thermostat. But notice that the explanation works, not as he says, because there is *as if* intentionality in the ascription, but because of the causal theory built into the word "thermostat." That is, thermostats are defined physically in terms of the production of certain physical effects. The difference between thermostats and computation, however, is that although thermostats are defined in terms of the production of physical effects, computation does not name a physical process. It names an abstract symbolic process that can be realized in an indefinitely large range of physical implementations.

To appreciate fully the failure of Objection 5, let us apply these lessons to the vestibular ocular reflex (VOR). In describing the VOR, Lisberger (1988) uses both computational and intentionalistic metaphors, but we all know that the real causal explanation is a brute hardware explanation in terms of the eighth cranial nerve, the semicircular canals, and all the rest of it. Now, McDermott reminds us that we can treat it as an *as if* intentional system and describe it computationally, as we can with the stomach or the thermostat or any other feedback mechanism. We can, in short, do a computational description of it as we can do a computational description of other natural processes. Which is giving the causal explanation? The account in terms of the eighth cranial nerve or the account in terms of 0's and 1's? Computational models are useful in studying the brain, but nobody really believes that the computational model is a substitute for a scientific account of how the system actually works, and they should stop pretending otherwise.

I think it is a profound mistake for the supporters of Objection 5 to think that we can give causal explanations of cognition at the computational level, while ignoring both the intentionalistic and the neurobiological levels, because the computational level, unlike the other two, is not intrinsic to the system. This brief argument has a simple logical structure and I will lay it out:

1. On the standard textbook definition, computation is defined syntactically in terms of symbol manipulation.

2. But syntax and symbols are not defined in terms of physics. Though symbol tokens are always physical tokens, "symbol" and "same symbol" are not defined in terms of physical features. Syntax, in short, is not intrinsic to physics.

3. So computation is not discovered in the physics, it is assigned to it. Certain physical phenomena are assigned or used or programmed or interpreted syntactically. Syntax and symbols are observer relative.

4. It follows that you could not *discover* that the brain or anything else was intrinsically a digital computer, although you could assign a computational interpretation to it as you could to anything else.

5. Some physical systems facilitate the computational use much better than others. That is why we build, program, and use them. In such cases we are the homunculus in the system interpreting the physics in both syntactical and semantic terms.

6. But the causal explanations we then give do not cite causal properties different from the physics of the implementation and the intentionality of the homunculus.

I present these points with some hesitation, because they depend on taking the standard definition of computation as symbol manipulation literally. Perhaps someone will come up with a better definition of computation that will avoid these difficulties. Also I am presenting here in a rather summary form arguments that deserve much more development. (For more, see Searle, 1990)

V. The best theory: Reply to Objection 6

I will reply very briefly to the objections of Chomsky, Matthews, Block, Freidin, Higginbotham, and Piattelli-Palmarini, that we have a good scientific theory of the deep unconscious and that it is hard to see how my views can accommodate this: First, most of the rules these commentators cite are not deep unconscious at all. It is ironic that Block gives as an example against me the linger-singer rule, because I cited that very rule in my favor more than 20 years ago (Searle 1969, p. 41–42). And the other examples of the split infinitive, the agreement in number between subject and verb, and so on, are all the stock in trade of school book grammar. No deep unconscious there. Similarly with Chomsky's example of the woman suffering from brain damage: Her case is not one of deep unconscious but is like the blind sight cases I discussed in the target article.

Still, there really are claims about the deep unconscious, and what about them? There is plenty of "evidence" for universal grammar, but it counts as evidence only if the mechanism postulated is legitimate to start with, otherwise it has the same status as the evidence for intentionality in plants.

This is a crucial point at issue, and as I believe that neither Chomsky nor any other commentator has met my argument, I will restate it: In the case of the language acquisition device, we find certain association patterns; for example, children can acquire certain sorts of languages and not others. Now the "best theory" says that the association patterns provide evidence that the children are following the deep unconscious rules of UG. But why then do we not say that association patterns also provide evidence that plants are following deep unconscious rules? Well, obviously because the plants don't meet the conditions for rule governed behavior. I have argued that as far as the deep unconscious is concerned humans don't meet those conditions either. Perhaps my argument can be answered, but it is no answer to say that we have a perfectly dandy theory that attributes deep unconscious rules. At one time, the best theory of plant behavior also attributed intentionality to plants, but the theory was still no good. Similarly, the theory of UG, if it is to be any good, has to meet certain ontological conditions that are external to the theory.

This is not an epistemic point. We have plenty of "evidence," but unless the Connection Principle is satisfied, evidence and theory just pass each other by. The association patterns provide the same amount of evidence for the deep unconscious rules of UG that they provide for the plant's mental life, namely zero.

I have argued that the discovery of the association patterns is important and useful, and that those patterns play a functional role in the life of the organism. This role can be properly appreciated once we have performed the Darwinian Inversion. Thus, contrary to Freidin's claim, I am not trying to throw out the past 35 years of linguistics, but rather to understand its achievements. Freidin also makes a point implicit in several other commentaries: It would be a terrible thing if our current forms of explanation did not work. But that is a bit like saying it would be a terrible thing if cancer existed. It does exist.

VI. The Future of science: Reply to Objection 7.

Several commentators, notably Taylor and Piattelli-Palmarini, accuse me of trying to legislate the future of science. Such was not my aim at all. No one can predict the future of science, and it is quite possible that we might have some incredible revolution in cognitive science comparable to the advent of quantum mechanics in theoretical physics that would render this entire discussion irrelevant. I was not attempting to legislate and certainly not to predict what the future of cognitive science might look like. I was talking, however, about a particular explanatory paradigm at a particular period in the history of the discipline. I have now been a practicing "cognitive scientist" for more than 10 years, and I have not seen anything in cognitive science remotely comparable to the intellectual revolutions of relativity theory or quantum mechanics in physics. I am not attempting to legislate the future but to describe some features of the present.

Freeman claims that "science deals with equations, not causes." I think the opposition is illusory. The germ theory of disease, the tectonic plate theory of geology, the DNA account of heredity, the Darwinian account of evolution by natural selection, and yes, even Newton's account of gravity are all causal accounts. But in any case, in cognitive science we are nowhere near the point of being able to state our explanations of cognition as a series of mathematical equations comparable to those of theoretical physics. Nonetheless, we are looking for accounts that are causally explanatory.

VII. Remaining issues

I now want to answer the major points that did not fall within the bounds of the seven most common objections:

Freeman and Skarda both raise questions concerning the role of causation in the account of brain functioning. I think some of the issues are too complex and too deep for me to try to settle in this short space, but I will say the following. They think that because mental states are just neurophysiological states at a higher level of description, the relation between the neurophysiological and the mental cannot be causal. But I see no reason to think that there cannot be causal relations between different levels within the brain. Consider pain as an obvious case. My conscious experience of pain is caused by, among other things, sequences of neuron firings in the thalamus and other regions of the brain. There is no doubt that that story of the production of pain is a causal story, even though the pain is a feature of the system at a level higher than that of individual neurons and synapses.

Now it is an empirical question whether or not the self-organizing, chaotic dynamics advocated by Skarda and Freeman is the correct account of how brain processes cause mental states. I have no opinions on the issue. It seems to me a factual scientific question. But the important thing to see is that both their account and the traditional neuronal account are causal. It is not an objection to this point to say that on chaotic dynamics you cannot get strict causal inferences. You often cannot get strict causal inference in nature anyway. Weather prediction is a classic case of the application of chaotic dynamics that does not use strict causal inferences, but, nonetheless, accounts of the weather can be causal.

Freeman raises the fascinating possibility that our very urge to seek causal explanations may be the result of a cycle of events in the limbic system, correlating sensory and motor systems. If he could demonstrate that, it would be one of the most important contributions to transcendental philosophy since Kant.

Czyzewska et al. think that I am attacking a straw man because, they say, the notion of the unconscious that I am discussing does not occur in "contemporary cognitive research." I believe they are mistaken about this point. I give examples from both linguistics and theories of vision in which the notion of the deep unconscious that I am discussing occurs. I find the postulation of deep unconscious processes to be quite common.

Schull, it seems to me, mistakes my position in two crucial respects. First, he thinks that I am advocating that functional explanations are not causal; that is not the case. The whole point of my discussion of the functional explanation of the plant turning its leaves to the sun is that this causes an increase in the plant's chances of survival and reproduction. There is no question that the account is causal. Second, when I say that the function is in the eye of the beholder, what I mean is that the only facts attributed by the functional explanation are causal facts. The normative component added by functional explanations is in the eye of the beholder.

Schull agrees with me that it is inappropriate to attribute intrinsic intentionality to an individual plant, but he seems to think that it is perfectly OK to attribute intrinsic intentionality to whole populations of plants. Frankly, I can't make any sense of this idea at all.

Taylor's commentary is thought provoking but it is hard to see how he is disagreeing with me. He claims that there might be a Durkheimian account of religion that appeals to, in principle, deep unconscious thoughts. I do not see how this hypothesis is really inconsistent with anything I have said. Either the unconscious thought here is supposed to be deep unconscious, in which case we are just talking about neurophysiological events, or it is a case of the shallow unconscious, in which case it is accessible to consciousness.

Taylor says I am making a "hard and fast distinction between hardware explanations and functional explanations," and he then considers the fascinating possibility that there might be larger field effects in the neurophysiology that are specified intentionalistically. Well, certainly I do not think that we can predict the future course of neurophysiology and there might well turn out to be intentionalistic criteria for the identification of neurophysiological processes. But given our present explanatory paradigms, there is a difference in evolutionary biology between hardware explanations and functional explanations. One might imagine a future course of science that runs them together, but I think that possibility is irrelevant to my present concerns.

Young points out that the mistakes that I am charging cognitive science with go back a long time, as early as Helmholtz in the nineteenth century. I am not competent enough in the history of psychology to say whether or not this is correct, but I am grateful for the historical information in his article.

Zelazo & Reznick wish to carry my arguments one step further and deny that there is any intentionality to shallow unconscious states. This would basically return us to

192

the Cartesian idea that the conscious and the mental are coextensive. This is a fascinating idea, but the difficulty I have with it is that the explanatory power of postulating unconscious states that continue even when unconscious seems to me too great. Zelazo & Reznick would be forced to say that the agent believes that Bush is president of the United States only when he is consciously thinking that Bush is president of the United States. And that seems to me not right. The issue is terminological, but the advantages of the common sense terminology seem to me much too great to be abandoned except in the face of very strong arguments.

Glymour begins his commentary by misrepresenting my argument. I do not know why he does not reproduce the argument as I gave it, instead of substituting something that is not equivalent to it. And in any case I do not derive the conclusion he attributes to me that no unconscious states are mental states. On the contrary, I am interested in the fact that I have plenty of unconscious mental states and I want to know exactly what facts correspond to that claim. It is perhaps revealing that Glymour says his formal argument is "something like" mine. Well, St. Thomas Aquinas is something like Ronald Reagan, I guess, but if I were discussing either St. Thomas or Reagan, I would not just want to get it "something like" right. I would want to try to get it exactly right. Throughout his commentary, Glymour invents arguments he thinks I might be making instead of looking at the actual arguments I presented.

Harman's argument is based entirely on using different and, I believe, unmotivated definitions of two of the crucial terms I was using, "intentionality" and "intrinsic." He is right in saying that "intentional" is a technical term in philosophy. It is a concept I have written about for a number of years, and though my usage does not conform with everybody's, certainly not with Harman's, it was my usage that I was using in my article, not surprisingly.

As I use the term, there is a distinction between intrinsic intentionality, as is reported by the sentence, "I am now thirsty," derived intentionality, as in the sentence, "'J'ai soif' means 'I am thirsty,'" and *as if* or metaphorical ascriptions of intentionality, as in, "My lawn is thirsty." On my usage, only intrinsic intentionality is mental. So there cannot be, on my usage, any question whether or not lawns have intrinsic intentionality, and in particular, the needs of a lawn are not intrinsically intentional.

My usage here is motivated by an important fact. The attribution of needs in such a sentence as, "The lawn needs water," is always observer relative. If you want the lawn to grow, it needs water. If you want the lawn to die of dehydration, it needs no water. In one sense this is a definitional question. **Harman** is entitled to his usage, but I do not see why he cannot see that mine is different.

"Intrinsic," unlike "intentional," is not, contrary to what **Harman** says, a technical term. Specifically, in this case, I am not contrasting it with "relational," as he does, but with "observer relative." On this usage, the type on the paper in front of me is intrinsically ink, but it is not intrinsically meaningful. Meaning is assigned to it by outside users and observers. Again, this is a matter of usage; he is welcome to his usage, but I do not see him giving any grounds for rejecting mine.

In support of his views Harman cites Putnam's twin

earth argument. I have answered this argument elsewhere (Intentionality, Chapter 8). He concludes his paper with the sentence: "But it is quite unclear that there is anything metaphorical in the claim that a lawn is thirsty." Well, if Harman believes that in the sense in which I now literally feel thirsty, in that very literal sense one can attribute thirst to a lawn, then I believe his philosophical difficulties exceed anything I could hope to correct in the space of this reply.

Rey's commentary gets off on the wrong foot at the beginning when he says that my distinction between intrinsic and *as if* intentionality is some sort of causal distinction. That is not the basis of the distinction. The basis is simply that real (subjective, inner, honest-to-john) mental states exist and should not be confused with things that have no mental reality but that can give rise to a lot of metaphorical talk.

Rey assumes both functionalism and the language of thought hypothesis, and I believe it is these assumptions that make my target article difficult for him to understand. I think both of these views are – at best – false; but this is not the place to refute them. He, like **Rosenthal**, thinks that the difference between wanting water and wanting H_2O is a matter of knowledge. But that is not the point at all. I know the identity, but though I often have a conscious desire for water, I hardly ever think about H_2O except when giving philosophical examples.

Rey says that I seem to want to believe in "materialism." In fact, I can't use words like "materialism" ("dualism," etc.) with a straight face, because I think this traditional vocabulary embodies massive intellectual confusions.

Holender and I are not, I believe, in serious disagreement; we are simply addressing different questions. I agree with what I take to be his main point, namely, that an experimental paradigm may survive after it has lost its theoretical basis.

VIII. Conclusion

I am grateful for all the work that went into the commentaries, but I am also anxious that in the discussion of the detailed points we should not lose sight of the larger issues. I have found cognitive science to be one of the most exciting developments of my intellectual lifetime. I believe that it has suffered from a persistent unclarity about its foundations, however. This emerges, for example, in the misunderstanding of the role of computer models of cognition and in the constant temptation to look over its shoulder for guidance to physics, or worse yet, to something abstractly called "science." Even more serious is the reluctance to face the precise ontology of its subject matter. Cognition is a matter of consciousness and intrinsic intentionality. Not all consciousness is intentional and not all intentionality is conscious, but there is an internal relation between the two, and I have tried in this discussion to characterize that relation. Furthermore, there is no way to study consciousness and intentionality without, directly or indirectly, studying the brain. But the brain is a biological organ like any other, and if we are to understand its functioning we need some way to specify where intentionality ends and brute neurobiology takes over. Until we are clear about that issue, we will be confusing

193

intentionalistic explanations with hardware and functional explanations. We will, in short, be making a pre-Darwinian anthropomorphic fallacy.

The two points I have been most concerned to establish are the Connection Principle and its consequence, Darwinian Inversion. None of these foundational issues is epistemic. The question is not, "How do you know?" but, "What is it that you know when you know?" The epistemic, methodological questions are relatively uninteresting because they always have the same answer: Use your ingenuity. Use any weapon at hand, and stick with any weapon that works.

ACKNOWLEDGMENT

I am indebted to Stevan Harnad, Jeannie Lum, and Dagmar Searle for helpful comments.

References

Allport, A. (1988) What concept of consciousness? In: *Consciousness in contemporary science*, ed. A. J. Marcel and E. Bisiach. Oxford University Press. [AWY]

Anderson, J. R. (1982) Acquisition of cognitive skill. *Psychological Review* 89:369–406. [RAC]

(1983) *The architecture of cognition*. Harvard University Press. [RAC]

Baars, B. (1988) *A cognitive theory of consciousness*. Cambridge University Press. [GU]

Balota, D. A. & Chumbley, J. I. (1984) Are lexical decisions a good measure of lexical access? The role of word frequency in the neglected decision stage. *Journal of Experimental Psychology: Human Perception and Performance* 10:340–57. [DH]

Bargh, J. A. (1989) Conditional automaticity: Varieties of automatic influence in social perception and cognition. In: *Unintended thought*, ed. J. S. Uleman & J. A. Bargh. Guilford Publications. [JSU]

Bealer, G. (1984) Mind and anti-mind. *Midwest Studies in Philosophy* 9:283–328. [GR]

Block, N. (1978) Troubles with functionalism. In: *Minnesota Studies in Philosophy of Science IX*, ed. C. W. Savage. [asp, GU]

(1980) Are absent qualia impossible? The Philosophical Review 89 257–74. [asp, GU]

(1986) Advertisement for a semantics for psychology. In: *Studies in the philosophy of mind*, vol. 10, Midwest Studies in Philosophy, ed. P. French, T. Euhling & H. Wettstein. University of Minnesota Press. [GR]

Bridgeman, B. (1988) *The biology of behavior and mind*. Wiley. [BB]

Carlson, R. A. & Dulany, D. E. (1985) Conscious attention and abstraction in concept learning. *Journal of Experimental Psychology: Learning, Memory, and Cognition* 11:45–58. [RAC]

(1988) Diagnostic reasoning with circumstantial evidence. *Cognitive Psychology* 20:463–92. [RAC]

Cheesman, J. & Merikle, P. M. (1985) Word recognition and consciousness. In: *Reading research*, vol. 5, ed. D. Besner, T. G. Waller & G. E. MacKinnon. Academic Press. [GU]

Chomsky, N. (1976) *Reflections on language*. Temple Smith. [aJRS]

(1980a) *Rules and representations*. Columbia University Press. [NC]

(1980b) Rules and representations. *Behavioral and Brain Sciences* 3:1–61. [DH]

(1986) *Knowledge of language. Its nature, origin, and use*. Praeger Special Studies. [aJRS, GR]

Cowie, A. (1979) Cortical maps and visual perception. *Quarterly Journal of Experimental Psychology* 31:1–17. [AIH]

Crutchfield, J., Farmer, J., Packard, N. & Shaw, R. (1987) Chaos. *Scientific American* 255:46–57. [CAS]

Dawkins, R. (1979) Twelve misunderstandings of kin selection. *Zeitschrift für Tierpsychologie* 51:184–200. [AIH]

De Groot, A. M. B. (1985) Word-context effects in word naming and lexical decision. *Quarterly Journal of Experimental Psychology* 37A:281–97. [DH]

Demopoulos, W. & Matthews, R. J. (1983) On the hypothesis that grammars

are mentally represented. *Behavioral and Brain Sciences* 6(3):423–86. [aJRS]

Dennett, D. (1981) *Brainstorms*. Bradford Books/MIT Press. [DM]

Devitt, M. (1981) *Designation*. Columbia University Press. [GR]

Dixon, N. (1981) *Preconscious processing*. John Wiley. [GU]

Dretske, F. (1981) *Knowledge and the flow of information*. MIT Press. [GR]

(1981) *Explaining behavior: Reasons in a world of causes*. MIT Press. [GR]

Dulany, D. E. (1968) Awareness, rules, and propositional control. A confrontation with S-R behavior theory. In: *Verbal behavior and general behavior theory*, ed. T. R. Dixon & D. L. Horton Prentice-Hall. [RAC]

(1984) A strategy for investigating consciousness. Paper presented at meetings of the Psychonomic Society, San Antonio, Texas. [RAC]

Dulany, D. E., Carlson, R. A. & Dewey, G. I. (1984) A case of syntactical learning and judgment: How conscious and how abstract? *Journal of Experimental Psychology: General* 114:25–32. [RAC]

(1985) On consciousness in syntactical learning and judgment: A reply to Reber, Allen, and Regan. *Journal of Experimental Psychology: General* 114:25–32. [RAC]

Eccles, J. C. (1970) *Facing reality. Adventures of a brain scientist*. Springer-Verlag. [WJF]

Ellenberger, H. F. (1970) *The discovery of the unconscious: The history and evolution of dynamic psychiatry*. Basic Books. [AWY]

Erdelyi, M. H. (1985) *Psychoanalysis: Freud's cognitive psychology*. W. H. Freeman. [DH]

(1986) Experimental indeterminacies in the dissociation paradigm of subliminal perception. *Behavioral and Brain Sciences* 9:30–31. [DH]

Fodor, J. A. (1975) The Language of thought. Thomas Y. Crowell. [DM]

(1983) *The modularity of mind*. MIT Press. [DH]

(1985) Précis on "The modularity of mind." *Behavioral and Brain Sciences* 8:1–42. [DH]

(1987) *Psychosemantics*. MIT Press. [GR]

(1990) Psychosemantics, or where do truth conditions come from? In: *Mind and cognition*, ed. W. Lycan. [GR]

Freeman, W. J. (1975) *Mass action in the nervous system*. Academic Press. [WJF]

(in press) Sobre el error de fijar origen del consciencia. *Interciencia*. [WJF]

Freeman, W. & Schneider, W. (1982) Changes in spatial patterns of rabbit olfactory EEG with conditioning to odors. *Psychophysiology* 19:44–56. [CAS]

Freeman, W. & Skarda, C. (1985) Spatial EEG patterns, nonlinear dynamics and perception: The neo-Sherringtonian view. *Brain Research Reviews* 10:147–75. [CAS]

Freidin, R. & Quicoli, A. C. (1989) Zero stimulation for parameter setting. *Behavioral and Brain Sciences* 12(2):338–39. [RF]

Freud, S. (1915) *The unconscious*, reprinted (1959) in Collected Papers, vol. 4. Basic Books. [aJRS]

(1949) Outline of psycho-analysis, trans. James Strachey. Hogarth Press. [aJRS]

Goldman-Rakic, P. (1987) Development of cortical circuitry and cognitive function. *Child Development* 58:601–22. [PDZ]

Gregory, R. L. (1963) Distortion of visual space as inappropriate constancy scaling. *Nature* 1939:678–80. [AWY]

Gregory, R. L. (1966) *Eye and brain: The psychology of seeing*. Weidenfeld and Nicolson. [AWY]

Halligan, P. W. & Marshall, J. C. (1988) Blindsight and insight in visuo-spatial neglect. *Nature* Dec. 22. [NC]

Harman, Gilbert (1990) The intrinsic quality of experience. *Philosophical Perspectives* 4:31–52. [GH]

Hayes, N. A. & Broadbent, D. E. (1988) Two modes of learning for interactive tasks. *Cognition* 28:249–76. [DH, GU]

Hill, T., Lewicki, P., Czyzewska, A. & Boss, A. (1989) Self-perpetuating development of encoding biases in person perception. *Journal of Personality and Social Psychology* 57:373–87. [MC]

Hille, B. (1983) Theories of anesthesia: General perturbations versus specific receptors. In: *Molecular mechanisms of anesthesia* (Progress in anesthesiology, vol. 2), ed. B. R. Fink. Raven Press. [JCK]

Hinton, G. (1985) Learning in parallel networks. *Byte* 10:265. [CAS]

Holender, D. (1986) Semantic activation without conscious identification in dichotic listening, parafoveal vision, and visual masking: A survey and appraisal. *Behavioral and Brain Sciences* 9:1–66. [DH]

(1987a) Semantic activation without conscious identification: Can progress be made? *Behavioral and Brain Sciences* 10:768–73. [DH]

(1987b) Is the unconscious amenable to scientific scrutiny? *Canadian Psychology* 28:120–24. [DH]

(in press) Comment: Writing systems and the modularity of language. In:

194

Modularity and the motor theory of speech perception, ed. I. G. Mattingly & M. Studdert-Kennedy. Erlbaum. [DH]

Hopfield, J. & Tank, D. (1986) Computing with neural circuits: A model. *Science* 233:625–33. [CAS]

Horn, B. K. P. (1977) Understanding image intensities. *Artificial Intelligence* 8:201–32. [AIH]

Huttenlocher, P. R. (1979) Synaptic density in the human frontal cortex – Developmental changes and the effects of aging. *Brain Research* 163:195–205. [PDZ]

James, W. (1890) *The principles of psychology.* Holt. [GU]
(1890/1950) *The principles of psychology.* Dover Publications. [JL]
(1896) *Principles of psychology.* Great Books edition, 1952. University of Chicago Press. [JS]

Kahneman, D., Slovic, P. & Tversky, A. (1982) *Judgment under uncertainty: Heuristics and biases.* Cambridge University Press. [MP-P]

Klatzky, R. L. (1984) *Memory and awareness: An information-processing perspective.* W. H. Freeman. [GU]

Kohonen, T. (1984) *Self-organization and associative memory.* Springer Verlag. [CAS]

Kushwaha, R., Williams, W. J., Shevrin, H. & Sachellares, C. (1989) An information flow technique in ERP application. *IEEE Engineering in Medicine and Biology Society 11th Annual Conference.* Session 2770:715–16. [HS]

Lehky, S. R. & Sejnowski, T. J. (1988) Network models of shape from shading. *Nature* 333:452–54. [AIH]

Lewicki, P. (1986) *Nonconscious social information processing.* Academic Press. [GU]

Lewicki, P. & Hill, T. (1989) On the status of nonconscious processes in human cognition. *Journal of Experimental Psychology: General* 118:239–41. [MC]

Limber, J. (1975) Goodbye behaviorism. *Behavioral and Brain Sciences* 1:581–83. [JL]
(1982) What can chimps tell us about the origins of language? In: *Language development,* vol. 2, ed. S. Kuczaj. Erlbaum. [JL]

Lisberger, S. G. (1988) The neural basis for learning of simple motor skills. *Science* 242:728–35. [aJRS]

Lisberger, S. G. & Pavelko, T. A. (1988) Brain stem neurons in modified pathways for motor learning in the primate vesnbulo-ocular reflex. *Science* 242:771–73. [aJRS]

Lloyd, D. (1989) *Simple minds.* Bradford Books/MIT Press. [DL]
(in press) Leaping to conclusions: Connectionism, consciousness, and the computational mind. In: *Connectionism and the philosophy of mind,* ed. T. Horgan & J. Tienson. Kluwer Academic Publishers. [DL]

Marcel, T. (1983) Conscious and unconscious perception: Experiments on visual masking and word recognition. *Cognitive Psychology* 15:197–237. [GU]

McKoon, G. & Ratcliff, R. (1986) Inferences about predictable events. *Journal of Experimental Psychology: Learning, Memory and Cognition* 12:82–91. [JSU]

Michenfelder, J. H. (1988) Assessing the brain. In: *Anesthesia and the brain,* ed. J. H. Michenfelder. Churchill Livingstone. [JCK]

Millikan, R. (1984) *Language, thought, and other biological categories.* MIT Press. [GR]
(1986) Thoughts without laws, cognitive science with content. *Philosophical Review* 95:47–80. [AC]

Neely, J. H. (in press) Semantic priming effect in visual word recognition: A selective review of current findings and theories. In: *Basic processes in reading: Visual word recognition,* ed. D. Besner & G. Humphreys. Erlbaum. [DH]

Neely, J. H., Keefe, D. E. & Ross, K. L. (1989) Semantic priming in the lexical decision task: Roles of prospective prime-generated expectancies and retrospective semantic matching. *Journal of Experimental Psychology: Learning, Memory, and Cognition* 15:1003–19. [DH]

Newman, L. A. & Uleman, J. S. (in press) Assimilation and contrast effects in spontaneous trait inferences. *Personality and Social Psychology Bulletin* 16. [JSU]

Nisbett, R. & Ross, L. (1980) *Human inference: Strategies and shortcomings of social judgment.* Prentice-Hall. [JSU]

Nisbett, R. & Wilson, T. (1977) On saying more than we can know. *Psychological Review* 84(3):231–59. [GR]

Pope, K. S. & Singer, J. L. (1978) Regulation of the stream of consciousness: Toward a theory of ongoing thought. In: *Consciousness and self-regulation,* vol. 2, ed. G. E. Schwartz & D. Shapiro. John Wiley. [GU]

Puccetti, R. (1981) The case for mental duality: Evidence from split-brain data and other considerations. *Behavioral and Brain Sciences* 4:93–123. [JCK]

Putnam, H. (1975) The meaning of meaning. In: *Mind, language, and reality:*

Philosophical papers, vol. 2, ed. H. Putnam. Cambridge University Press. [GH]

Quine, W. V. O. (1960) *Word and object.* Technology Press of MIT and John Wiley & Sons. [aJRS]

Reber, A. S. (1989) Implicit learning and tacit knowledge. *Journal of Experimental Psychology: General* 118:219–35. [DH]

Reber, A. S., Kassim, S. M., Lewis, S. & Cantor, S. (1980) On the relationship between implicit and explicit modes of learning a complex rule structure. *Journal of Experimental Psychology: Human Learning and Performance* 6:492–502. [GU]

Reeves, J. W. (1965) *Thinking about thinking: Studies in the background of some psychological approaches.* Secker and Warburg. [AWY]

Reingold, E. M. & Merikle, P. M. (1988) Using direct and indirect measures to study perception without awareness. *Perception and Psychophysics* 44:563–75. [DH]

Rey, G. (1986) What's really going on in Searle's "Chinese Room." *Philosophical Studies* 50:169–85. [GR]

Rips, L. J. (1983) Cognitive processes in propositional reasoning. *Psychological Review* 90:38–71. [RAC]

Rock, I. (1984) *Perception.* W. H. Freeman. [aJRS, GR]

Rogers, R. L., Papanicolaou, A. C., Baumann, S. B., Eisenberg, H. M. & Savdjari, C. (1990) Spatially distributed cortical excitation patterns of auditory processing during contralateral and ipsilateral stimulation. *Journal of Cognitive Neuroscience* 2:44–50. [JCK]

Rose, S. (1987) *Molecules and minds.* Open University Press. [CAS]

Rosenthal, David M. (1985) Intentionality. Midwest Studies. *Philosophy X.* 151–84 (reprinted, with postscript [1989] in Rerepresentation: Readings in the philosophy of mental representation, ed. S. Silvers. D. Reidel Publishing Co.) [DMR]
(1986) Two concepts of consciousness. *Philosophical Studies* 49:3: 329–59. [DMR]
(1990) A theory of consciousness. Report, Center for Interdisciplinary Research (ZiF), Research Group on Mind and Brain. University of Bielefeld, West Germany. [DMR]

Ross, L., Lepper, M. R. & Hubbard, M. (1975) Perseverance in self-perception and social perception: Biased attribution processes in the debriefing paradigm. *Journal of Personality and Social Psychology* 32:880–92. [JSU]

Rozin, P. & Schull, J. (1988) The adaptive-evolutionary point of view in experimental psychology. In: S. S. *Stevens' handbook of experimental psychology,* 2d ed., ed. R. Atkinson, R. Herrnstein, G. Lindzey & R. D. Luce. Wiley. [JS]

Sarna, S. K. & Otterson, M. F. Gastrointestinal motility: Some basic concepts. *Pharmacology Supplement* 36:7–14. [aJRS]

Schacter, D. L. (1987) Implicit memory: History and current status. *Journal of Experimental Psychology: Learning, Memory, and Cognition* 13:501–18. [DH, JSU]

Schacter, D. L., McAndrews, M. P. & Moscovitch, M. (1988) Access to consciousness: Dissociations between implicit and explicit knowledge in neuropsychological syndromes. In: *Thought without language,* ed. L. Weiskrantz. Clarendon Press. [DH]

Schull, J. (1990) Are species intelligent? *Behavioral and Brain Sciences* 13(1):63–75. [JS]
(in press) Evolution and learning: Analogies and interactions. In: *The evolution paradigm,* ed. Ervin Laszlo. Gordon and Breach Science Publishers. [JS]
(in preparation) William James and the nature of selection. [JS]

Searle, J. R. (1969) *Speech acts: An essay in the philosophy of language.* Cambridge University Press. [rJRS]
(1976) The rules of the language game (review of Noam Chomsky, "Reflections on Language"). In: *The Times Literary Supplement,* Sept. 10. [aJRS]
(1980a) Minds, brains, and programs. *Behavioral and Brain Sciences* 3:417–57. [aJRS]
(1980b) Intrinsic intentionality. Reply to criticisms of Minds, brains and programs. *Behavioral and Brain Sciences* 3:450–6. [aJRS]
(1980c) Rules and causation. *Behavioral and Brain Sciences* 3:37–38. [DH]
(1983) *Intentionality. An essay in the philosophy of mind.* Cambridge University Press. [aJRS, RAC, DMR, PDZ]
(1984a) *Minds, Brains and Science.* Harvard University Press.
(1984b) Intentionality and its place in nature. *Synthese* 61:3–16. [aJRS]
(1987) Indeterminacy, empiricism and the first person. *Journal of Philosophy* March:123–46. [aJRS]
(1989) Consciousness, unconsciousness, and intentionality. *Philosophical Tendencies* 17(1):269–284. [JRS]
(1990) Is the brain a digital computer? Presidential address to the Pacific

195

Division of the American Philosophical Association, Los Angeles, March. [rJRS]

Shallice, T. (1972) Dual functions of consciousness. *Psychological Review* 79:383–93. [GU]

Sherrington, C. (1906) *The integrative action of the nervous system*. Yale University Press. [CAS]

Shevrin, H. (1988) Unconscious conflict: A convergent psychodynamic and electrophysiological approach. In: *Psychodynamics and cognition*, ed. M. S. Horowitz. University of Chicago Press. [HS]

Shevrin, H., Williams, W. J., Marshall, R. E., Hertel, R. K., Bond, J. A. & Brakel, L. A. (1988) Event-related potential indicators of the dynamic unconscious. International Conference on Psychophysiology, Prague. [HS]

Skarda, C. (1986) Explaining behavior: Bringing the brain back in. *Inquiry* 29:187–202. [CAS]

Skarda, C. & Freeman, W. (1987) How brains make chaos in order to make sense of the world. *Behavioral and Brain Sciences* 10:161–73. [CAS] (1988) EEG research of neural dynamics: Implications for models of learning and memory. In: *Systems with learning and memory*, ed. J. Delacour & J. C. S. Levy. Elsevier. [CAS]

Stampe, D. (1977) Towards a causal theory of linguistic representation. *Midwest Studies in Philosophy*, ed. P. French, T. Euhling & H. Wettstein. University of Minnesota Press. [GR]

Taylor, S. E. & Fiske, S. T. (1975) Point-of-view and perceptions of causality. *Journal of Personality and Social Psychology* 32:439–45. [JSU]

Thach, W. T. (1978) The cerebellum. In: *Medical physiology*, ed. V. B. Mountcastle. Mosby Press. [CAS]

Titchener, E. B. (1910) *A textbook of psychology*. Macmillan. [JCK]

Treisman, A. (1985) Preattentive processing in vision. *Computer Vision, Graphics and Image Processing* 31:156–77. [AIH] (1988) Features and objects. *Quarterly Journal of Experimental Psychology* 40:201–238. [AIH]

Treisman, A. & Gelade, G. (1980) A feature-integration theory of attention. *Cognitive Psychology* 12:97–136. [AIH]

Treisman, A. & Gormican, S. (1988) Feature analysis in early vision: Evidence from search asymmetries. *Psychological Review* 95:15–48. [AIH]

Uleman, J. S. (1987) Consciousness and control: The case of spontaneous trait inferences. *Personality and Social Psychology Bulletin* 13:337–54. [JSU] (1989) A framework for thinking intentionally about unintended thoughts. In: *Unintended thought*, ed. J. S. Uleman & J. A. Bargh. Guilford Publications. [JSU]

Underwood, G. (1978) Attentional selectivity and behavioural control. In: *Strategies of information processing*, ed. G. Underwood. Academic Press. [GU] (1981) Lexical recognition of embedded unattended words: Some implications for reading processes. *Acta Psychologia* 47:267–83. [GU] (1982) Attention and awareness in cognitive and motor skills. In: *Aspects of consciousness*, vol. 3, ed. G. Underwood. Academic Press. [GU]

Van Riemsdijk, H. & Williams, E. (1986) *Introduction to the theory of grammar*. MIT Press. [MP-P]

Vickers, M. D. A. (1987) Detecting consciousness by clinical means. In: *Consciousness, awareness and pain in general anaesthesia*, ed. M. Rosen & J. N. Lunn. Butterworths. [JCK]

Von Helmholtz, H. L. F. (1856–1867) *Treatise on physiological optics*. Trans. from German pub. 1924–1925. Optical Society of America. [AWY]

Weiskrantz, L. (1982) A follow-up study of blindsight. Paper presented at the Fifth INS European Conference. Deauville, France. June 16–18. [aJRS]

Williams, W. J., Shevrin, H. & Marshall, R. E. (1987) Information modeling and analysis of event-related potentials. *IEEE Transactions on Biomedical Engineering* U-BME-34: 12.928–37. [HS]

Williams, W. J. & Jeong, J. (1989) New time-frequency distributions: Theory and applications. Transactions of International Symposium on Circuits and Systems, session 1692-2, VS-1243–47. [HS]

Wittgenstein, L. (1953) *Philosophical investigations*. Macmillan. [GR]

Zeki, S. & Shipp, S. (1988) The functional logic of cortical connections. *Nature* 335:311–17. [AIH]

196

EPIPHENOMENAL QUALIA

By Frank Jackson

It is undeniable that the physical, chemical and biological sciences have provided a great deal of information about the world we live in and about ourselves. I will use the label 'physical information' for this kind of information, and also for information that automatically comes along with it. For example, if a medical scientist tells me enough about the processes that go on in my nervous system, and about how they relate to happenings in the world around me, to what has happened in the past and is likely to happen in the future, to what happens to other similar and dissimilar organisms, and the like, he or she tells me — if I am clever enough to fit it together appropriately — about what is often called the functional role of those states in me (and in organisms in general in similar cases). This information, and its kin, I also label 'physical'.

I do not mean these sketchy remarks to constitute a definition of 'physical information', and of the correlative notions of physical property, process, and so on, but to indicate what I have in mind here. It is well known that there are problems with giving a precise definition of these notions, and so of the thesis of Physicalism that all (correct) information is physical information.[1] But — unlike some — I take the question of definition to cut across the central problems I want to discuss in this paper.

I am what is sometimes known as a "qualia freak". I think that there are certain features of the bodily sensations especially, but also of certain perceptual experiences, which no amount of purely physical information includes. Tell me everything physical there is to tell about what is going on in a living brain, the kind of states, their functional role, their relation to what goes on at other times and in other brains, and so on and so forth, and be I as clever as can be in fitting it all together, you won't have told me about the hurtfulness of pains, the itchiness of itches, pangs of jealousy, or about the characteristic experience of tasting a lemon, smelling a rose, hearing a loud noise or seeing the sky.

There are many qualia freaks, and some of them say that their rejection of Physicalism is an unargued intuition.[2] I think that they are being unfair to themselves. They have the following argument. Nothing you could tell of a physical sort captures the smell of a rose, for instance. Therefore, Physicalism is false. By our lights this is a perfectly good argument. It is

[1]See, e.g., D. H. Mellor, "Materialism and Phenomenal Qualities", *Aristotelian Society Supp. Vol.* 47 (1973), 107-19; and J. W. Cornman, *Materialism and Sensations* (New Haven and London, 1971).

[2]Particularly in discussion, but see, e.g., Keith Campbell, *Metaphysics* (Belmont, 1976), p. 67.

obviously not to the point to question its validity, and the premise is intuitively obviously true both to them and to me.

I must, however, admit that it is weak from a polemical point of view. There are, unfortunately for us, many who do not find the premise intuitively obvious. The task then is to present an argument whose premises are obvious to all, or at least to as many as possible. This I try to do in §I with what I will call "the Knowledge argument". In §II I contrast the Knowledge argument with the Modal argument and in §III with the "What is it like to be" argument. In §IV I tackle the question of the causal role of qualia. The major factor in stopping people from admitting qualia is the belief that they would have to be given a causal role with respect to the physical world and especially the brain;[3] and it is hard to do this without sounding like someone who believes in fairies. I seek in §IV to turn this objection by arguing that the view that qualia are epiphenomenal is a perfectly possible one.

I. The Knowledge Argument for Qualia

People vary considerably in their ability to discriminate colours. Suppose that in an experiment to catalogue this variation Fred is discovered. Fred has better colour vision than anyone else on record; he makes every discrimination that anyone has ever made, and moreover he makes one that we cannot even begin to make. Show him a batch of ripe tomatoes and he sorts them into two roughly equal groups and does so with complete consistency. That is, if you blindfold him, shuffle the tomatoes up, and then remove the blindfold and ask him to sort them out again, he sorts them into exactly the same two groups.

We ask Fred how he does it. He explains that all ripe tomatoes do not look the same colour to him, and in fact that this is true of a great many objects that we classify together as red. He sees two colours where we see one, and he has in consequence developed for his own use two words 'red$_1$' and 'red$_2$' to mark the difference. Perhaps he tells us that he has often tried to teach the difference between red$_1$ and red$_2$ to his friends but has got nowhere and has concluded that the rest of the world is red$_1$-red$_2$ colour-blind — or perhaps he has had partial success with his children, it doesn't matter. In any case he explains to us that it would be quite wrong to think that because 'red' appears in both 'red$_1$' and 'red$_2$' that the two colours are shades of the one colour. He only uses the common term 'red' to fit more easily into our restricted usage. To him red$_1$ and red$_2$ are as different from each other and all the other colours as yellow is from blue. And his discriminatory behaviour bears this out: he sorts red$_1$ from red$_2$ tomatoes with the greatest of ease in a wide variety of viewing circumstances. Moreover, an investigation of the physiological basis of Fred's exceptional ability reveals that Fred's optical system is able to separate out two groups of wave-

[3]See, e.g., D. C. Dennett, "Current Issues in the Philosophy of Mind", *American Philosophical Quarterly*, 15 (1978), 249-61.

lengths in the red spectrum as sharply as we are able to sort out yellow from blue.[4]

I think that we should admit that Fred can see, really see, at least one more colour than we can; red_1 is a different colour from red_2. We are to Fred as a totally red-green colour-blind person is to us. H. G. Wells' story "The Country of the Blind" is about a sighted person in a totally blind community.[5] This person never manages to convince them that he can see, that he has an extra sense. They ridicule this sense as quite inconceivable, and treat his capacity to avoid falling into ditches, to win fights and so on as precisely that capacity and nothing more. We would be making their mistake if we refused to allow that Fred can see one more colour than we can.

What kind of experience does Fred have when he sees red_1 and red_2? What is the new colour or colours like? We would dearly like to know but do not; and it seems that no amount of physical information about Fred's brain and optical system tells us. We find out perhaps that Fred's cones respond differentially to certain light waves in the red section of the spectrum that make no difference to ours (or perhaps he has an extra cone) and that this leads in Fred to a wider range of those brain states responsible for visual discriminatory behaviour. But none of this tells us what we really want to know about his colour experience. There is something about it we don't know. But we know, we may suppose, everything about Fred's body, his behaviour and dispositions to behaviour and about his internal physiology, and everything about his history and relation to others that can be given in physical accounts of persons. We have all the physical information. Therefore, knowing all this is *not* knowing everything about Fred. It follows that Physicalism leaves something out.

To reinforce this conclusion, imagine that as a result of our investigations into the internal workings of Fred we find out how to make everyone's physiology like Fred's in the relevant respects; or perhaps Fred donates his body to science and on his death we are able to transplant his optical system into someone else — again the fine detail doesn't matter. The important point is that such a happening would create enormous interest. People would say, "At last we will know what it is like to see the extra colour, at last we will know how Fred has differed from us in the way he has struggled to tell us about for so long". Then it cannot be that we knew all along all about Fred. But *ex hypothesi* we did know all along everything about Fred that features in the physicalist scheme; hence the physicalist scheme leaves something out.

Put it this way. *After* the operation, we will know *more* about Fred and especially about his colour experiences. But beforehand we had all the physical information we could desire about his body and brain, and indeed

[4]Put this, and similar simplifications below, in terms of Land's theory if you prefer. See, e.g., Edwin H. Land, "Experiments in Color Vision", *Scientific American*, 200 (5 May 1959), 84-99.

[5]H. G. Wells, *The Country of the Blind and Other Stories* (London, n.d.).

everything that has ever featured in physicalist accounts of mind and consciousness. Hence there is more to know than all that. Hence Physicalism is incomplete.

Fred and the new colour(s) are of course essentially rhetorical devices. The same point can be made with normal people and familiar colours. Mary is a brilliant scientist who is, for whatever reason, forced to investigate the world from a black and white room *via* a black and white television monitor. She specialises in the neurophysiology of vision and acquires, let us suppose, all the physical information there is to obtain about what goes on when we see ripe tomatoes, or the sky, and use terms like 'red', 'blue', and so on. She discovers, for example, just which wave-length combinations from the sky stimulate the retina, and exactly how this produces *via* the central nervous system the contraction of the vocal chords and expulsion of air from the lungs that results in the uttering of the sentence 'The sky is blue'. (It can hardly be denied that it is in principle possible to obtain all this physical information from black and white television, otherwise the Open University would *of necessity* need to use colour television.)

What will happen when Mary is released from her black and white room or is given a colour television monitor? Will she *learn* anything or not? It seems just obvious that she will learn something about the world and our visual experience of it. But then it is inescapable that her previous knowledge was incomplete. But she had *all* the physical information. *Ergo* there is more to have than that, and Physicalism is false.

Clearly the same style of Knowledge argument could be deployed for taste, hearing, the bodily sensations and generally speaking for the various mental states which are said to have (as it is variously put) raw feels, phenomenal features or qualia. The conclusion in each case is that the qualia are left out of the physicalist story. And the polemical strength of the Knowledge argument is that it is so hard to deny the central claim that one can have all the physical information without having all the information there is to have.

II. THE MODAL ARGUMENT

By the Modal Argument I mean an argument of the following style.[6] Sceptics about other minds are not making a mistake in deductive logic, whatever else may be wrong with their position. No amount of physical information about another *logically entails* that he or she is conscious or feels anything at all. Consequently there is a possible world with organisms exactly like us in every physical respect (and remember that includes functional states, physical history, *et al.*) but which differ from us profoundly in that they have no conscious mental life at all. But then what is it that we have and they lack? Not anything physical *ex hypothesi*. In all physical

[6]See, e.g., Keith Campbell, *Body and Mind* (New York, 1970); and Robert Kirk, "Sentience and Behaviour", *Mind*, 83 (1974), 43-60.

regards we and they are exactly alike. Consequently there is more to us than the purely physical. Thus Physicalism is false.[7]

It is sometimes objected that the Modal argument misconceives Physicalism on the ground that that doctrine is advanced as a *contingent* truth.[8] But to say this is only to say that physicalists restrict their claim to *some* possible worlds, including especially ours; and the Modal argument is only directed against this lesser claim. If we in *our* world, let alone beings in any others, have features additional to those of our physical replicas in other possible worlds, then we have non-physical features or qualia.

The trouble rather with the Modal argument is that it rests on a disputable modal intuition. Disputable because it is disputed. Some sincerely deny that there can be physical replicas of us in other possible worlds which nevertheless lack consciousness. Moreover, at least one person who once had the intuition now has doubts.[9]

Head-counting may seem a poor approach to a discussion of the Modal argument. But frequently we can do no better when modal intuitions are in question, and remember our initial goal was to find the argument with the greatest polemical utility.

Of course, *qua* protagonists of the Knowledge argument we may well accept the modal intuition in question; but this will be a *consequence* of our already having an argument to the conclusion that qualia are left out of the physicalist story, not our ground for that conclusion. Moreover, the matter is complicated by the possibility that the connection between matters physical and qualia is like that sometimes held to obtain between aesthetic qualities and natural ones. Two possible worlds which agree in all "natural" respects (including the experiences of sentient creatures) must agree in all aesthetic qualities also, but it is plausibly held that the aesthetic qualities cannot be reduced to the natural.

III. The "What is it like to be" Argument

In "What is it like to be a bat?" Thomas Nagel argues that no amount of physical information can tell us what it is like to be a bat, and indeed that we, human beings, cannot imagine what it is like to be a bat.[10] His

---

[7] I have presented the argument in an inter-world rather than the more usual intra-world fashion to avoid inessential complications to do with supervenience, causal anomalies and the like.

[8] See, e.g., W. G. Lycan, "A New Lilliputian Argument Against Machine Functionalism", *Philosophical Studies*, 35 (1979), 279-87, p. 280; and Don Locke, "Zombies, Schizophrenics and Purely Physical Objects", *Mind*, 85 (1976), 97-9.

[9] See R. Kirk, "From Physical Explicability to Full-Blooded Materialism", *The Philosophical Quarterly*, 29 (1979), 229-37. See also the arguments against the modal intuition in, e.g., Sydney Shoemaker, "Functionalism and Qualia", *Philosophical Studies*, 27 (1975), 291-315.

[10] *The Philosophical Review*, 83 (1974), 435-50. Two things need to be said about this article. One is that, despite my dissociations to come, I am much indebted to it. The other is that the emphasis changes through the article, and by the end Nagel is objecting not so much to Physicalism as to all extant theories of mind for ignoring points of view, including those that admit (irreducible) qualia.

reason is that what this is like can only be understood from a bat's point of view, which is not our point of view and is not something capturable in physical terms which are essentially terms understandable equally from many points of view.

It is important to distinguish this argument from the Knowledge argument. When I complained that all the physical knowledge about Fred was not enough to tell us what his special colour experience was like, I was not complaining that we weren't finding out what it is like to *be* Fred. I was complaining that there is something *about* his experience, a property of it, of which we were left ignorant. And if and when we come to know what this property is we still will not know what it is like to *be* Fred, but we will know more *about* him. No amount of knowledge about Fred, be it physical or not, amounts to knowledge "from the inside" concerning Fred. We are not Fred. There is thus a whole set of items of knowledge expressed by forms of words like 'that it is *I myself* who is . . .' which Fred has and we simply cannot have because we are not him.[11]

When Fred sees the colour he alone can see, one thing he knows is the way his experience of it differs from his experience of seeing red and so on, *another* is that he himself is seeing it. Physicalist and qualia freaks alike should acknowledge that no amount of information of whatever kind that *others* have *about* Fred amounts to knowledge of the second. My complaint though concerned the first and was that the special quality of his experience is certainly a fact about it, and one which Physicalism leaves out because no amount of physical information told us what it is.

Nagel speaks as if the problem he is raising is one of extrapolating from knowledge of one experience to another, of imagining what an unfamiliar experience would be like on the basis of familiar ones. In terms of Hume's example, from knowledge of some shades of blue we can work out what it would be like to see other shades of blue. Nagel argues that the trouble with bats *et al.* is that they are too unlike us. It is hard to see an objection to Physicalism here. Physicalism makes no special claims about the imaginative or extrapolative powers of human beings, and it is hard to see why it need do so.[12]

Anyway, our Knowledge argument makes no assumptions on this point. If Physicalism were true, enough physical information about Fred would obviate any need to extrapolate or to perform special feats of imagination or understanding in order to know all about his special colour experience. *The information would already be in our possession.* But it clearly isn't. That was the nub of the argument.

---

[11]Knowledge *de se* in the terms of David Lewis, "Attitudes De Dicto and De Se", *The Philosophical Review*, 88 (1979), 513-43.

[12]See Laurence Nemirow's comments on "What is it . . ." in his review of T. Nagel, *Mortal Questions*, in *The Philosophical Review*, 89 (1980), 473-7. I am indebted here in particular to a discussion with David Lewis.

IV. The Bogey of Epiphenomenalism

Is there any really *good* reason for refusing to countenance the idea that qualia are causally impotent with respect to the physical world? I will argue for the answer no, but in doing this I will say nothing about two views associated with the classical epiphenomenalist position. The first is that mental *states* are inefficacious with respect to the physical world. All I will be concerned to defend is that it is possible to hold that certain *properties* of certain mental states, namely those I've called qualia, are such that their possession or absence makes no difference to the physical world. The second is that the mental is *totally* causally inefficacious. For all I will say it may be that you have to hold that the instantiation of *qualia* makes a difference to *other mental states* though not to anything physical. Indeed general considerations to do with how you could come to be aware of the instantiation of qualia suggest such a position.[13]

Three reasons are standardly given for holding that a quale like the hurtfulness of a pain must be causally efficacious in the physical world, and so, for instance, that its instantiation must sometimes make a difference to what happens in the brain. None, I will argue, has any real force. (I am much indebted to Alec Hyslop and John Lucas for convincing me of this.)

(i) It is supposed to be just obvious that the hurtfulness of pain is partly responsible for the subject seeking to avoid pain, saying 'It hurts' and so on. But, to reverse Hume, anything can fail to cause anything. No matter how often *B* follows *A*, and no matter how initially obvious the causality of the connection seems, the hypothesis that *A* causes *B* can be overturned by an over-arching theory which shows the two as distinct effects of a common underlying causal process.

To the untutored the image on the screen of Lee Marvin's fist moving from left to right immediately followed by the image of John Wayne's head moving in the same general direction looks as causal as anything.[14] And of course throughout countless Westerns images similar to the first are followed by images similar to the second. All this counts for precisely nothing when we know the over-arching theory concerning how the relevant images are both effects of an underlying causal process involving the projector and the film. The epiphenomenalist can say exactly the same about the connection between, for example, hurtfulness and behaviour. It is simply a consequence of the fact that certain happenings in the brain cause both.

(ii) The second objection relates to Darwin's Theory of Evolution. According to natural selection the traits that evolve over time are those conducive to physical survival. We may assume that qualia evolved over time — we have them, the earliest forms of life do not — and so we should

[13]See my review of K. Campbell, *Body and Mind*, in *Australasian Journal of Philosophy*, 50 (1972), 77-80.

[14]Cf. Jean Piaget, "The Child's Conception of Physical Causality", reprinted in *The Essential Piaget* (London, 1977).

expect qualia to be conducive to survival. The objection is that they could hardly help us to survive if they do nothing to the physical world.

The appeal of this argument is undeniable, but there is a good reply to it. Polar bears have particularly thick, warm coats. The Theory of Evolution explains this (we suppose) by pointing out that having a thick, warm coat is conducive to survival in the Arctic. But having a thick coat goes along with having a heavy coat, and having a heavy coat is *not* conducive to survival. It slows the animal down.

Does this mean that we have refuted Darwin because we have found an evolved trait — having a heavy coat — which is not conducive to survival? Clearly not. Having a heavy coat is an unavoidable concomitant of having a warm coat (in the context, modern insulation was not available), and the advantages for survival of having a warm coat outweighed the disadvantages of having a heavy one. The point is that all we can extract from Darwin's theory is that we should expect any evolved characteristic to be *either* conducive to survival *or* a by-product of one that is so conducive. The epiphenomenalist holds that qualia fall into the latter category. They are a by-product of certain brain processes that are highly conducive to survival.

(iii) The third objection is based on a point about how we come to know about other minds. We know about other minds by knowing about other behaviour, at least in part. The nature of the inference is a matter of some controversy, but it is not a matter of controversy that it proceeds from behaviour. That is why we think that stones do not feel and dogs do feel. But, runs the objection, how can a person's behaviour provide any reason for believing he has qualia like mine, or indeed any qualia at all, unless this behaviour can be regarded as the *outcome* of the qualia. Man Friday's footprint was evidence of Man Friday because footprints are causal outcomes of feet attached to people. And an epiphenomenalist cannot regard behaviour, or indeed anything physical, as an outcome of qualia.

But consider my reading in *The Times* that Spurs won. This provides excellent evidence that *The Telegraph* has also reported that Spurs won, despite the fact that (I trust) *The Telegraph* does not get the results from *The Times*. They each send their own reporters to the game. *The Telegraph*'s report is in no sense an outcome of *The Times*', but the latter provides good evidence for the former nevertheless.

The reasoning involved can be reconstructed thus. I read in *The Times* that Spurs won. This gives me reason to think that Spurs won because I know that Spurs' winning is the most likely candidate to be what caused the report in *The Times*. But I also know that Spurs' winning would have had many effects, including almost certainly a report in *The Telegraph*.

I am arguing from one effect back to its cause and out again to another effect. The fact that neither effect causes the other is irrelevant. Now the epiphenomenalist allows that qualia are effects of what goes on in the brain. Qualia cause nothing physical but are caused by something physical. Hence the epiphenomenalist can argue from the behaviour of others to the qualia of others by arguing from the behaviour of others back to its causes in the brains of others and out again to their qualia.

You may well feel for one reason or another that this is a more dubious chain of reasoning than its model in the case of newspaper reports. You are right. The problem of other minds is a major philosophical problem, the problem of other newspaper reports is not. But there is no special problem of Epiphenomenalism as opposed to, say, Interactionism here.

There is a very understandable response to the three replies I have just made. "All right, there is no knockdown refutation of the existence of epiphenomenal qualia. But the fact remains that they are an excrescence. They *do* nothing, they *explain* nothing, they serve merely to soothe the intuitions of dualists, and it is left a total mystery how they fit into the world view of science. In short we do not and cannot understand the how and why of them."

This is perfectly true; but is no objection to qualia, for it rests on an overly optimistic view of the human animal, and its powers. We are the products of Evolution. We understand and sense what we need to understand and sense in order to survive. Epiphenomenal qualia are totally irrelevant to survival. At no stage of our evolution did natural selection favour those who could make sense of how they are caused and the laws governing them, or in fact why they exist at all. And that is why we can't.

It is not sufficiently appreciated that Physicalism is an extremely optimistic view of our powers. If it is true, we have, in very broad outline admittedly, a grasp of our place in the scheme of things. Certain matters of sheer complexity defeat us — there are an awful lot of neurons — but in principle we have it all. But consider the antecedent probability that everything in the Universe be of a kind that is relevant in some way or other to the survival of *homo sapiens*. It is very low surely. But then one must admit that it is very likely that there is a part of the whole scheme of things, maybe a big part, which no amount of evolution will ever bring us near to knowledge about or understanding. For the simple reason that such knowledge and understanding is irrelevant to survival.

Physicalists typically emphasise that we are a part of nature on their view, which is fair enough. But if we are a part of nature, we are as nature has left us after however many years of evolution it is, and each step in that evolutionary progression has been a matter of chance constrained just by the need to preserve or increase survival value. The wonder is that we understand as much as we do, and there is no wonder that there should be matters which fall quite outside our comprehension. Perhaps exactly how epiphenomenal qualia fit into the scheme of things is one such.

This may seem an unduly pessimistic view of our capacity to articulate a truly comprehensive picture of our world and our place in it. But suppose we discovered living on the bottom of the deepest oceans a sort of sea slug which manifested intelligence. Perhaps survival in the conditions required rational powers. Despite their intelligence, these sea slugs have only a very restricted conception of the world by comparison with ours, the explanation for this being the nature of their immediate environment. Nevertheless they have developed sciences which work surprisingly well in these restricted terms. They also have philosophers, called slugists. Some call themselves tough-minded slugists, others confess to being soft-minded slugists.

The tough-minded slugists hold that the restricted terms (or ones pretty like them which may be introduced as their sciences progress) suffice in principle to describe everything without remainder. These tough-minded slugists admit in moments of weakness to a feeling that their theory leaves something out. They resist this feeling and their opponents, the soft-minded slugists, by pointing out — absolutely correctly — that no slugist has ever succeeded in spelling out how this mysterious residue fits into the highly successful view that their sciences have and are developing of how their world works.

Our sea slugs don't exist, but they might. And there might also exist super beings which stand to us as we stand to the sea slugs. We cannot adopt the perspective of these super beings, because we are not them, but the possibility of such a perspective is, I think, an antidote to excessive optimism.[15]

Monash University

[15]I am indebted to Robert Pargetter for a number of comments and, despite his dissent, to §IV of Paul E. Meehl, "The Compleat Autocerebroscopist" in *Mind, Matter, and Method,* ed. Paul Feyerabend and Grover Maxwell (Minneapolis, 1966).

David J. Chalmers

Facing Up to the Problem of Consciousness

I: Introduction*

Consciousness poses the most baffling problems in the science of the mind. There is nothing that we know more intimately than conscious experience, but there is nothing that is harder to explain. All sorts of mental phenomena have yielded to scientific investigation in recent years, but consciousness has stubbornly resisted. Many have tried to explain it, but the explanations always seem to fall short of the target. Some have been led to suppose that the problem is intractable, and that no good explanation can be given.

To make progress on the problem of consciousness, we have to confront it directly. In this paper, I first isolate the truly hard part of the problem, separating it from more tractable parts and giving an account of why it is so difficult to explain. I critique some recent work that uses reductive methods to address consciousness, and argue that these methods inevitably fail to come to grips with the hardest part of the problem. Once this failure is recognized, the door to further progress is opened. In the second half of the paper, I argue that if we move to a new kind of nonreductive explanation, a naturalistic account of consciousness can be given. I put forward my own candidate for such an account: a nonreductive theory based on principles of structural coherence and organizational invariance and a double-aspect view of information.

II: The Easy Problems and the Hard Problem

There is not just one problem of consciousness. 'Consciousness' is an ambiguous term, referring to many different phenomena. Each of these phenomena needs to be explained, but some are easier to explain than others. At the start, it is useful to divide the associated problems of consciousness into 'hard' and 'easy' problems. The easy problems of consciousness are those that seem directly susceptible to the standard methods of cognitive science, whereby a phenomenon is explained in terms of computational or neural mechanisms. The hard problems are those that seem to resist those methods.

The easy problems of consciousness include those of explaining the following phenomena:

* This paper was originally published in the *Journal of Consciousness Studies*, 2, No.3 (1995), pp. 200–19.

- the ability to discriminate, categorize, and react to environmental stimuli;
- the integration of information by a cognitive system;
- the reportability of mental states;
- the ability of a system to access its own internal states;
- the focus of attention;
- the deliberate control of behaviour;
- the difference between wakefulness and sleep.

All of these phenomena are associated with the notion of consciousness. For example, one sometimes says that a mental state is conscious when it is verbally reportable, or when it is internally accessible. Sometimes a system is said to be conscious of some information when it has the ability to react on the basis of that information, or, more strongly, when it attends to that information, or when it can integrate that information and exploit it in the sophisticated control of behaviour. We sometimes say that an action is conscious precisely when it is deliberate. Often, we say that an organism is conscious as another way of saying that it is awake.

There is no real issue about whether *these* phenomena can be explained scientifically. All of them are straightforwardly vulnerable to explanation in terms of computational or neural mechanisms. To explain access and reportability, for example, we need only specify the mechanism by which information about internal states is retrieved and made available for verbal report. To explain the integration of information, we need only exhibit mechanisms by which information is brought together and exploited by later processes. For an account of sleep and wakefulness, an appropriate neurophysiological account of the processes responsible for organisms' contrasting behaviour in those states will suffice. In each case, an appropriate cognitive or neurophysiological model can clearly do the explanatory work.

If these phenomena were all there was to consciousness, then consciousness would not be much of a problem. Although we do not yet have anything close to a complete explanation of these phenomena, we have a clear idea of how we might go about explaining them. This is why I call these problems the easy problems. Of course, 'easy' is a relative term. Getting the details right will probably take a century or two of difficult empirical work. Still, there is every reason to believe that the methods of cognitive science and neuroscience will succeed.

The really hard problem of consciousness is the problem of *experience*. When we think and perceive, there is a whir of information-processing, but there is also a subjective aspect. As Nagel (1974) has put it, there is *something it is like* to be a conscious organism. This subjective aspect is experience. When we see, for example, we *experience* visual sensations: the felt quality of redness, the experience of dark and light, the quality of depth in a visual field. Other experiences go along with perception in different modalities: the sound of a clarinet, the smell of mothballs. Then there are bodily sensations, from pains to orgasms; mental images that are conjured up internally; the felt quality of emotion, and the experience of a stream of conscious thought. What unites all of these states is that there is something it is like to be in them. All of them are states of experience.

It is undeniable that some organisms are subjects of experience. But the question of how it is that these systems are subjects of experience is perplexing. Why is it that when our cognitive systems engage in visual and auditory information-processing, we have visual or auditory experience: the quality of deep blue, the sensation of middle C? How can we explain why there is something it is like to entertain a mental image, or to experience an emotion? It is widely agreed that experience arises from a physical basis, but we have no good explanation of why and how it so arises. Why should physical processing give rise to a rich inner life at all? It seems objectively unreasonable that it should, and yet it does.

If any problem qualifies as *the* problem of consciousness, it is this one. In this central sense of 'consciousness', an organism is conscious if there is something it is like to be that organism, and a mental state is conscious if there is something it is like to be in that state. Sometimes terms such as 'phenomenal consciousness' and 'qualia' are also used here, but I find it more natural to speak of 'conscious experience' or simply 'experience'. Another useful way to avoid confusion (used by e.g. Newell 1990, Chalmers 1996) is to reserve the term 'consciousness' for the phenomena of experience, using the less loaded term 'awareness' for the more straightforward phenomena described earlier. If such a convention were widely adopted, communication would be much easier. As things stand, those who talk about 'consciousness' are frequently talking past each other.

The ambiguity of the term 'consciousness' is often exploited by both philosophers and scientists writing on the subject. It is common to see a paper on consciousness begin with an invocation of the mystery of consciousness, noting the strange intangibility and ineffability of subjectivity, and worrying that so far we have no theory of the phenomenon. Here, the topic is clearly the hard problem — the problem of experience. In the second half of the paper, the tone becomes more optimistic, and the author's own theory of consciousness is outlined. Upon examination, this theory turns out to be a theory of one of the more straightforward phenomena — of reportability, of introspective access, or whatever. At the close, the author declares that consciousness has turned out to be tractable after all, but the reader is left feeling like the victim of a bait-and-switch. The hard problem remains untouched.

III: Functional Explanation

Why are the easy problems easy, and why is the hard problem hard? The easy problems are easy precisely because they concern the explanation of cognitive *abilities* and *functions*. To explain a cognitive function, we need only specify a mechanism that can perform the function. The methods of cognitive science are well-suited for this sort of explanation, and so are well-suited to the easy problems of consciousness. By contrast, the hard problem is hard precisely because it is not a problem about the performance of functions. The problem persists even when the performance of all the relevant functions is explained.[1]

To explain reportability, for instance, is just to explain how a system could perform the function of producing reports on internal states. To explain internal access, we need to explain how a system could be appropriately affected by its internal states and use information about those states in directing later processes. To explain integration and

[1] Here 'function' is not used in the narrow teleological sense of something that a system is designed to do, but in the broader sense of any causal role in the production of behaviour that a system might perform.

control, we need to explain how a system's central processes can bring information contents together and use them in the facilitation of various behaviours. These are all problems about the explanation of functions.

How do we explain the performance of a function? By specifying a *mechanism* that performs the function. Here, neurophysiological and cognitive modelling are perfect for the task. If we want a detailed low-level explanation, we can specify the neural mechanism that is responsible for the function. If we want a more abstract explanation, we can specify a mechanism in computational terms. Either way, a full and satisfying explanation will result. Once we have specified the neural or computational mechanism that performs the function of verbal report, for example, the bulk of our work in explaining reportability is over.

In a way, the point is trivial. It is a *conceptual* fact about these phenomena that their explanation only involves the explanation of various functions, as the phenomena are *functionally definable*. All it *means* for reportability to be instantiated in a system is that the system has the capacity for verbal reports of internal information. All it means for a system to be awake is for it to be appropriately receptive to information from the environment and for it to be able to use this information in directing behaviour in an appropriate way. To see that this sort of thing is a conceptual fact, note that someone who says 'you have explained the performance of the verbal report function, but you have not explained reportability' is making a trivial conceptual mistake about reportability. All it could *possibly* take to explain reportability is an explanation of how the relevant function is performed; the same goes for the other phenomena in question.

Throughout the higher-level sciences, reductive explanation works in just this way. To explain the gene, for instance, we needed to specify the mechanism that stores and transmits hereditary information from one generation to the next. It turns out that DNA performs this function; once we explain how the function is performed, we have explained the gene. To explain life, we ultimately need to explain how a system can reproduce, adapt to its environment, metabolize, and so on. All of these are questions about the performance of functions, and so are well-suited to reductive explanation. The same holds for most problems in cognitive science. To explain learning, we need to explain the way in which a system's behavioural capacities are modified in light of environmental information, and the way in which new information can be brought to bear in adapting a system's actions to its environment. If we show how a neural or computational mechanism does the job, we have explained learning. We can say the same for other cognitive phenomena, such as perception, memory, and language. Sometimes the relevant functions need to be characterized quite subtly, but it is clear that insofar as cognitive science explains these phenomena at all, it does so by explaining the performance of functions.

When it comes to conscious experience, this sort of explanation fails. What makes the hard problem hard and almost unique is that it goes *beyond* problems about the performance of functions. To see this, note that even when we have explained the performance of all the cognitive and behavioural functions in the vicinity of experience — perceptual discrimination, categorization, internal access, verbal report — there may still remain a further unanswered question: *Why is the performance of these functions accompanied by experience?* A simple explanation of the functions leaves this question open.

There is no analogous further question in the explanation of genes, or of life, or of learning. If someone says 'I can see that you have explained how DNA stores and transmits hereditary information from one generation to the next, but you have not explained how it is a *gene*,' then they are making a conceptual mistake. All it means to be a gene is to be an entity that performs the relevant storage and transmission function. But if someone says 'I can see that you have explained how information is discriminated, integrated, and reported, but you have not explained how it is *experienced*,' they are not making a conceptual mistake. This is a nontrivial further question.

This further question is the key question in the problem of consciousness. Why doesn't all this information-processing go on 'in the dark', free of any inner feel? Why is it that when electromagnetic waveforms impinge on a retina and are discriminated and categorized by a visual system, this discrimination and categorization is experienced as a sensation of vivid red? We know that conscious experience *does* arise when these functions are performed, but the very fact that it arises is the central mystery. There is an *explanatory gap* (a term due to Levine 1983) between the functions and experience, and we need an explanatory bridge to cross it. A mere account of the functions stays on one side of the gap, so the materials for the bridge must be found elsewhere.

This is not to say that experience *has* no function. Perhaps it will turn out to play an important cognitive role. But for any role it might play, there will be more to the explanation of experience than a simple explanation of the function. Perhaps it will even turn out that in the course of explaining a function, we will be led to the key insight that allows an explanation of experience. If this happens, though, the discovery will be an *extra* explanatory reward. There is no cognitive function such that we can say in advance that explanation of that function will *automatically* explain experience.

To explain experience, we need a new approach. The usual explanatory methods of cognitive science and neuroscience do not suffice. These methods have been developed precisely to explain the performance of cognitive functions, and they do a good job of it. But as these methods stand, they are *only* equipped to explain the performance of functions. When it comes to the hard problem, the standard approach has nothing to say.

IV: Some Case-Studies

In the last few years, a number of works have addressed the problems of consciousness within the framework of cognitive science and neuroscience. This might suggest that the analysis above is faulty, but in fact a close examination of the relevant work only lends the analysis further support. When we investigate just which aspects of consciousness these studies are aimed at, and which aspects they end up explaining, we find that the ultimate target of explanation is always one of the easy problems. I will illustrate this with two representative examples.

The first is the 'neurobiological theory of consciousness' outlined by Francis Crick and Christof Koch (1990; see also Crick 1994). This theory centers on certain 35–75 hertz neural oscillations in the cerebral cortex; Crick and Koch hypothesize that these oscillations are the basis of consciousness. This is partly because the oscillations seem to be correlated with awareness in a number of different modalities — within the visual and olfactory systems, for example — and also because they suggest a mechanism by which the *binding* of information contents might be achieved. Binding is the process whereby separately represented pieces of information about a single entity are brought together to be used by later processing, as when information about the colour and shape of a perceived object is integrated from separate visual pathways. Following others (e.g. Eckhorn *et al.* 1988), Crick and Koch hypothesize that binding may be achieved by the synchronized oscillations of neuronal groups representing the relevant contents. When two pieces of information are to be bound together, the relevant neural groups will oscillate with the same frequency and phase.

The details of how this binding might be achieved are still poorly understood, but suppose that they can be worked out. What might the resulting theory explain? Clearly it might explain the binding of information contents, and perhaps it might yield a more general account of the integration of information in the brain. Crick and Koch also suggest that these oscillations activate the mechanisms of working memory, so that there may be an account of this and perhaps other forms of memory in the distance. The theory might eventually lead to a general account of how perceived information is bound and stored in memory, for use by later processing.

Such a theory would be valuable, but it would tell us nothing about why the relevant contents are experienced. Crick and Koch suggest that these oscillations are the neural *correlates* of experience. This claim is arguable — does not binding also take place in the processing of unconscious information? — but even if it is accepted, the *explanatory* question remains: Why do the oscillations give rise to experience? The only basis for an explanatory connection is the role they play in binding and storage, but the question of why binding and storage should themselves be accompanied by experience is never addressed. If we do not know why binding and storage should give rise to experience, telling a story about the oscillations cannot help us. Conversely, if we *knew* why binding and storage gave rise to experience, the neurophysiological details would be just the icing on the cake. Crick and Koch's theory gains its purchase by *assuming* a connection between binding and experience, and so can do nothing to explain that link.

I do not think that Crick and Koch are ultimately claiming to address the hard problem, although some have interpreted them otherwise. A published interview with Koch gives a clear statement of the limitations on the theory's ambitions.

> Well, let's first forget about the really difficult aspects, like subjective feelings, for they may not have a scientific solution. The subjective state of play, of pain, of pleasure, of seeing blue, of smelling a rose — there seems to be a huge jump between the materialistic level, of explaining molecules and neurons, and the subjective level. Let's focus on things that are easier to study — like visual awareness. You're now talking to me, but you're not looking at me, you're looking at the cappuccino, and so you are aware of it. You can say, 'It's a cup and there's some liquid in it.' If I give it to you, you'll move your arm and you'll take it — you'll respond in a meaningful manner. That's what I call awareness. ('What is Consciousness?', *Discover*, November 1992, p. 96.)

The second example is an approach at the level of cognitive psychology. This is Bernard Baars' global workspace theory of consciousness, presented in his book *A Cognitive Theory of Consciousness* (1988). According to this theory, the contents of consciousness

212

are contained in a *global workspace*, a central processor used to mediate communication between a host of specialized nonconscious processors. When these specialized processors need to broadcast information to the rest of the system, they do so by sending this information to the workspace, which acts as a kind of communal blackboard for the rest of the system, accessible to all the other processors.

Baars uses this model to address many aspects of human cognition, and to explain a number of contrasts between conscious and unconscious cognitive functioning. Ultimately, however, it is a theory of *cognitive accessibility*, explaining how it is that certain information contents are widely accessible within a system, as well as a theory of informational integration and reportability. The theory shows promise as a theory of awareness, the functional correlate of conscious experience, but an explanation of experience itself is not on offer.

One might suppose that according to this theory, the contents of experience are precisely the contents of the workspace. But even if this is so, nothing internal to the theory *explains* why the information within the global workspace is experienced. The best the theory can do is to say that the information is experienced because it is *globally accessible*. But now the question arises in a different form: why should global accessibility give rise to conscious experience? As always, this bridging question is unanswered.

Almost all work taking a cognitive or neuroscientific approach to consciousness in recent years could be subjected to a similar critique. The 'Neural Darwinism' model of Edelman (1989), for instance, addresses questions about perceptual awareness and the self-concept, but says nothing about why there should also be experience. The 'multiple drafts' model of Dennett (1991) is largely directed at explaining the reportability of certain mental contents. The 'intermediate level' theory of Jackendoff (1987) provides an account of some computational processes that underlie consciousness, but Jackendoff stresses that the question of how these 'project' into conscious experience remains mysterious.

Researchers using these methods are often inexplicit about their attitudes to the problem of conscious experience, although sometimes they take a clear stand. Even among those who are clear about it, attitudes differ widely. In placing this sort of work with respect to the problem of experience, a number of different strategies are available. It would be useful if these strategic choices were more often made explicit.

The first strategy is simply to *explain something else*. Some researchers are explicit that the problem of experience is too difficult for now, and perhaps even outside the domain of science altogether. These researchers instead choose to address one of the more tractable problems such as reportability or the self-concept. Although I have called these problems the 'easy' problems, they are among the most interesting unsolved problems in cognitive science, so this work is certainly worthwhile. The worst that can be said of this choice is that in the context of research on consciousness it is relatively unambitious, and the work can sometimes be misinterpreted.

The second choice is to take a harder line and *deny the phenomenon*. (Variations on this approach are taken by Allport 1988; Dennett 1991; Wilkes 1988.) According to this line, once we have explained the functions such as accessibility, reportability, and the like, there is no further phenomenon called 'experience' to explain. Some explicitly deny the phenomenon, holding for example that what is not externally verifiable cannot be real.

Others achieve the same effect by allowing that experience exists, but only if we equate 'experience' with something like the capacity to discriminate and report. These approaches lead to a simpler theory, but are ultimately unsatisfactory. Experience is the most central and manifest aspect of our mental lives, and indeed is perhaps the key explanandum in the science of the mind. Because of this status as an explanandum, experience cannot be discarded like the vital spirit when a new theory comes along. Rather, it is the central fact that any theory of consciousness must explain. A theory that denies the phenomenon 'solves' the problem by ducking the question.

In a third option, some researchers *claim to be explaining experience* in the full sense. These researchers (unlike those above) wish to take experience very seriously; they lay out their functional model or theory, and claim that it explains the full subjective quality of experience (e.g. Flohr 1992; Humphrey 1992). The relevant step in the explanation is usually passed over quickly, however, and usually ends up looking something like magic. After some details about information processing are given, experience suddenly enters the picture, but it is left obscure *how* these processes should suddenly give rise to experience. Perhaps it is simply taken for granted that it does, but then we have an incomplete explanation and a version of the fifth strategy below.

A fourth, more promising approach appeals to these methods to *explain the structure of experience*. For example, it is arguable that an account of the discriminations made by the visual system can account for the structural relations between different colour experiences, as well as for the geometric structure of the visual field (see e.g. Clark 1992; Hardin 1992). In general, certain facts about structures found in processing will correspond to and arguably explain facts about the structure of experience. This strategy is plausible but limited. At best, it takes the existence of experience for granted and accounts for some facts about its structure, providing a sort of nonreductive explanation of the structural aspects of experience (I will say more on this later). This is useful for many purposes, but it tells us nothing about why there should be experience in the first place.

A fifth and reasonable strategy is to *isolate the substrate of experience*. After all, almost everyone allows that experience *arises* one way or another from brain processes, and it makes sense to identify the sort of process from which it arises. Crick and Koch put their work forward as isolating the neural correlate of consciousness, for example, and Edelman (1989) and Jackendoff (1987) make related claims. Justification of these claims requires a careful theoretical analysis, especially as experience is not directly observable in experimental contexts, but when applied judiciously this strategy can shed indirect light on the problem of experience. Nevertheless, the strategy is clearly incomplete. For a satisfactory theory, we need to know more than *which* processes give rise to experience; we need an account of why and how. A full theory of consciousness must build an explanatory bridge.

V: The Extra Ingredient

We have seen that there are systematic reasons why the usual methods of cognitive science and neuroscience fail to account for conscious experience. These are simply the wrong sort of methods: nothing that they give to us can yield an explanation. To account for conscious experience, we need an *extra ingredient* in the explanation. This makes for a challenge to those who are serious about the hard problem of consciousness: What is your extra ingredient, and why should *that* account for conscious experience?

There is no shortage of extra ingredients to be had. Some propose an injection of chaos and nonlinear dynamics. Some think that the key lies in nonalgorithmic processing. Some appeal to future discoveries in neurophysiology. Some suppose that the key to the mystery will lie at the level of quantum mechanics. It is easy to see why all these suggestions are put forward. None of the old methods work, so the solution must lie with *something* new. Unfortunately, these suggestions all suffer from the same old problems.

Nonalgorithmic processing, for example, is put forward by Penrose (1989; 1994) because of the role it might play in the process of conscious mathematical insight. The arguments about mathematics are controversial, but even if they succeed and an account of nonalgorithmic processing in the human brain is given, it will still only be an account of the *functions* involved in mathematical reasoning and the like. For a nonalgorithmic process as much as an algorithmic process, the question is left unanswered: why should this process give rise to experience? In answering *this* question, there is no special role for nonalgorithmic processing.

The same goes for nonlinear and chaotic dynamics. These might provide a novel account of the dynamics of cognitive functioning, quite different from that given by standard methods in cognitive science. But from dynamics, one only gets more dynamics. The question about experience here is as mysterious as ever. The point is even clearer for new discoveries in neurophysiology. These new discoveries may help us make significant progress in understanding brain function, but for any neural process we isolate, the same question will always arise. It is difficult to imagine what a proponent of new neurophysiology expects to happen, over and above the explanation of further cognitive functions. It is not as if we will suddenly discover a phenomenal glow inside a neuron!

Perhaps the most popular 'extra ingredient' of all is quantum mechanics (e.g. Hameroff 1994). The attractiveness of quantum theories of consciousness may stem from a Law of Minimization of Mystery: consciousness is mysterious and quantum mechanics is mysterious, so maybe the two mysteries have a common source. Nevertheless, quantum theories of consciousness suffer from the same difficulties as neural or computational theories. Quantum phenomena have some remarkable functional properties, such as nondeterminism and nonlocality. It is natural to speculate that these properties may play some role in the explanation of cognitive functions, such as random choice and the integration of information, and this hypothesis cannot be ruled out *a priori*. But when it comes to the explanation of experience, quantum processes are in the same boat as any other. The question of why these processes should give rise to experience is entirely unanswered.[2]

---

[2] One special attraction of quantum theories is the fact that on some interpretations of quantum mechanics, consciousness plays an active role in 'collapsing' the quantum wave function. Such interpretations are controversial, but in any case they offer no hope of *explaining* consciousness in terms of quantum processes. Rather, these theories *assume* the existence of consciousness, and use it in the explanation of quantum processes. At best, these theories tell us something about a physical role that consciousness may play. They tell us nothing about how it arises.

At the end of the day, the same criticism applies to *any* purely physical account of consciousness. For any physical process we specify there will be an unanswered question: Why should this process give rise to experience? Given any such process, it is conceptually coherent that it could be instantiated in the absence of experience. It follows that no mere account of the physical process will tell us why experience arises. The emergence of experience goes beyond what can be derived from physical theory.

Purely physical explanation is well-suited to the explanation of physical *structures*, explaining macroscopic structures in terms of detailed microstructural constituents; and it provides a satisfying explanation of the performance of *functions*, accounting for these functions in terms of the physical mechanisms that perform them. This is because a physical account can *entail* the facts about structures and functions: once the internal details of the physical account are given, the structural and functional properties fall out as an automatic consequence. But the structure and dynamics of physical processes yield only more structure and dynamics, so structures and functions are all we can expect these processes to explain. The facts about experience cannot be an automatic consequence of any physical account, as it is conceptually coherent that any given process could exist without experience. Experience may *arise* from the physical, but it is not *entailed* by the physical.

The moral of all this is that *you can't explain conscious experience on the cheap*. It is a remarkable fact that reductive methods — methods that explain a high-level phenomenon wholly in terms of more basic physical processes — work well in so many domains. In a sense, one *can* explain most biological and cognitive phenomena on the cheap, in that these phenomena are seen as automatic consequences of more fundamental processes. It would be wonderful if reductive methods could explain experience, too; I hoped for a long time that they might. Unfortunately, there are systematic reasons why these methods must fail. Reductive methods are successful in most domains because what needs explaining in those domains are structures and functions, and these are the kind of thing that a physical account can entail. When it comes to a problem over and above the explanation of structures and functions, these methods are impotent.

This might seem reminiscent of the vitalist claim that no physical account could explain life, but the cases are disanalogous. What drove vitalist scepticism was doubt about whether physical mechanisms could perform the many remarkable functions associated with life, such as complex adaptive behaviour and reproduction. The conceptual claim that explanation of functions is what is needed was implicitly accepted, but lacking detailed knowledge of biochemical mechanisms, vitalists doubted whether any physical process could do the job and put forward the hypothesis of the vital spirit as an alternative explanation. Once it turned out that physical processes could perform the relevant functions, vitalist doubts melted away.

With experience, on the other hand, physical explanation of the functions is not in question. The key is instead the *conceptual* point that the explanation of functions does not suffice for the explanation of experience. This basic conceptual point is not something that further neuroscientific investigation will affect. In a similar way, experience is disanalogous to the *elan vital*. The vital spirit was put forward as an explanatory posit, in order to explain the relevant functions, and could therefore be discarded when those functions were explained without it. Experience is not an explanatory posit but an explanandum in its own right, and so is not a candidate for this sort of elimination.

It is tempting to note that all sorts of puzzling phenomena have eventually turned out to be explainable in physical terms. But each of these were problems about the observable behaviour of physical objects, coming down to problems in the explanation of structures and functions. Because of this, these phenomena have always been the kind of thing that a physical account *might* explain, even if at some points there have been good reasons to suspect that no such explanation would be forthcoming. The tempting induction from these cases fails in the case of consciousness, which is not a problem about physical structures and functions. The problem of consciousness is puzzling in an entirely different way. An analysis of the problem shows us that conscious experience is just not the kind of thing that a wholly reductive account could succeed in explaining.

VI: Nonreductive Explanation

At this point some are tempted to give up, holding that we will never have a theory of conscious experience. McGinn (1989), for example, argues that the problem is too hard for our limited minds; we are 'cognitively closed' with respect to the phenomenon. Others have argued that conscious experience lies outside the domain of scientific theory altogether.

I think this pessimism is premature. This is not the place to give up; it is the place where things get interesting. When simple methods of explanation are ruled out, we need to investigate the alternatives. Given that reductive explanation fails, *nonreductive* explanation is the natural choice.

Although a remarkable number of phenomena have turned out to be explicable wholly in terms of entities simpler than themselves, this is not universal. In physics, it occasionally happens that an entity has to be taken as *fundamental*. Fundamental entities are not explained in terms of anything simpler. Instead, one takes them as basic, and gives a theory of how they relate to everything else in the world. For example, in the nineteenth century it turned out that electromagnetic processes could not be explained in terms of the wholly mechanical processes that previous physical theories appealed to, so Maxwell and others introduced electromagnetic charge and electromagnetic forces as new fundamental components of a physical theory. To explain electromagnetism, the ontology of physics had to be expanded. New basic properties and basic laws were needed to give a satisfactory account of the phenomena.

Other features that physical theory takes as fundamental include mass and space-time. No attempt is made to explain these features in terms of anything simpler. But this does not rule out the possibility of a theory of mass or of space-time. There is an intricate theory of how these features interrelate, and of the basic laws they enter into. These basic principles are used to explain many familiar phenomena concerning mass, space, and time at a higher level.

I suggest that a theory of consciousness should take experience as fundamental. We know that a theory of consciousness requires the addition of *something* fundamental to our ontology, as everything in physical theory is compatible with the absence of consciousness. We might add some entirely new nonphysical feature, from which experience can be derived, but it is hard to see what such a feature would be like. More likely, we will take experience itself as a fundamental feature of the world, alongside mass, charge, and space-time. If we take experience as fundamental, then we can go about the business of constructing a theory of experience.

Where there is a fundamental property, there are fundamental laws. A nonreductive theory of experience will add new principles to the furniture of the basic laws of nature. These basic principles will ultimately carry the explanatory burden in a theory of consciousness. Just as we explain familiar high-level phenomena involving mass in terms of more basic principles involving mass and other entities, we might explain familiar phenomena involving experience in terms of more basic principles involving experience and other entities.

In particular, a nonreductive theory of experience will specify basic principles telling us how experience depends on physical features of the world. These *psychophysical* principles will not interfere with physical laws, as it seems that physical laws already form a closed system. Rather, they will be a supplement to a physical theory. A physical theory gives a theory of physical processes, and a psychophysical theory tells us how those processes give rise to experience. We know that experience depends on physical processes, but we also know that this dependence cannot be derived from physical laws alone. The new basic principles postulated by a nonreductive theory give us the extra ingredient that we need to build an explanatory bridge.

Of course, by taking experience as fundamental, there is a sense in which this approach does not tell us why there is experience in the first place. But this is the same for any fundamental theory. Nothing in physics tells us why there is matter in the first place, but we do not count this against theories of matter. Certain features of the world need to be taken as fundamental by any scientific theory. A theory of matter can still explain all sorts of facts about matter, by showing how they are consequences of the basic laws. The same goes for a theory of experience.

This position qualifies as a variety of dualism, as it postulates basic properties over and above the properties invoked by physics. But it is an innocent version of dualism, entirely compatible with the scientific view of the world. Nothing in this approach contradicts anything in physical theory; we simply need to add further *bridging* principles to explain how experience arises from physical processes. There is nothing particularly spiritual or mystical about this theory — its overall shape is like that of a physical theory, with a few fundamental entities connected by fundamental laws. It expands the ontology slightly, to be sure, but Maxwell did the same thing. Indeed, the overall structure of this position is entirely naturalistic, allowing that ultimately the universe comes down to a network of basic entities obeying simple laws, and allowing that there may ultimately be a theory of consciousness cast in terms of such laws. If the position is to have a name, a good choice might be *naturalistic dualism.*

If this view is right, then in some ways a theory of consciousness will have more in common with a theory in physics than a theory in biology. Biological theories involve no principles that are fundamental in this way, so biological theory has a certain complexity and messiness to it; but theories in physics, insofar as they deal with fundamental principles, aspire to simplicity and elegance. The fundamental laws of nature are part of the basic furniture of the world, and physical theories are telling us that this basic furniture is remarkably simple. If a theory of consciousness also involves fundamental principles, then we should expect the same. The principles of simplicity, elegance, and even beauty that drive physicists' search for a fundamental theory will also apply to a theory of consciousness.[3]

VII: Toward of a Theory of Consciousness

It is not too soon to begin work on a theory. We are already in a position to understand some key facts about the relationship between physical processes and experience, and about the regularities that connect them. Once reductive explanation is set aside, we can lay those facts on the table so that they can play their proper role as the initial pieces in a nonreductive theory of consciousness, and as constraints on the basic laws that constitute an ultimate theory.

There is an obvious problem that plagues the development of a theory of consciousness, and that is the paucity of objective data. Conscious experience is not directly observable in an experimental context, so we cannot generate data about the relationship between physical processes and experience at will. Nevertheless, we all have access to a rich source of data in our own case. Many important regularities between experience and processing can be inferred from considerations about one's own experience. There are also good indirect sources of data from observable cases, as when one relies on the verbal report of a subject as an indication of experience. These methods have their limitations, but we have more than enough data to get a theory off the ground.

Philosophical analysis is also useful in getting value for money out of the data we have. This sort of analysis can yield a number of principles relating consciousness and cognition, thereby strongly constraining the shape of an ultimate theory. The method of thought-experimentation can also yield significant rewards, as we will see. Finally, the fact that we are searching for a *fundamental* theory means that we can appeal to such nonempirical constraints as simplicity, homogeneity, and the like in developing a theory. We must seek to systematize the information we have, to extend it as far as possible by careful analysis, and then make the inference to the simplest possible theory that explains the data while remaining a plausible candidate to be part of the fundamental furniture of the world.

Such theories will always retain an element of speculation that is not present in other scientific theories, because of the impossibility of conclusive intersubjective experimental tests. Still, we can certainly construct theories that are compatible with the data that

---

[3] Some philosophers argue that even though there is a *conceptual* gap between physical processes and experience, there need be no metaphysical gap, so that experience might in a certain sense still be physical (e.g. Hill 1991; Levine 1983; Loar 1990). Usually this line of argument is supported by an appeal to the notion of *a posteriori* necessity (Kripke 1980). I think that this position rests on a misunderstanding of *a posteriori* necessity, however, or else requires an entirely new sort of necessity that we have no reason to believe in; see Chalmers 1996 (also Jackson 1994; Lewis 1994) for details. In any case, this position still concedes an *explanatory* gap between physical processes and experience. For example, the principles connecting the physical and the experiential will not be derivable from the laws of physics, so such principles must be taken as *explanatorily* fundamental. So even on this sort of view, the explanatory structure of a theory of consciousness will be much as I have described.

we have, and evaluate them in comparison to each other. Even in the absence of intersubjective observation, there are numerous criteria available for the evaluation of such theories: simplicity, internal coherence, coherence with theories in other domains, the ability to reproduce the properties of experience that are familiar from our own case, and even an overall fit with the dictates of common sense. Perhaps there will be significant indeterminacies remaining even when all these constraints are applied, but we can at least develop plausible candidates. Only when candidate theories have been developed will we be able to evaluate them.

A nonreductive theory of consciousness will consist of a number of *psychophysical principles*, principles connecting the properties of physical processes to the properties of experience. We can think of these principles as encapsulating the way in which experience arises from the physical. Ultimately, these principles should tell us what sort of physical systems will have associated experiences, and for the systems that do, they should tell us what sort of physical properties are relevant to the emergence of experience, and just what sort of experience we should expect any given physical system to yield. This is a tall order, but there is no reason why we should not get started.

In what follows, I present my own candidates for the psychophysical principles that might go into a theory of consciousness. The first two of these are *nonbasic principles* — systematic connections between processing and experience at a relatively high level. These principles can play a significant role in developing and constraining a theory of consciousness, but they are not cast at a sufficiently fundamental level to qualify as truly basic laws. The final principle is a candidate for a *basic principle* that might form the cornerstone of a fundamental theory of consciousness. This principle is particularly speculative, but it is the kind of speculation that is required if we are ever to have a satisfying theory of consciousness. I can present these principles only briefly here; I argue for them at much greater length in Chalmers 1996.

1. The principle of structural coherence

This is a principle of coherence between the *structure of consciousness* and the *structure of awareness*. Recall that 'awareness' was used earlier to refer to the various functional phenomena that are associated with consciousness. I am now using it to refer to a somewhat more specific process in the cognitive underpinnings of experience. In particular, the contents of awareness are to be understood as those information contents that are accessible to central systems, and brought to bear in a widespread way in the control of behaviour. Briefly put, we can think of awareness as *direct availability for global control*. To a first approximation, the contents of awareness are the contents that are directly accessible and potentially reportable, at least in a language-using system.

Awareness is a purely functional notion, but it is nevertheless intimately linked to conscious experience. In familiar cases, wherever we find consciousness, we find awareness. Wherever there is conscious experience, there is some corresponding information in the cognitive system that is available in the control of behaviour, and available for verbal report. Conversely, it seems that whenever information is available for report and for global control, there is a corresponding conscious experience. Thus, there is a direct correspondence between consciousness and awareness.

The correspondence can be taken further. It is a central fact about experience that it has a complex structure. The visual field has a complex geometry, for instance. There are also relations of similarity and difference between experiences, and relations in such things as relative intensity. Every subject's experience can be at least partly characterized and decomposed in terms of these structural properties: similarity and difference relations, perceived location, relative intensity, geometric structure, and so on. It is also a central fact that to each of these structural features, there is a corresponding feature in the information-processing structure of awareness.

Take colour sensations as an example. For every distinction between colour experiences, there is a corresponding distinction in processing. The different phenomenal colours that we experience form a complex three-dimensional space, varying in hue, saturation, and intensity. The properties of this space can be recovered from information-processing considerations: examination of the visual systems shows that waveforms of light are discriminated and analysed along three different axes, and it is this three-dimensional information that is relevant to later processing. The three-dimensional structure of phenomenal colour space therefore corresponds directly to the three dimensional structure of visual awareness. This is precisely what we would expect. After all, every colour distinction corresponds to some reportable information, and therefore to a distinction that is represented in the structure of processing.

In a more straightforward way, the geometric structure of the visual field is directly reflected in a structure that can be recovered from visual processing. Every geometric relation corresponds to something that can be reported and is therefore cognitively represented. If we were given only the story about information-processing in an agent's visual and cognitive system, we could not *directly* observe that agent's visual experiences, but we could nevertheless infer those experiences' structural properties.

In general, any information that is consciously experienced will also be cognitively represented. The fine-grained structure of the visual field will correspond to some fine-grained structure in visual processing. The same goes for experiences in other modalities, and even for nonsensory experiences. Internal mental images have geometric properties that are represented in processing. Even emotions have structural properties, such as relative intensity, that correspond directly to a structural property of processing; where there is greater intensity, we find a greater effect on later processes. In general, precisely because the structural properties of experience are accessible and reportable, those properties will be directly represented in the structure of awareness.

It is this isomorphism between the structures of consciousness and awareness that constitutes the principle of structural coherence. This principle reflects the central fact that even though cognitive processes do not conceptually entail facts about conscious experience, consciousness and cognition do not float free of one another but cohere in an intimate way.

This principle has its limits. It allows us to recover structural properties of experience from information-processing properties, but not all properties of experience are structural properties. There are properties of experience, such as the intrinsic nature of a sensation of red, that cannot be fully captured in a structural description. The very intelligibility of inverted spectrum scenarios, where experiences of red and green are inverted but all structural properties remain the same, show that structural properties constrain experience without exhausting it. Nevertheless, the very fact that we feel compelled to leave structural properties unaltered when we imagine experiences inverted between functionally identical systems shows how central the principle of structural coherence is to our conception of our mental lives. It is not a *logically* necessary principle, as after all we can imagine all the information processing occurring without any experience at all, but it is nevertheless a strong and familiar constraint on the psychophysical connection.

The principle of structural coherence allows for a very useful kind of indirect explanation of experience in terms of physical processes. For example, we can use facts about neural processing of visual information to indirectly explain the structure of colour space. The facts about neural processing can entail and explain the structure of awareness; if we take the coherence principle for granted, the structure of experience will also be explained. Empirical investigation might even lead us to better understand the structure of awareness within animals, shedding indirect light on Nagel's vexing question of what it is like to be a bat. This principle provides a natural interpretation of much existing work on the explanation of consciousness (e.g. Clark 1992, Hardin 1992 on colours; Akins 1993 on bats), although it is often appealed to inexplicitly. It is so familiar that it is taken for granted by almost everybody, and is a central plank in the cognitive explanation of consciousness.

The coherence between consciousness and awareness also allows a natural interpretation of work in neuroscience directed at isolating the *substrate* (or the *neural correlate*) of consciousness. Various specific hypotheses have been put forward. For example, Crick and Koch (1990) suggest that 40-hertz oscillations may be the neural correlate of consciousness, whereas Libet (1993) suggests that temporally-extended neural activity is central. If we accept the principle of coherence, the most *direct* physical correlate of consciousness is awareness: the process whereby information is made directly available for global control. The different specific hypotheses can be interpreted as empirical suggestions about how awareness might be achieved. For example, Crick and Koch suggest that 40-Hz oscillations are the gateway by which information is integrated into working memory and thereby made available to later processes. Similarly, it is natural to suppose that Libet's temporally extended activity is relevant precisely because only that sort of activity achieves global availability. The same applies to other suggested correlates such as the 'global workspace' of Baars (1988), the 'high-quality representations' of Farah (1994), and the 'selector inputs to action systems' of Shallice (1972). All these can be seen as hypotheses about the *mechanisms of awareness*: the mechanisms that perform the function of making information directly available for global control.

Given the coherence between consciousness and awareness, it follows that a mechanism of awareness will itself be a correlate of conscious experience. The question of just *which* mechanisms in the brain govern global availability is an empirical one; perhaps there are many such mechanisms. But if we accept the coherence principle, we have reason to believe that the processes that *explain* awareness will at the same time be part of the *basis* of consciousness.

2. The principle of organizational invariance

This principle states that any two systems with the same fine-grained *functional organization* will have qualitatively identical experiences. If the causal patterns of neural organization were duplicated in silicon, for example, with a silicon chip for every neuron and the same patterns of interaction, then the same experiences would arise. According to this principle, what matters for the emergence of experience is not the specific physical makeup of a system, but the abstract pattern of causal interaction between its components. This principle is controversial, of course. Some (e.g. Searle 1980) have thought that consciousness is tied to a specific biology, so that a silicon isomorph of a human need not be conscious. I believe that the principle can be given significant support by the analysis of thought-experiments, however.

Very briefly: suppose (for the purposes of a *reductio ad absurdum*) that the principle is false, and that there could be two functionally isomorphic systems with different experiences. Perhaps only one of the systems is conscious, or perhaps both are conscious but they have different experiences. For the purposes of illustration, let us say that one system is made of neurons and the other of silicon, and that one experiences red where the other experiences blue. The two systems have the same organization, so we can imagine gradually transforming one into the other, perhaps replacing neurons one at a time by silicon chips with the same local function. We thus gain a spectrum of intermediate cases, each with the same organization, but with slightly different physical makeup and slightly different experiences. Along this spectrum, there must be two systems A and B between which we replace less than one tenth of the system, but whose experiences differ. These two systems are physically identical, except that a small neural circuit in A has been replaced by a silicon circuit in B.

The key step in the thought-experiment is to take the relevant neural circuit in A, and install alongside it a causally isomorphic silicon circuit, with a switch between the two. What happens when we flip the switch? By hypothesis, the system's conscious experiences will change; from red to blue, say, for the purposes of illustration. This follows from the fact that the system after the change is essentially a version of B, whereas before the change it is just A.

But given the assumptions, there is no way for the system to *notice* the changes! Its causal organization stays constant, so that all of its functional states and behavioural dispositions stay fixed. As far as the system is concerned, nothing unusual has happened. There is no room for the thought, 'Hmm! Something strange just happened!' In general, the structure of any such thought must be reflected in processing, but the structure of processing remains constant here. If there were to be such a thought it must float entirely free of the system and would be utterly impotent to affect later processing. (If it affected later processing, the systems would be functionally distinct, contrary to hypothesis.) We might even flip the switch a number of times, so that experiences of red and blue dance back and forth before the system's 'inner eye'. According to hypothesis, the system can never notice these 'dancing qualia'.

This I take to be a *reductio* of the original assumption. It is a central fact about experience, very familiar from our own case, that whenever experiences change significantly and we are paying attention, we can notice the change; if this were not to be the case, we would be led to the sceptical possibility that our experiences are dancing before our eyes all the time. This hypothesis has the same status as the possibility that the world was created five minutes ago: perhaps it is logically coherent, but it is not plausible. Given the extremely plausible assumption that changes in experience correspond to changes in processing, we are led to the conclusion that the original hypothesis is impossible, and that any two functionally isomorphic systems must have the same sort of experiences. To put it in technical terms, the philosophical hypotheses of 'absent qualia' and 'inverted qualia', while logically possible, are empirically and nomologically impossible.[4]

There is more to be said here, but this gives the basic flavour. Once again, this thought experiment draws on familiar facts about the coherence between consciousness and cognitive processing to yield a strong conclusion about the relation between physical structure and experience. If the argument goes through, we know that the only physical properties directly relevant to the emergence of experience are *organizational* properties. This acts as a further strong constraint on a theory of consciousness.

3. The double-aspect theory of information

The two preceding principles have been *nonbasic* principles. They involve high-level notions such as 'awareness' and 'organization', and therefore lie at the wrong level to constitute the fundamental laws in a theory of consciousness. Nevertheless, they act as strong constraints. What is further needed are *basic* principles that fit these constraints and that might ultimately explain them.

The basic principle that I suggest centrally involves the notion of *information*. I understand information in more or less the sense of Shannon (1948). Where there is information, there are *information states* embedded in an *information space*. An information space has a basic structure of *difference* relations between its elements, characterizing the ways in which different elements in a space are similar or different, possibly in complex ways. An information space is an abstract object, but following Shannon we can see information as *physically embodied* when there is a space of distinct physical states, the differences between which can be transmitted down some causal pathway. The states that are transmitted can be seen as themselves constituting an information space. To borrow a phrase from Bateson (1972), physical information is a *difference that makes a difference*.

The double-aspect principle stems from the observation that there is a direct isomorphism between certain physically embodied information spaces and certain *phenomenal* (or experiential) information spaces. From the same sort of observations that went into the principle of structural coherence, we can note that the differences between phenomenal states have a structure that corresponds directly to the differences embedded in physical processes; in particular, to those differences that make a difference down certain causal pathways implicated in global availability and control. That is, we can find the

---

[4] Some may worry that a silicon isomorph of a neural system might be impossible for technical reasons. That question is open. The invariance principle says only that *if* an isomorph is possible,

same abstract information space embedded in physical processing and in conscious experience.

This leads to a natural hypothesis: that information (or at least some information) has two basic aspects, a physical aspect and a phenomenal aspect. This has the status of a basic principle that might underlie and explain the emergence of experience from the physical. Experience arises by virtue of its status as one aspect of information, when the other aspect is found embodied in physical processing.

This principle is lent support by a number of considerations, which I can only outline briefly here. First, consideration of the sort of physical changes that correspond to changes in conscious experience suggests that such changes are always relevant by virtue of their role in constituting *informational changes* — differences within an abstract space of states that are divided up precisely according to their causal differences along certain causal pathways. Second, if the principle of organizational invariance is to hold, then we need to find some fundamental *organizational* property for experience to be linked to, and information is an organizational property *par excellence*. Third, this principle offers some hope of explaining the principle of structural coherence in terms of the structure present within information spaces. Fourth, analysis of the cognitive explanation of our *judgments* and *claims* about conscious experience — judgments that are functionally explainable but nevertheless deeply tied to experience itself — suggests that explanation centrally involves the information states embedded in cognitive processing. It follows that a theory based on information allows a deep coherence between the explanation of experience and the explanation of our judgments and claims about it.

Wheeler (1990) has suggested that information is fundamental to the physics of the universe. According to this 'it from bit' doctrine, the laws of physics can be cast in terms of information, postulating different states that give rise to different effects without actually saying what those states *are*. It is only their position in an information space that counts. If so, then information is a natural candidate to also play a role in a fundamental theory of consciousness. We are led to a conception of the world on which information is truly fundamental, and on which it has two basic aspects, corresponding to the physical and the phenomenal features of the world.

Of course, the double-aspect principle is extremely speculative and is also underdetermined, leaving a number of key questions unanswered. An obvious question is whether *all* information has a phenomenal aspect. One possibility is that we need a further constraint on the fundamental theory, indicating just what *sort* of information has a phenomenal aspect. The other possibility is that there is no such constraint. If not, then experience is much more widespread than we might have believed, as information is everywhere. This is counterintuitive at first, but on reflection I think the position gains a certain plausibility and elegance. Where there is simple information processing, there is simple experience, and where there is complex information processing, there is complex experience. A mouse has a simpler information-processing structure than a human, and has correspondingly simpler experience; perhaps a thermostat, a maximally simple information processing structure, might have maximally simple experience? Indeed, if experience is truly a fundamental property, it would be surprising for it to arise only every now and then; most fundamental properties are more evenly spread. In any case, this is very much an open question, but I believe that the position is not as implausible as it is often thought to be.

Once a fundamental link between information and experience is on the table, the door is opened to some grander metaphysical speculation concerning the nature of the world. For example, it is often noted that physics characterizes its basic entities only *extrinsically*, in terms of their relations to other entities, which are themselves characterized extrinsically, and so on. The intrinsic nature of physical entities is left aside. Some argue that no such intrinsic properties exist, but then one is left with a world that is pure causal flux (a pure flow of information) with no properties for the causation to relate. If one allows that intrinsic properties exist, a natural speculation given the above is that the intrinsic properties of the physical — the properties that causation ultimately relates — are themselves phenomenal properties. We might say that phenomenal properties are the internal aspect of information. This could answer a concern about the causal relevance of experience — a natural worry, given a picture on which the physical domain is causally closed, and on which experience is supplementary to the physical. The informational view allows us to understand how experience might have a subtle kind of causal relevance in virtue of its status as the intrinsic aspect of the physical. This metaphysical speculation is probably best ignored for the purposes of developing a scientific theory, but in addressing some philosophical issues it is quite suggestive.

VIII: Conclusion

The theory I have presented is speculative, but it is a candidate theory. I suspect that the principles of structural coherence and organizational invariance will be planks in any satisfactory theory of consciousness; the status of the double-aspect theory of information is much less certain. Indeed, right now it is more of an idea than a theory. To have any hope of eventual explanatory success, it will have to be specified more fully and fleshed out into a more powerful form. Still, reflection on just what is plausible and implausible about it, on where it works and where it fails, can only lead to a better theory.

Most existing theories of consciousness either deny the phenomenon, explain something else, or elevate the problem to an eternal mystery. I hope to have shown that it is possible to make progress on the problem even while taking it seriously. To make further progress, we will need further investigation, more refined theories, and more careful analysis. The hard problem is a hard problem, but there is no reason to believe that it will remain permanently unsolved.

Further Reading

The problems of consciousness have been widely discussed in the recent philosophical literature. For some conceptual clarification of the various problems of consciousness, see Block 1995, Nelkin 1993 and Tye 1995. Those who have stressed the difficulties of explaining experience in physical terms include Hodgson 1988, Jackson 1982, Levine 1983, Lockwood 1989, McGinn 1989, Nagel 1974, Seager 1991, Searle 1992, Strawson 1994 and Velmans 1991, among others. Those who take a reductive approach include Churchland 1995, Clark 1992, Dennett 1991, Dretske 1995, Kirk 1994, Rosenthal 1996 and Tye 1995. There have not been many attempts to build detailed nonreductive theories in the literature, but see Hodgson 1988 and Lockwood 1989 for some thoughts in that direction. Two excellent collections of recent articles on consciousness are Block, Flanagan and Güzeldere 1996 and Metzinger 1995.

References

Akins, K. (1993), 'What is it like to be boring and myopic?' in *Dennett and his Critics*, ed. B. Dahlbom (Oxford: Blackwell).

Allport, A. (1988), 'What concept of consciousness?' in (eds.) *Consciousness in Contemporary Science*, ed. A. Marcel and E. Bisiach (Oxford: Oxford University Press).

Baars, B.J. (1988), *A Cognitive Theory of Consciousness* (Cambridge: Cambridge University Press).

Bateson, G. (1972), *Steps to an Ecology of Mind* (Chandler Publishing).

Block, N. (1995), 'On a confusion about the function of consciousness', *Behavioral and Brain Sciences*, in press.

Block, N, Flanagan, O. and Güzeldere, G. (eds. 1996), *The Nature of Consciousness: Philosophical and Scientific Debates* (Cambridge, MA: MIT Press).

Chalmers, D.J. (1996), *The Conscious Mind* (New York: Oxford University Press).

Churchland, P.M. (1995), *The Engine of Reason, The Seat of the Soul: A Philosophical Journey into the Brain* (Cambridge, MA: MIT Press).

Clark, A. (1992), *Sensory Qualities* (Oxford: Oxford University Press).

Crick, F. and Koch, C. (1990), 'Toward a neurobiological theory of consciousness', *Seminars in the Neurosciences*, 2, pp. 263–75.

Crick, F. (1994), *The Astonishing Hypothesis: The Scientific Search for the Soul* (New York: Scribners).

Dennett, D.C. (1991), *Consciousness Explained* (Boston: Little, Brown).

Dretske, F.I. (1995), *Naturalizing the Mind* (Cambridge, MA: MIT Press).

Edelman, G. (1989), *The Remembered Present: A Biological Theory of Consciousness* (New York: Basic Books).

Farah, M.J. (1994), 'Visual perception and visual awareness after brain damage: a tutorial overview', in *Consciousness and Unconscious Information Processing: Attention and Performance 15*, ed. C. Umilta and M. Moscovitch (Cambridge, MA: MIT Press).

Flohr, H. (1992), 'Qualia and brain processes', in *Emergence or Reduction?: Prospects for Nonreductive Physicalism*, ed. A. Beckermann, H. Flohr, and J. Kim (Berlin: De Gruyter).

Hameroff, S.R. (1994), 'Quantum coherence in microtubules: a neural basis for emergent consciousness?', *Journal of Consciousness Studies*, 1, pp. 91–118.

Hardin, C.L. (1992), 'Physiology, phenomenology, and Spinoza's true colors', in *Emergence or Reduction?: Prospects for Nonreductive Physicalism*, ed. A. Beckermann, H. Flohr, and J. Kim (Berlin: De Gruyter).

Hill, C.S. (1991), *Sensations: A Defense of Type Materialism* (Cambridge: Cambridge University Press).

Hodgson, D. (1988), *The Mind Matters: Consciousness and Choice in a Quantum World* (Oxford: Oxford University Press).

Humphrey, N. (1992), *A History of the Mind* (New York: Simon and Schuster).

Jackendoff, R. (1987), *Consciousness and the Computational Mind* (Cambridge, MA: MIT Press).

Jackson, F. (1982), 'Epiphenomenal qualia', *Philosophical Quarterly*, **32**, pp. 127–36.

Jackson, F. (1994), 'Finding the mind in the natural world', in *Philosophy and the Cognitive Sciences*, ed. R. Casati, B. Smith, and S. White (Vienna: Hölder-Pichler-Tempsky).

Kirk, R. (1994), *Raw Feeling: A Philosophical Account of the Essence of Consciousness* (Oxford: Oxford University Press).

Kripke, S. (1980), *Naming and Necessity* (Cambridge, MA: Harvard University Press).

Levine, J. (1983), 'Materialism and qualia: the explanatory gap', *Pacific Philosophical Quarterly*, **64**, pp. 354–61.

Lewis, D. (1994), 'Reduction of mind', in *A Companion to the Philosophy of Mind*, ed. S. Guttenplan (Oxford: Blackwell).

Libet, B. (1993), 'The neural time factor in conscious and unconscious events', in *Experimental and Theoretical Studies of Consciousness* (Ciba Foundation Symposium 174), ed. G.R. Block and J. Marsh (Chichester: John Wiley and Sons).

Loar, B. (1990), 'Phenomenal states', *Philosophical Perspectives*, **4**, pp. 81–108.

Lockwood, M. (1989), *Mind, Brai, and the Quantum* (Oxford: Blackwell).

McGinn, C. (1989), 'Can we solve the mind–body problem?', *Mind*, **98**, pp. 349–66.

Metzinger, T. (ed. 1995), *Conscious Experience* (Exeter: Imprint Academic).

Nagel, T. (1974), 'What is it like to be a bat?', *Philosophical Review*, **4**, pp. 435–50.

Nelkin, N. (1993), 'What is consciousness?', *Philosophy of Science*, **60**, pp. 419–34.

Newell, A. (1990), *Unified Theories of Cognition* (Cambridge, MA: Harvard University Press).

Penrose, R. (1989), *The Emperor's New Mind* (Oxford: Oxford University Press).

Penrose, R. (1994), *Shadows of the Mind* (Oxford: Oxford University Press).

Rosenthal, D.M. (1996), 'A theory of consciousness', in *The Nature of Consciousness*, ed. N. Block, O. Flanagan, and G. Güzeldere (Cambridge, MA: MIT Press).

Seager, W.E. (1991), *Metaphysics of Consciousness* (London: Routledge).

Searle, J.R. (1980), 'Minds, brains and programs', *Behavioral and Brain Sciences*, **3**, pp. 417–57.

Searle, J.R. (1992), *The Rediscovery of the Mind* (Cambridge, MA: MIT Press).

Shallice, T. (1972), 'Dual functions of consciousness', *Psychological Review*, **79**, pp. 383–93.

Shannon, C.E. (1948), 'A mathematical theory of communication', *Bell Systems Technical Journal*, **27**, pp. 379–423.

Strawson, G. (1994), *Mental Reality* (Cambridge, MA: MIT Press).

Tye, M. (1995), *Ten Problems of Consciousness* (Cambridge, MA: MIT Press).

Velmans, M. (1991), 'Is human information-processing conscious?' *Behavioral and Brain Sciences*, **14**, pp. 651–69.

Wheeler, J.A. (1990), 'Information, physics, quantum: the search for links', in *Complexity, Entropy, and the Physics of Information*, ed. W. Zurek (Redwood City, CA: Addison-Wesley).

Wilkes, K.V. (1988), '— , Yishi, Duh, Um and consciousness', in *Consciousness in Contemporary Science*, ed. A. Marcel and E. Bisiach (Oxford: Oxford University Press).

COGNITIVE SCIENCE **18,** 283–323 (1994)

Explaining Emotions

PAUL O'RORKE

University of California, Irvine

ANDREW ORTONY

Northwestern University

Emotions and cognition are inextricably intertwined. Feelings influence thoughts and actions, which in turn can give rise to new emotional reactions. We claim that people infer emotional states in others using commonsense psychological theories of the interactions among emotions, cognition, and action. We present a situation calculus theory of emotion elicitation representing knowledge underlying commonsense causal reasoning involving emotions, and show how the theory can be used to construct explanations for emotional states. The method for constructing explanations is based on the notion of abduction. This method has been implemented in a computer program called AbMaL. The results of computational experiments using AbMaL to construct explanations of examples based on cases taken from a diary study of emotions indicate that the abductive approach to explanatory reasoning about emotions offers significant advantages. We found that the majority of the diary study examples cannot be explained using deduction alone, but they can be explained by making abjuctive inferences. These inferences provide useful information relevant to emotional states.

1. INTRODUCTION

Explaining people's actions often requires reasoning about emotions. This is because experiences frequently give rise to emotional states, which in turn make some actions more likely than others. For example, if we see someone striking another person, we may explain the aggression as being a result of

An earlier abbreviated version of this article appeared in the *Proceedings of the 14th Annual Conference of the Cognitive Science Society.* Paul O'Rorke's work on this project was supported in part by grant number IRI-8813048 from the National Science Foundation. Andrew Ortony was supported in part by grant number IRI-8812699 from the National Science Foundation and in part by Andersen Consulting through Northwestern University's Institute for the Learning Sciences. Terry Turner kindly provided diary study data. Steven Morris, Clark Elliott, Tim Cain, David Aha, Tony Wieser, Stephanie Sage, Patrick Murphy, and Milton Epstein participated in various stages of this work. Anonymous reviewers provided useful criticism and suggestions.

Correspondence and requests for reprints should be sent to Paul O'Rorke, Department of Information and Computer Science, University of California, Irvine, CA 92717-3425.

229

anger. As well as reasoning about actions induced by emotional states, we can reason about emotional states themselves. In the right context, we might reason that a person was angry because we knew that he or she had been insulted. Explaining emotional states often requires reasoning about the cognitive antecedents of emotions. This article focuses on explanations of this kind.

Although people appear to generate explanations involving emotions effortlessly, the question of how one might compute such explanations remains a difficult open question, just as the more general question of how to automate commonsense reasoning remains open. We present a computational model of the construction of explanations of emotions. The model is comprised of two main components. The first component is a method for constructing explanations. The second component is a situation calculus representation of a theory of emotion elicitation. The representation of emotion-eliciting conditions is inspired by a theory of the cognitive structure of emotions proposed by Ortony, Clore, and Collins (1988). In addition to codifying a set of general rules of emotion elicitation, we have also codified a large collection of cases based on diary study data provided by Turner (1985). We have implemented a computer program that constructs explanations of emotions arising in these scenarios. The program constructs explanations using abduction. We describe the representation in some detail in later sections. In the remainder of the introduction, we provide some background information on the reasoning component, the theory of emotions, and the diary study.

1.1 Abductive Explanation

Our approach to constructing explanations is based on work in artificial intelligence and cognitive science on computational methods employing Charles Sanders Peirces's notion of abduction (Peirce, 1931–1958). Peirce used the term abduction as a name for a particular form of explanatory hypothesis generation. His description was basically:

> The surprising fact C is observed;
> But if A were true, C would be a matter of course,
> hence there is reason to suspect that A is true.

In other words, if there is a causal or logical reason A for C, and C is observed, then one might conjecture that A is true in an effort to explain C.

Since Peirce's original formulation, many variants of this form of reasoning have also come to be referred to as abduction. We focus on a view of abduction advocated by Poole (e.g., Poole, Goebel, & Aleliunas, 1987). In this approach, observations O are explained given some background knowledge expressed as a logical theory T by finding some hypotheses H such that

$$H \wedge T \vdash O.$$

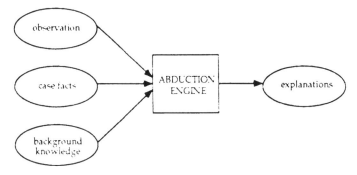

Figure 1. Inputs and Outputs of the Abduction Engine

In other words, if the hypotheses are assumed, the observation follows by way of general laws and other facts given in the background knowledge. Consistency is also desirable so it is usually required that

$H \wedge T \nvdash false.$

We use an abduction engine: a computer program that automatically constructs explanations. The explanations of observations are based on general background knowledge and knowledge of particular cases provided to the program in a machine-readable form.

The particular abduction machinery that we use is based on an early approach to mechanizing abduction described in Pople (1973). The abduction engine is part of an explanation-based learning system called AbMaL (Abductive Macro Learner). It is implemented in PROLOG.[1]

An input/output characterization of the program is given in Figure 1. AbMaL takes as input a collection of PROLOG clauses encoding theories. One theory represents background knowledge, another captures the facts of the case at hand. An observation to be explained is given as a query. AbMaL's output includes an explanation of the given observation. AbMaL generates one explanation at a time: It will search for an alternative explanation only if the user rejects the first one found. An explanation can include assumptions made in order to complete the explanation.

In addition to background knowledge, case facts, and a query, AbMaL is also given an operationality criterion and an assumability criterion. The operationality criterion is used to flag queries that should be turned over to the underlying PROLOG interpreter. The intuition is that AbMaL performs explanatory reasoning, whereas the PROLOG interpreter performs lower level reasoning in a more efficient manner without keeping track of explanatory relationships. Separate theories are provided: Explanatory rules

---

[1] PROLOG was chosen because the basic operation involved in constructing explanations, abductive inference, is related to backward chaining. PROLOG provides basic operations such as unification that are essential parts of backward chaining.

are used to construct explanations and nonexplanatory rules are used for operational inferences. The two types of rules will be distinguished in this article by using the symbol "—" in explanatory rules and the symbol ': -" in nonexplanatory rules (implemented as PROLOG clauses).

The assumability criterion determines whether a hypothesis (a query that could not be proved or disproved) may be assumed. The query may or may not be operational.[2]

A simplified sketch of the procedure followed by the abduction engine is shown in Figure 2.[3] The abduction engine attempts to construct explanations of given observations using general laws and specific facts. In the implementation, explanations are proofs represented as AND trees. Observations to be explained correspond to conclusions of the proofs (roots of the trees). General laws are encoded as rules and these are used to generate the proofs through a process based upon backward chaining.

The mechanization of abduction is comprised of three steps (Figure 2). The first step corresponds to backward chaining as it is implemented in PROLOG interpreters. The observation is treated as a query. Initially, there is only one query but, in general, there may be a number of open questions in the query list Q. The search process attempts to ground the explanation tree in known facts. If a query is operational, AbMaL attempts to identify it with a fact in the database or in its deductive closure. In attempting to prove operational queries, AbMaL does not keep track of an explanation and it does not use "explanatory" clauses. However, it does allow for the possibility that a query may be operational and/or provable in several ways. If one operationalization of the query fails to pan out, backtracking is used to search for another. If the query is not operational, or no direct operational explanation is possible, then explanatory rules may be used to extend the partial explanation, replacing existing queries with new queries. Before queries are allowed to generate new queries in this manner, a test is applied and those deemed inadmissible are disallowed. Several explanatory rules may apply to a single query. In this case, the alternatives are tried if the first rule fails to lead to an explanation.

The second step begins when backward chaining fails. This "identification" or "merging" step is based on the *synthesis* operator advocated by Pople (1973) and justified by him in terms of Occam's razor. In this step, the remaining unexplained queries are examined and some of them are assumed to be "the same" (identical in the sense that they are coreferential

---

[2] Assumptions involving operational hypotheses are allowed (see the example in Section 3.2).

[3] For more details of the explanation system, see O'Rorke (in press). That article focuses on abduction and explanation-based learning and on lessons learned in case studies involving several other domains.

Given: a first-in-first-out queue of queries Q containing observations to be explained;
Find: explanations for the queries, using abductive inferences.

1. BACKWARD CHAINING:

While the query list Q is not empty, do:

  (a) Select the first query q and remove it from Q.

  (b) If q is operational then compute an answer directly if possible, else

  (c) If q is an admissible goal and is indirectly explainable using a rule, then use the rule to generate new queries and add them to Q, else

    If q is an admissible hypothesis

      then add q to the list U of unexplained queries,

      else fail and backtrack.

2. IDENTIFICATION/MERGING:

While there are unifiable pairs of queries in U, unify and replace pairs.

3. ASSUMPTION:

While there are unexplained queries in U,

  (a) Select the first query u and remove it from U.

  (b) If the truth of u is not known, and it is an admissible hypothesis, and it is ratified by the user (optional), then assume u, else fail and backtrack.

Figure 2. A Procedural Sketch of the Abduction Engine

statements). The idea is that some of the open questions may actually be the same question even though they may have been generated by separate lines of inquiry. Merging them simplifies explanations, reduces the number of assumptions that must be made, and increases the plausibility of the explanation. Another advantage is that identification assumptions often introduce new information. The identification of two previously unrelated statements in different parts of an explanation often causes sharing of information between the separate branches of the explanation. In the implementation, statements are identified by unifying two well-formed formulae. This can cause variables in both formulae to take on new bindings. The new bindings then propagate to other statements that share the same variables.

Merging is implemented using unification. In terms of Figure 2, at the beginning of this stage Q is the empty list, U is a non-nil list of unexplained statements, and the explanation is incomplete. The algorithm continues by first selecting an arbitrary unexplained statement u from U. If u can be identified (unified) with any other statement in U, then the pair is replaced in U with their identification. The identification step ends when no more queries in U are pairwise identifiable. Unlike the previous step, this step is not deductively sound. The assumption that corresponding variables mentioned in merged queries refer to the same individual may be incorrect. When erroneous assumptions of this type are detected at explanation time, they are recoverable through backtracking.

The third abduction step tests whether remaining queries can be assumed. The queries are tested to ensure that they are not known to be true (or false). Nonexplanatory theorem proving is allowed in testing whether a hypothesis is known to be true. AbMaL calls PROLOG and if the hypothesis is proven true, then it is not allowed as an assumption. A test against stored negative assertions is used to determine whether a hypothesis is false (we do not use negation as failure). This test is a limited form of the consistency check called for in the formal specification of abduction. Together, these two tests ensure that a hypothesis is not known true or false. Next, an "assumability" test is used to decide whether to assume that a hypothesis is true. The test includes a domain-independent component and a hook that takes advantage of domain-dependent information about admissible hypotheses. This test is applied to each of the queries u in list U. If u is not assumable, then the current attempt to find an explanation is aborted and backtracking is invoked[4] in order to continue the search for acceptable explanations.

Like all methods for constructing explanations, the one just described can spend substantial time searching large spaces of potential explanations. The amount of computation required depends on the domain, the task, and the specific problem at hand. One aspect of the method that favors efficiency

---

[4] The most recent choice that may be changed is the choice of goals merged in "identication."

is that it does not attempt to compute multiple candidates simultaneously. It does not try to find all possible explanations, it only tries for one. Even so, the depth-first search tends to run away. We keep this tendency in check using depth bounds. We also use other forms of search control that take advantage of both general and domain-specific information. Tests for admissibility are applied to reject inadmissible queries and hypotheses arising in partial explanations. With these constraints on search, the algorithm finds explanations for most examples in a few seconds.[5]

1.2 A Theory of Emotions

The theory of the cognitive structure of emotions employed in the research we describe views emotions as valenced reactions to events, agents and their actions, and objects. It specifies a total of 22 emotion types summarized in an abbreviated form in Table 1. We provide only a brief sketch here. A full description can be found in Ortony, Clore, and Collins (1988).

The emotion types are essentially just classes of eliciting conditions, but each emotion type is labeled with a word or phrase, generally an English emotion word corresponding to a relatively neutral example of an emotion fitting the type. The simplest emotions are the well-being emotions *joy* and *distress*. These are an individual's positive and negative reactions to desirable or undesirable events.

The fortunes-of-others group covers four emotion types: *happy-for*, *gloating, resentment,* and *sorry-for.* Each type in this group is a combination of pleasure or displeasure over an event further specialized as being presumed to be desirable or undesirable for another person.

The prospect-based group includes six emotion types: *hope, satisfaction, relief, fear, fears-confirmed,* and *disappointment.* Each type is a reaction to a desirable or undesirable event that is still pending or that has been confirmed or disconfirmed.

The attribution group covers four types: *pride, admiration, shame,* and *reproach.* Each attribution emotion type is a (positive or negative) reaction to either one's own or another's action.

The attraction group is a structureless group of reactions to objects. The two emotions in this group are the momentary feelings (as opposed to stable dispositions) of liking or disliking.

The final group is comprised of four compounds of well being × attribution emotion types. These compound emotions do not correspond to the co-occurrence of their component emotions. Rather, each compound's eliciting conditions are the union of the component's eliciting conditions. For example, the eliciting conditions for anger combine the eliciting conditions for reproach with those for distress.

---

[5] The examples run in LPA MacPROLOG on a MacIIci.

TABLE 1
Emotion Types

Group	Specification	Types (name)
Well-being	Appraisal of event	pleased (joy) displeased (distress)
Fortunes-of-others	Presumed value of an event affecting another	pleased about an event desirable for another (happy-for) pleased about an event undesirable for another (gloating) displeased about an event desirable for another (resentment) displeased about an event undesirable for another (sorry for)
Prospect-based	Appraisal of a prospective event	pleased about a prospective desirable event (hope) pleased about a confirmed desirable event (satisfaction) pleased about a disconfirmed undesirable event (relief) displeased about a prospective undesirable event (fear) displeased about a confirmed undesirable event (fears-confirmed) displeased aobut a disconfirmed desirable event (disappointment)
Attribution	Appraisal of an agent's action	approving of one's own action (pride) approving of another's action (admiration) disapproving of one's own action (shame) disapproving of another's action (reproach)
Attraction	Appraisal of an object	liking an appealing object (love) disliking an unappealing object (hate)
Well-being Attribution	Compound emotions	admiration + joy → gratitude reproach + distress → anger pride + joy → gratification shame + distress → remorse

In general, eliciting conditions are specified in terms of variables that contribute toward increasing the intensity of emotions. The theory specifies global variables that affect all emotions, and local variables that affect subsets of emotions. The variables have values and weights associated with them, and the theory claims that an emotion is experienced only when certain levels, the emotion thresholds, are exceeded.

For anger, the variables affecting intensity are:

236

- the degree of judged blameworthiness;
- the degree of deviation from personal or role-based expectations, and
- the degree to which the event is undesirable.

The first variable, blameworthiness, is the evaluation of an action against the standards of the judger. The second variable, deviations from expectations, gauges the extent to which the action is unexpected of the agent. The third variable reflects an evaluation of the event (perpetrated by the agent) in terms of its impact upon personal goals.

1.3 A Diary Study of Emotions

We use data taken from a diary study of emotions (Turner, 1985) as the source of examples. Most of the subjects who participated in the study were sophomores at the University of Illinois at Champaign-Urbana. They were asked to describe emotional experiences that occurred within the previous 24 hours. They typed answers to a computerized questionnaire containing questions about which emotion they felt, the event giving rise to the emotion, the people involved, the goals affected, and so on. Over 1,000 descriptions of emotion episodes were collected, compiled, and recorded on magnetic media.

We chose to use this diary study as a source of examples because, although nearly every emotion type is represented, the situations and events described in the entries tend to cluster into a relatively small number of stereotypical scenarios. This is a natural consequence of the importance of examinations, dating, and so on, in the emotional lives of undergraduate students. We were thus able to focus on aspects of the theory and computational model most relevant to emotions, rather than being distracted by problems having to do with representing a wide range of domain-specific knowledge.

2. A SITUATION CALCULUS THEORY OF EMOTION ELICITATION

In this section, we describe a representation language designed to support the construction of explanations involving emotions. The language is based on work on two major contributions to knowledge representation, the situation calculus (McCarthy, 1968) and conceptual dependency (Schank, 1972).

Before we proceed, a few words on methodology may be in order. We use logic-based methods for reasoning and for representing knowledge in this article. There is some controversy over methodological issues associated with whether and how to use logic in artificial intelligence. The logicist approach is presented in Genesereth and Nilsson (1987) and the use of logic in representing commonsense knowledge is discussed in Davis (1990). Good examples of debates about the role of logic are McDermott (1987), Nilsson

(1991), and Birnbaum (1991). We use logic in this article in an attempt to make the presentation of the relevant commonsense knowledge and inference techniques clear, complete, and comprehensible. However, the use of logic-based representations and reasoning methods in this article does not represent a commitment on our part to logicism. Charniak's (1987) quip also applies to us: we are not now, nor have we ever been, logicists.

2.1 Situation Calculus

The situation calculus provides us with a language for expressing causal laws relating actions and physical situations. This first-order logical language was originally invented by McCarthy (1968, 1977). We employ a version of the situation calculus incorporating improvements by Green (1969), McCarthy and Hayes (1969), Fikes and Nilsson (1971), Kowalski (1979), and Reiter (1991).

Situations are represented by terms. Fluents are statements that may or may not be true in a given situation. The negation of a fluent F is (also) a fluent. Partial descriptions of the state of affairs in a given situation S state that fluents such as P hold in S: *holds(P, S)*.

Actions are functions that map situations representing the state of the world before the execution of the action into situations representing the state of the world afterward. The situation resulting from applying action A to state S is designated by the term *do(A, S)*. We treat the negation of an action A the same as the action of not executing A.

A number of examples that we have studied suggest that people do not make a strong distinction between actions and fluents in the sense that they often want an action to be done without focusing on any explicit effect caused by the action. Actions that are done for their own sake because they are intrinsically enjoyable rather than to achieve other goals are good examples (e.g., chatting on the phone, skiing, watching one's favorite sports team). In response to this observation, we introduced a function *did* that maps actions to fluents and we added a causal law relating corresponding actions and fluents:

  causes (A, did(A), S).

The intuition behind this fluent for actions is that, if nothing else, doing an action at least causes it to be done. [If action A is executed in situation S, then it causes the fluent *did(A)* to be true in the resulting situation.] This law allows us to refer to actions through fluents and not just as mappings between situations. This is useful for actions that are done for their own sake but more generally, the *did* fluent is useful whenever we do not wish to focus on a specific effect of an action but rather on the action itself. The importance of this extension in reasoning about emotions is discussed in Section 4.3.

Actions are defined by specifying their preconditions and effects. Preconditions are divided into *action* and *fluent* preconditions following Reiter

(1991). Action preconditions are fluents that imply that the action is possible in a given situation. Fluent preconditions imply that individual effects follow upon execution of an action.

The fact that an action A is possible in a situation S is represented as $poss(A, S)$. If P_1 and P_n are action preconditions for doing A, this is represented:

$poss(A, S)$ ― $holds(P_1, S) / \ldots \land holds(P_n, S)$.

The effects of actions are represented using something like the "add" and "delete" statements of STRIPS (Fikes & Nilsson, 1971). These statements specify fluents added or deleted upon execution of an action. Both positive and negative effects are encoded in conditional "cause" statements of the form:

$causes(A, F, S)$ ― $holds(F_1, S) \land \ldots \land holds(F_n, S)$.

where each F_i is a fluent condition for action A causing fluent F in situation S. Positive and negative effects are inferred through the following law of direct effects:

$holds(F, do(A, S))$ ― $causes(A, F, S) \cdot poss(A, S)$.

This law states that a fluent holds in the situation resulting from the execution of an action if the action was possible to begin with and if the action causes the fluent.

Our "causes" statements are similar to the "causal associations" of Peng and Reggia (1990) as opposed to their "causation events." A causal association specifies a possible causal relationship whereas a causation event is said to hold iff both the cause and the effect hold and the effect is *actually caused* by the cause. Our statements and rules involving "causes" only capture associations between actions and their effects and the fluent preconditions under which the action, if executed, would lead to the effects. The action preconditions still must be satisfied in order for the action to be possible and actually cause the relevant effect.

Frame axioms state that nothing changes unless it is explicitly stated that it changes as a result of some action. We use the following frame axiom schema:

$holds(P, do(A, S))$ ― $causes(A, \overline{P}, S) \land holds(P, S) \land poss(A, S)$.

This frame axiom states that the fluent P will hold after execution of an action if it held before and the action did not cancel it.[6]

---

[6] In our implementation, queries of the form *not P* are considered to be operational, so the explanatory machinery turns *not(causes (A, not P, S))* over to PROLOG, which attempts to show that *causes(A, not P, S)*. If the attempt to prove this goal fails, negation as failure is used to "prove" the negation. Attempts to prove *causes(A, not P, S)* can match this literal against facts and rules about causes.

The advantage of the frame axiom we have adapted is that there is no need to have a separate frame axiom for every relation. Instead one only needs a single frame axiom. Kowalski (1979) pointed out that earlier versions of frame axioms can be had by forming macros from the very general frame axiom and specific statements about what is deleted.[7]

We provided the operational metalevel predicates *agent*, *precedes*, and *precondition*. They express important constraints and help to control inference. The *agent* predicate is used to identify or extract the agent of a given action. The *precedes* predicate applies to two arguments:

$$precedes[S, do(A, S)].$$

The situation S precedes the situation resulting from the execution of action A in situation S. The situation S_0 precedes the result of doing A in S if it precedes S:

$$precedes[S_t, do(A, S)] :\!-precedes(S_0, S).$$

These rules give sufficient conditions for one situation to precede another.[8] The *precondition* predicate applies to two arguments, an action and a fluent:

$$precondition(A, C).$$

This statement is true if C is an action precondition of A. To determine whether it is true, domain-level rules about when actions are possible are consulted. To be exact, *precondition*(A, C) is true if the system finds an explanatory clause with a conclusion of the form $poss(A, S)$ and a condition of the form $holds(C, S)$.

The following general laws facilitate reasoning about goals:

$$wants[P, did(A), S] \leftarrow causes(A, F, S) \wedge diff[F, did(A)] \wedge wants(P, F, S);$$

$$wants(P, F, S) \leftarrow precondition(A, F) \wedge causes[A, G, S] \wedge wants(P, G, S).$$

The first rule states that a person may want an action to be done if some effect caused by the action is desired. Note the use of the operational metapredicate *diff*, which ensures that the effect of the action A is different from $did(A)$. (This is needed here in order to avoid useless recursion.) The second statement allows for the fact that an action may be directed at satisfying goals that contribute to the eventual achievement of longer term goals. In particular, a person may want a fluent to hold if it is a precondition of an action that causes another desired fluent.

---

[7] Interestingly, this can be done by explanation-based learning (DeJong & Mooney, 1986; Mitchell, Keller, & Kedar-Cabelli, 1986).

[8] This amounts to something like an induction schema for the *precedes* relation. Because assumptions about this relation are disallowed (see Section 2.4), negation as failure ensures that only the instances covered by these rules are allowed.

2.2 Conceptual Dependency

Primitive actions provided by conceptual dependency (*ptrans, move, atrans, propel, grasp, ingest, expel, mtrans,* and *attend*) are encoded in our situation calculus representation as functions mapping situations and CD roles such as "agent" into new situations. Variants of the functional representations are used when the value of some argument is unknown or unimportant.

For example, the function *atrans* is used to represent transfers of ownership. The most explicit form of *atrans* has arguments for the agent responsible for the transaction, the object in question, the new owner, and the previous owner. In many cases, the previous owner is the agent. We use a three-argument version of *atrans* in such cases. A two-argument version is used when the agent is the recipient and the previous owner is irrelevant.

We show how knowledge about actions is encoded using an example of *ptrans.* It is possible for an agent P to move an object T from one location *From* to a destination *To* if the thing T is initially *at* the location *From*:

$poss(ptrans(P, To, From, T), S) \leftarrow holds(at(T, From), S).$

This is an example of an action precondition. The effects of *ptrans* are encoded as follows. A physical transfer of a thing T to a destination *To* causes the thing to be *at* the destination:

$causes(ptrans(P, To, From, T), at(T, To), S).$

This is a positive effect of the transfer. A negative effect is that the thing is no longer at its original location:

$causes(ptrans(P, To, From, T), \overline{at(T, From)}, S) :-diff(To, From).$

Whereas the positive effect is unconditional, the negative effect has a fluent precondition in this formulation of *ptrans*. The operational metapredicate *diff* is used to ensure that the destination and origin are different locations. This predicate uses PROLOG to ensure that its two arguments cannot be unified. Note that the predicate "causes" should be interpreted with care (see the discussion of "causal associations" vs. "causation events" in Section 2.1). For example, *causes[ptrans(P, D, F, T), at(T, D), S]* may be true even when *holds[at(T, F), S]*, and *poss[ptrans(P, D, F, T), S]* are false so that *do[ptrans(P, D, F, T), S]* is impossible.

Additional actions required by the examples that we have encoded include *abuse, attack, breakup, call, close, die, excel, fight, gossip, hurt, insult, kill, open,* and *score*. Preconditions and effects of actions are encoded using fluents such as *alive, dead, did, have, rested, single,* and *unfaithful*. These actions and their preconditions and effects are represented using general laws similar to those shown previously in the example of *ptrans*.

241

2.3 Emotion-Eliciting Conditions

Eliciting conditions for emotions are encoded in a collection of rules for all emotion types except likes and dislikes. The rules are an initial attempt to represent the emotion-eliciting conditions proposed by Ortony et al. (1988), and sketched in Section 1.2. The elicitation rules correspond to explanatory rules in the computational implementation.[9] Simplifying assumptions and limitations of this initial representation are discussed in Section 4.4. We present the rules in pairs corresponding to complementary emotions.

People may experience joy over a fluent[10] in a situation if they want it and it holds, but they may experience distress if they want a fluent that holds not to hold:

$$joy(P, F, S) \leftarrow wants(P, F, S) \wedge holds(F, S);$$

$$distress(P, F, S) \leftarrow wants(P, \overline{F}, S) \wedge holds(F, S).$$

A person may experience neither joy nor distress in the event that a fluent holds, if he or she desires neither the fluent nor its negation. Even if we grant the law of the excluded middle for a fluent, it is still possible that a person is indifferent to it.

A person may be happy for another if he or she experiences joy over a fluent presumed to be desirable for the other. We express this in terms of joy over the other's joy:

$$happy\_for(P_1, P_2, F, S) \leftarrow joy[P_1, joy(P_2, F, S_0), S].$$

Note that the desire for the fluent is implicit in the embedded joy. Although the rule does not encode the fact that a person is usually happy for another before or while the other is happy, the rule does reflect the fact that they may be happy in different situations (at different times).

A person may be sorry for another if he or she experiences distress over a fluent presumably undesirable for the other. We express this in terms of distress over the other's distress:

$$sorry\_for(P_1, P_2, F, S) \leftarrow distress[P_1, distress(P_2, F, S_0), S].$$

The undesirability of the fluent is implicit in the embedded distress. The two people may be distressed in different situations and no temporal constraints are placed on these situations in the present formalization.

---

[9] For the distinction between explanatory and nonexplanatory rules, see Section 1.1.

[10] Note that in this formalization of the theory sketched in Table 2 fluents are used in place of events. In many cases, emotional reactions to events are actual reactions to effects of the events, rather than to the event itself. Even when the focus is on an event, we can cover this case by representing the fact that the event occurred as a fluent. For example, in the case of actions we use fluents of the form *did(action)*.

A person may gloat over a fluent that gives them joy that (they believe) is not wanted by another. We express this in terms of joy over another's distress:

$gloats(P_1, P_2, F, S) \leftarrow joy[P_1, distress(P_2, F, S_0), S]$.

People may resent another person if they are distressed about an event that pleases the other person. We express this in terms of distress over another's joy:

$resents(P_1, P_2, F, S) \leftarrow distress[P_1, joy(P_2, F, S_0), S]$.

Again, the desirability of the event for the other is implicit in the embedded distress and joy and we currently do not require any particular temporal order for the relevant situations.

The *hopes* rule captures the idea that people may experience hope if they want a fluent and anticipate it:

$hopes(P, F, S) \leftarrow wants(P, F, S) \wedge anticipates(P, F, S)$.

People may experience fear if they want an anticipated fluent not to obtain:

$fears(P, F, S) \leftarrow wants(P, \overline{F}, S) \wedge anticipates(P, F, S)$.

These rules allow for hopes and fears even if, unknown to the person, the hoped-for or feared fluent in fact already holds.

Although many examples of hopes and fears involve expectations, we use the predicate *anticipates* in order to suggest the notion of "entertaining the prospect of" a state of affairs. The purpose of this is to avoid suggesting that hoped-for and feared events necessarily have a high subjective probability. We also want to avoid suggesting that they always occur in the future.

Satisfaction occurs when a hoped-for fluent holds:

$satisfied(P, F, S) \leftarrow precedes(S_0, S) \wedge hopes(P, F, S_0) \wedge holds(F, S)$.

Fears are confirmed when feared fluents hold:

$fears\_confirmed(P, F, S) \leftarrow precedes(S_0, S) \wedge fears(P, F, S_0) \wedge holds(F, S)$.

We require the fear to precede its confirmation and we expect the hope to occur before it is satisfied.

Relief may be experienced when the negation of a feared fluent holds:

$relieved(P, \overline{F}, S) \leftarrow precedes(S_0, S) \wedge fears(P, F, S_0) \wedge holds(\overline{F}, S)$;

$relieved(P, \overline{F}, S) \leftarrow fears(P, F, S_0) \wedge holds(\overline{F}, S)$.

We give two rules for relief because fear usually occurs before the fluent holds, but sometimes relief occurs in the absence of prior fear (as when a person discovers that a missed flight crashed).

Disappointment occurs when the negation of a hoped-for fluent holds:

$disappointed(P, \overline{F}, S) \leftarrow precedes(S_0, S) \wedge hopes(P, F, S_c) \wedge holds(\overline{F}, S).$

$disappointed(P, \overline{F}, S) \leftarrow hopes(P, F, S_0) \wedge holds(\overline{F}, S).$

The fluent is usually hoped-for in a situation that occurs in advance of the present situation but disappointment (e.g., at a missed opportunity) may occur in the absence of prior hope.

Pride and shame can occur for individuals or groups. An agent experiences pride over praiseworthy actions executed either by the agent or by another member of a "cognitive unit" (Heider, 1958) containing the agent. Agents may experience shame if they do a blameworthy act or if another member of their cognitive unit does a blameworthy act:

$proud(P, A, S) \leftarrow agent(A, P) \wedge holds[did(A), S] \wedge praiseworthy(A);$

$proud(P_1, P_2, A, S) \leftarrow agent(A, P_2) \wedge holds[did(A), S] \wedge praiseworthy(A)$
$\wedge cognitive\_unit(P_1, P_2);$

$shame(P, A, S) \leftarrow agent(A, P) \wedge holds[did(A), S] \wedge blameworthy(A);$

$shame(P_1, P_2, A, S) \leftarrow agent(A, P_2) \wedge holds[did(A), S] \wedge blameworthy(A)$
$\wedge cognitive\_unit(P_1, P_2).$

The predicates *praiseworthy* and *blameworthy* are intended to reflect personal standards rather than normative or social standards, except insofar as the judging person subscribes to such standards.

A person may admire another person if the other person does something praiseworthy, but a person may feel reproach toward another if the other does something blameworthy:

$admire(P_1, P_2, A, S) \leftarrow agent(A, P_2) \wedge holds[did(A), S] \wedge praiseworthy(A);$

$reproach(P_1, P_2, A, S) \leftarrow agent(A, P_2) \wedge holds[did(A), S] \wedge blameworthy(A).$

Compound emotion types are comprised of the eliciting conditions of components taken from the well-being and attribution groups. We do not include the component emotions in the eliciting conditions in order to avoid suggesting that the component emotions are necessarily felt as part of feeling the compound. Instead, we collect the eliciting conditions of the components and simplify them, eliminating redundancies.

Gratitude is a compound of the eliciting conditions of joy and admiration. A person may be grateful toward another person if the other person does a praiseworthy action that causes a desirable fluent to hold:

$grateful(P_1, P_2, A, S_1) \leftarrow agent(A, P_2) \wedge holds[did(A), S_1] \wedge precedes(S_0, S_1)$
$\wedge causes(A, F, S_0) \wedge praiseworthy(A) \wedge wants(P_1, F, S_1) \wedge holds(F, S_1).$

The *angry_at* emotion type focuses on anger at other agents. It is a compound comprised of the eliciting conditions of reproach and distress. A person may be angry at another if an undesirable fluent is caused by a blameworthy action taken by the other person:

$$angry\_at(P_1, P_2, A, S1) \leftarrow agent(A, P_2) \land holds[did(A), S_1] \cdot precedes(S_0, S_1)$$
$$\land causes(A, F, S_0) \land blameworthy(A) \land wants(P_1, \bar{F}, S_1) \land holds(F, S_1).$$

We distinguish this from angry about, which because it focuses on the undesirability of the situation, is better thought of as a special case of *distress:* frustration (typically at goal blockage).

Gratification is a compound emotion comprised of the eliciting conditions of pride and joy. A person may be gratified if he or she does a praiseworthy action that results in a desirable fluent:

$$gratified(P, A, S_1) \leftarrow agent(A, P_2) \land holds[did(A), S_1] \cdot precedes(S_1, S_1)$$
$$causes(A, F, S_0) \land wants(P, F, S_1) \cdot holds(F, S_1) \cdot praiseworthy(A);$$

$$gratified(P_1, P_2, A, S_1) \leftarrow agent(A, P_2) \land holds[did(A), S_1] \cdot precedes(S_2, S_1)$$
$$causes(A, F, S_0) \land cognitive\_unit(P_1, P_2) \land wants(P_1, F, S_1)$$
$$holds(F, S_1) \land praiseworthy(A).$$

Because there is a cognitive unit variant of pride, there is also a variant of gratification. This variant of *gratified* is closely related to *grateful*.

Remorse is a compound emotion comprised of the eliciting conditions of shame and distress. People may be remorseful if they do a blameworthy action that results in an undesirable fluent:

$$remorseful(P, A, S_1) \leftarrow agent(A, P_2) \land holds[did(A), S_1] \cdot precedes(S_0, S_1)$$
$$causes(A, F, S_0) \land wants(P_1, \bar{F}, S_1) \cdot holds(F, S_1) \cdot blameworthy(A);$$

$$remorseful(P_1, P_2, A, S_1) \leftarrow agent(A, P_2) \land holds[did(A), S_1] \cdot precedes(S_0, S_1)$$
$$causes(A, F, S_0) \land cognitive\_unit(P_1, P_2) \land wants(P_1, \bar{F}, S_1)$$
$$holds(F, S_1) \land blameworthy(A);$$

The second rule here provides the eliciting conditions of a cognitive-unit variant of *remorseful*. These conditions are derived from the corresponding variant of *shame*. This variant is closely related to *angry_at*.

The eliciting conditions given in this section represent a first attempt at formalizing necessary, but not necessarily sufficient, conditions for the elicitation of the corresponding emotions. (This is why we say "a person *may* experience emotion *x* under conditions *y*.") In some cases, additional conditions may be required to capture more fully commonsense knowledge about emotion elicitation. It may be that a person's disposition towards another person should play a role in explaining the elicitation of "fortunes of others" emotions. For example, perhaps the eliciting conditions of *happy_for* should include a requirement that a person may be happy for another if that person does not dislike the other.

Cognitive units play an important role in some emotions and are provided for in the situation calculus theory of emotions. A special predicate is used for cognitive units in the eliciting conditions of the *attribution* emotions *pride* and *shame*. This predicate is also used in the given "background knowledge" to encode groups that may form cognitive units. This is an attempt to capture the idea that people can form (relatively stable) cognitive units with their family members and close friends.

Studies of a number of examples suggest that many goals are shared by members of the same cognitive unit. People want good things to happen not just to themselves but also to others in their cognitive unit. They want to avoid bad things and they do not want bad things to happen to others in their cognitive unit. For example, everyone wants to excel, and they want people in their cognitive unit to excel, too. People generally do not want to experience harm, and they do not want other members of their group to be harmed either. This sort of general law is represented using conditional *wants:*

$$wants[P, \overline{harmed(Q)}] \leftarrow cognitive\_unit(P, Q).$$

2.4 Finding Plausible Explanations of Emotions Efficiently
In this section, we describe three additional sources of knowledge that contribute to efficiency. The first source, the assumability criterion, specifies which queries are admissible as hypotheses. The second source, the operationality criterion, specifies which queries can be answered without explanation. The third source, a set of rewrite rules, specifies transformations that map alternative representations into a canonical form. In addition to improving efficiency, these three knowledge sources also help limit the search for explanations for emotions to plausible candidates.

In general, many explanations are possible and it is important to constrain the search to avoid large numbers of implausible hypotheses and explanations. In early experiments without constraints we found that the abduction engine conjectured large numbers of implausible causal relationships. This problem was addressed by disallowing assumptions involving metapredicates like *diff* and instances of the following:

$$preconditions(A, F);$$

$$causes(A, F, S).$$

In other words, the abduction engine was not allowed to assume arbitrary preconditions for actions, nor was it allowed to assume unprovable causal relationships between actions and effects.

The operationality criterion provides a second source of knowledge that enables efficient recognition of the truth or falsity of queries. Operational queries can be answered relatively efficiently because they do not involve the additional overhead associated with constructing explanations. Such queries are turned over to the base-level interpreter (PROLOG). In most cases, they are answered by simply attempting to find a matching statement in the database, although in general backward chaining is allowed. The following predicates are considered to be operational:

$$diff(X, Y);$$
$$member(X, Y);$$
$$opposite(X, Y)'$$
$$action(X);$$
$$precondition(A, C);$$
$$agent(A, P);$$
$$precedes(S_1, S_2);$$
$$cognitive\_unit(P, Q);$$
$$d\_likes(P, Q).$$

We assume that no explanatory reasoning is involved in answering these queries. For example, the dispositional attitude *d_likes* is operational because we assume that likes and dislikes are not explainable. In addition to these predicates, simple queries present in the database as facts are also considered to be operational.

Rewrite rules enable the system to recognize alternative ways of representing the same expression. For example, since the negation of doing an action is considered to be identical to the execution of the negation of that action, the term $\overline{do(A)}$ is considered to be an alias for $do(\overline{A})$. At key points in the computation, the system uses rewrite rules to transform such expressions into a canonical form. Note that this use of canonical forms does not put us in danger of subscribing to Woods's (1975) "canonical form myth." Woods pointed out that it is provably impossible to find canonical form functions for many formal systems and that such functions should not be expected to map all internal representations of sentences with the same "meaning" into a single canonical form. But our rewrite rules do not carry much inferential burden; instead, they provide representational flexibility.

This allows us to use two ways of associating fluents and situations. One advantage of the *holds* relation introduced by Kowalski (1979) is that it avoids the need for an extra state parameter for all relations. The disadvantage of the use of *holds* (as opposed to including extra arguments for situations) is that it requires an extra predicate in the sense that it requires us to embed fluents in *holds* statements. Besides being aesthetically undesirable in some situations, the use of *holds* can increase the branching factor of some explanatory searches because facts and clauses tend to be indexed and fetched by the top-level predicate. Sometimes it is more convenient to use situations as arguments of fluents. In other cases, it is preferable to associate fluents with situations using the *holds* predicate. The rewrite rules listed here facilitate explanations involving chains of fluents by linking equivalent representations:

$$holds[F(Args), S] \Rightarrow F(Args, S);$$
$$holds\{holds[F(Args), S_1], S_2\} \Rightarrow holds[F(Args), S_1];$$
$$holds[F(Args, S_1), S_2] \Rightarrow F(Args, S_1).$$

The first rule establishes the equivalence between a fluent with a situation as an argument and the corresponding situationless fluent holding in the same situation. The second rule "unwraps" embedded *holds* statements. The truth of a *holds* statement depends only on the situation it applies to. The third rule is a consequence of the first two.

In the emotion-eliciting conditions, the emotion types occurring in the heads of the rules are expressed as fluents with situational arguments. The bodies of the rules contain *holds* statements whose arguments are fluents eliciting emotional reactions. It is important to ensure that the alternative representations are viewed as equivalent so that inferences are not lost. We enforce the equivalence by mapping complex representations to relatively simple canonical forms using rewrite rules. This facilitates explanations involving emotional chains (as shown in Sections 3.2 and 4.3).

3. EXPLAINING EMOTIONS

In this section, we show how our situation calculus representation of emotion elicitation can be used to explain emotional states. We describe the process of codifying examples. We show how explanations are produced by the abduction engine sketched earlier. We provide assumptions and explanations produced by the computer program and show how its outputs are interpreted. The examples discussed in this section were chosen to show some of the breadth of the reasoning and representation methods. In a later section, we will use these examples to illustrate strengths and weaknesses of the methods.

3.1 An Explanation for Joy

The first example is based on the following "case," a simplified version of a scenario taken from Turner's (1985) diary study:

Mary went home to see her family.
She ate a home-cooked meal.
She wanted to stay home.
Mary was happy.

We hand-coded the case into the inputs shown in Figure 3. The first input states that Mary wants to be home in the situation that follows after she went home and ingested a home-cooked meal. Abbreviations at the bottom of the table are used for the relevant situations. Note that this codification of the example is crude in the sense that we have not attempted to capture

Case Facts

    wants(mary, at(mary, home), s2).

Query

    why(joy(mary, -, s2))

Explanation

    joy(mary, at(mary, home), s2)
        wants(mary, at(mary, home), s2)
        holds(at(mary, home), s2)
        not causes(ingest(mary, home_cooked_meal), not at(mary, home), s1)
        holds(at(mary, home), s1)
            causes(ptrans(mary, home), at(mary, home), s0)
            poss(ptrans(mary, home), s0)
        poss(ingest(mary, home_cooked_meal), s1)
            [holds(have(mary, home_cooked_meal), s1)]

Abbreviations

    s1 = do(ptrans(mary, home), s0)
    s2 = do(ingest(mary, home_cooked_meal), s1)

Figure 3. An Explanation for Joy

249

much of the information associated with Mary visiting her family. The underscores and the use of a constant for "home_cooked_meal" also signify simplification aimed at avoiding having to deal with issues related to quality of food and different methods of food preparation. Such subtleties are lost. We strive only to capture basic facts of the case. The case specifies that Mary is happy in a situation following certain actions. It specifies that she wants to be at home in this situation but it does not specify the fluent that she is happy about. In the query about why Mary is experiencing joy, the situation is specified but the fluent is left blank (using a "don't care" variable designated in PROLOG as an underscore "_").

Figure 3 also shows an explanation produced by the abduction engine. The explanation is in the form of a tree. The first line is the root of the tree; indented lines are branches. The first level of indentation shows the propositions immediately supporting the root. The second level of indentation shows their supporters, and so on. The deepest levels of indentation correspond to leaves of branches of the explanation tree.

This explanation for Mary's happiness is interpreted as follows. In the first line, we see that the fluent she is happy about has been identified as the fact that she is at home in the given situation. Mary's location prior to going home was not specified in the given case fact and neither is it determined during the construction of the explanation. Her joy is supported by the second line, which was part of the input. A case fact stating that Mary wanted to be at home was given. The fat that she is at home is explained by the remainder of the tree. Mary is a home because she went there and the fact that she ingested a meal did nothing to cancel this result. The explanation that Mary is at home after she ate the meal rests on an assumption that it was possible to eat because she had it after she went home. Abductive inferences (assumptions) are distinguished from leaves of explanation trees that are known to be true by enclosing them in boxes as in Figure 3.

The explanation was constructed by the abduction engine using the situation calculus theory of emotion elicitation described earlier. The system is not allowed to explain the initial query (why was Mary happy?) directly, even though it is a known fact. Instead, it finds reasons by backward chaining on rules of situation calculus and emotion elicitation rules. In this case, the rule specifying the eliciting conditions for joy applied. Backward chaining on this rule generated two new queries: Does Mary want something—something that holds in the given situation? A given case fact stated that Mary wants to be at home. This fact was used to "ground out" one of the queries. The query about how the desired state of affairs came to pass (how Mary came to be at home) was answered by backward chaining on a law of situation calculus. The frame axiom stating that effects of earlier actions persist unless canceled by later actions was applied to infer that Mary was at home because she went there earlier and nothing she did in the meantime canceled this effect. Most of the remaining queries grounded out in known

250

facts provided as facts of situation calculus and conceptual dependency (e.g., Mary was at home because she did a *ptrans* and physically transferred herself home). But an assumption was made in order to complete the explanation. An action precondition of *ingest* states that it is possible to ingest something if one has it first. It was not stated in the input whether Mary had possession of the meal, but the abduction engine assumed that she did.[11]

3.2 An Explanation for Happy_For

This example illustrates the fortunes-of-others emotion, *happy_for:*

Mary's roommate is going to Europe.
Mary is happy for her.

Here, Mary's roommate is going to Europe and thus Mary is happy for her. The explanation constructed by the abduction engine is shown in Figure 4. The first reason translates the emotion to be explained into joy over another's joy. The explanation states that Mary is happy for her because she will be happy in Europe and Mary wants her roommate to be happy because she likes her. In the construction of this part of the explanation, a new query is generated in an effort to explain Mary's roommate's joyful reaction to being in Europe. The query is initially in the form:

$holds(joy(roommate(mary), at(roommate(mary), europe), s2), s1)$;

but this is immediately simplified, using rewrite rules mapping fluents into a canonical form (Section 2.4), to strip off a superfluous *holds* predicate and an unnecessary situation $s1$:

$joy(roommate(mary), at(roommate(mary), europe), s2)$.

This simplified version of the query invokes the emotion eliciting rule for *joy* producing the explanation shown.

Note that the predicate *d_likes* is to be interpreted as dispositional liking as opposed to momentary liking. Also, it is *assumed* that Mary likes her roommate.[12] Her roommate will be happy in Europe because she is there, assuming that she wants to be there. She will be in Europe as a result of going there.

3.3 An Explanation for Gloating

In our third example (Figure 5), the parenthetical comments were not encoded, but they are included to retain fidelity to the original report from the diary study:

---

[11] See Section 4.1 for a discussion of several issues associated with the abductive inference of preconditions.

[12] This is an example where an operational predicate (dispositional liking) is assumed in spite of the fact that an attempt to prove it fails because there is nothing in the list of known facts about whether Mary likes her roommate.

Case Facts

NONE;

Query

why(happy_for(mary, roommate(mary)), at(roommate(mary), europe), s1))

Explanation

happy_for(mary, roommate(mary)), at(roommate(mary), europe), s1

joy(mary, joy(roommate(mary), at(roommate(mary), europe), s2), s1

wants(mary, joy(roommate(mary), at(roommate(mary), europe), s2), s1

d_likes(mary, roommate(mary))

joy(roommate(mary), at(roommate(mary), europe), s2)

wants(roommate(mary), at(roommate(mary), europe), s2)

holds(at(roommate(mary), europe), s2)

causes(ptrans(roommate(mary), europe), at(roommate(mary), europe), _29530)

poss(ptrans(roommate(mary), europe), _29530)

Abbreviations

s1 do(ptrans(roommate(mary), europe), s0)

s2=do(ptrans(roommate(mary), europe), _29530)

Figure 4. An Explanation for Happy_For

252

Case Facts
  NONE

Query
  why(gloats(john, guy, did(expel(guy,vomit)), s2))

Explanation
  gloats(john, guy, did(expel(guy, vomit)), s2)
  joy(john, distress(guy, did(expel(guy, vomit)), do(expel(guy, vomit), _21375)), s2)
    wants(john, distress(guy, did(expel(guy, vomit)), do(expel(guy, vomit), _21375)), s2)
    distress(guy, did(expel(guy, vomit)), do(expel(guy, vomit), _21375))
    wants(guy, did(not expel(guy, vomit)), do(expel(guy, vomit), _21375))
    holds(did(expel(guy, vomit)), do(expel(guy, vomit), _21375))
      causes(expel(guy, vomit), did(expel(guy, vomit), _21375))
      poss(expel(guy, vomit), _21375)

Abbreviations
  s1 = do(expel(guy,vomit), s0)
  s2   do(hear(john, expel(guy,vomit)), s1)

Figure 5. An Explanation for Gloating

253

John heard one of the guys who lived below him throwing up
(the morning after a party).
(The guy got what he deserved.)
John gloated.

The explanation illustrating explanatory reasoning involved *gloating* shown in Figure 5 states that John gloats over his neighbor's retching because John takes pleasure in his neighbor's distress. The system is forced to assume that John wanted his neighbor to experience distress. John's neighbor felt distress over vomiting because it is an undesirable event. Deeper inferences might explain how this unpleasant action occurred. These would require additional causal connections (e.g., between the party, excessive drinking, and retching). Another line of inference currently outside the scope of our implementation is: John may wish his neighbor ill because the neighbor's party had been noisy and disturbed John. This explanation would require the addition of knowledge about parties and noise, and about needs associated with sleep.

3.4 Explanations for Relief and Fear
The following case provides examples of *relief* and *fear:*

Mary wanted to go to sleep.
Karen returned.
T.C. finally left her place.
Mary was relieved.

The case is encoded as shown in Figure 6. The case facts say Mary wants sleep. The query asks why Mary is relieved that T.C. is not at her home in the situation that results after T.C.'s departure. T.C.'s departure occurred in the situation resulting from Karen's return.

The automatically constructed explanation assumes that Mary fears T.C.'s presence in her home because she does not want T.C. to be in her home but she anticipates that he will be there. A deeper explanation connecting this desire and anticipation to Mary's desire for restful sleep should be possible. For example, the presence of T.C. might interfere with Mary's sleep.

The explanation in Figure 6 states that Mary is relieved because T.C. is no longer at her home. The explanation of his absence does not include the possibility that he may have been driven away by Karen's return. But it is interesting for another reason. It illustrates the use of causal laws to infer negative fluents relevant to emotional reactions. In this case, because T.C. moved from Mary's home to another location, he is no longer *at* Mary's home.

Case Facts

wants(mary, sleep(mary), _)

Query

why(relieved(mary, not at(tc, home(mary)), s2))

Explanation

relieved(mary, not at(tc, home(mary)), s2)
  precedes(s1, s2)
  fears(mary, at(tc, home(mary)), s1
    wants(mary, not at(tc, home(mary)), s1
    anticipates(mary, at(tc, home(mary)), s1
  holds(not at(tc, home(mary)), s2)
    causes(ptrans(tc, _29887, home(mary), tc), not at(tc, home(mary)), s1)
    poss(ptrans(tc, _29887, home(mary), tc), s1)
    • holds(at(tc, home(mary)), s1)
      not causes(ptrans(karen, home(mary)), not at(tc, home(mary)), s0)
      holds(at(tc, home(mary)), s0)
      poss(ptrans(karen, home(mary)), s0)

Abbreviations

s1=do(ptrans(karen, home(mary)), s0)
s2=do(ptrans(tc, _29887, home(mary), tc), s1)

Figure 6. Explanations for Relief and Fear

255

3.5 An Explanation for Anger

The following example, based on a diary study case involving a dating scenario, illustrates the *angry_at* emotion type. The example involves the breakup of a couple of college students, Kim and John:

> *Kim wanted to break up with John.*
> *John didn't want to break up with Kim.*
> *They broke up.*
> *John is angry at Kim.*

The example is encoded as shown in Figure 7. The query encodes the question "Why is John angry with Kim over the breakup?" The explanation states that John is angry with Kim because Kim initiated the breakup and it caused John to be single. The fact that John doesn't want to be single was given but the remainder of the explanation involves two assumptions. The first is that Kim and John were a couple prior to the breakup. The second assumption is that John views Kim's breaking up with him to be *blameworthy*.

4. DISCUSSION

The previous sections described a representation for knowledge about emotion elicitation and a computer program that constructs explanations based on cases taken from a diary study about emotion episodes. In this section, we discuss some of the strengths and weaknesses of our explanatory reasoner and our representation of knowledge about emotion elicitation.

4.1 Advantages of Abductive Reasoning about Emotions

We claim that abduction is superior to deduction as a basis for explanatory reasoning about emotions because it subsumes deduction which, on its own, will fail when a proof cannot be derived from a given set of facts. The primary advantage of abduction is that it allows for the possibility that assumptions may be required to complete explanations, so that an explanation of a given observation can be proposed even when it does not follow logically from given facts.

It is unreasonable to expect all the information needed to construct explanations involving emotions to be provided in advance. It is particularly unlikely that all the relevant information about mental states will be provided. Indeed, we would like to acquire this sort of information by inference, and abductive inference allows us to do so during the construction of explanations. Abduction may be viewed as a search for plausible hypotheses that help explain observations.

The majority of the cases in the diary study data require assumptions. In this regard, the examples that we have discussed are representative. The kinds of assumptions needed included missing preconditions, goals, prospects, and judgments.

Case Facts

    wants(kim,single(kim),_)
    wants(john,not(single(john)),_)

Query

    why(angry_at(john,kim,breakup(kim,kim,john),s1))

Explanation

    angry_at(john, kim, breakup(kim, kim, john), s1)
      agent(breakup(kim, kim, john), kim)
      holds(did(breakup(kim, kim, john)), s1)
        causes(breakup(kim, kim, john), did(breakup(kim, kim, john)), s)
        poss(breakup(kim, kim, john), s)
          ┌─────────────────────────────┐
          │ holds(couple(kim, john), s) │
          └─────────────────────────────┘
      precedes(s, s1)
      causes(breakup(kim, kim, john), single(john)), s)
        ┌──────────────────────────────────────┐
        │ blameworthy(breakup(kim, kim, john)) │
        └──────────────────────────────────────┘
        wants(john, not single(john)), s1)
        holds(single(john), s1)
          causes(breakup(kim, kim, john), single(john)), single(john), s)
          poss(breakup(kim, kim, john), s)
            ┌─────────────────────────────┐
            │ holds(couple(kim, john), s) │
            └─────────────────────────────┘

Abbreviations

    s1 = do(breakup(kim, kim, john), s)

Figure 7. An Explanation for Anger

257

Examples of preconditions inferred by abductive inference included the following. In the example of *joy* (Section 3.1) the explanation was completed with an assumption that Mary had possession of a home-cooked meal. This explained how it was possible for her to ingest it. In the example of *gloating* (Section 3.3) it was assumed that it was possible for John's neighbor to vomit. In Section 3.5 it was assumed that Kim and John were a couple before she broke up with John and he became angry.[13]

Examples of abductive assumptions about goals occur frequently. The example in Section 3.4 required an assumption that Mary wanted T.C. to go somewhere else in order to explain Mary's fear that T.C. would be at her home. Assumptions about others' goals occur in explaining fortunes-of-others emotions. In the example for *happy_for* (Section 3.2), an assumption was made about Mary's roommate's desire to be in Europe. The example of *gloating* (Section 3.3) required an assumption that John wanted his neighbor to be distressed.

Abductive assumptions about other mental states include assumptions about whether agents anticipate events. In the example of *relief* (Section 3.4), it was necessary to assume that Mary anticipated T.C.'s continued (unwelcome) presence in her home.

Assumptions about judgments of blameworthiness and praiseworthiness are important in explaining attribution emotions and compound emotions. For example, in the *anger* case, the assumption that Kim's breaking up with John was blameworthy was made in order to explain why John was angry at Kim.

None of the explanations constructed in these examples could have been constructed by the abduction engine without its abductive inference capability, given the background knowledge and codifications of the cases provided with the observations to be explained. Given the same information, a purely deductive PROLOG-style interpreter would have failed to find an explanation.

Admittedly, the knowledge base could conceivably be extended so that some assumptions could be eliminated and replaced by deductive inferences. Knowledge of ethics and standards of behavior could be provided, reducing the number of assumptions in explanations requiring judgments of blameworthiness and praiseworthiness. Some additional inferences could be made deductively, rather than abductively. For example, we saw several instances of necessary preconditions in the emotion cases: It is necessary for

---

[13] The reader may wonder how it is possible to make these assumptions given that the abduction engine is not allowed to assume that an arbitrary fluent might be a precondition for anaction (Section 2.4). The reason is that assumptions of the form *preconditions(A, F)* are prohibited, whereas assumptions of the form *holds(F, S)* are permitted. In other words, we allow a conjecture that a condition is true in a given situation but we disallow a conjecture that the condition is a precondition of an action.

two people to form a couple in order to break up, it is necessary to have food prior to eating it, and so on. If we provide the reasoning system with facts stating that such preconditions are necessary, it should be possible to replace these abductive inferences with deductive inferences.

This is special case of a more general idea. Recent research on the relationship between abduction and other forms of reasoning shows that there is a close relationship between abduction and an alternative deductive approach based on closure and minimization (Konolige, 1992). It is possible to translate abduction into the alternative approach by rewriting a logical theory and adding "closure statements," for example, statements to the effect that the known causes or preconditions are the only ones. This is obviously appropriate if the known preconditions are necessary and not just sufficient. However, it is not likely that all relevant preconditions, causes, desires, prospects, and judgments can be provided in advance.

The abductive approach is well suited to the domain of emotion-relevant reasoning because it does not require complete knowledge of causation, and causal closures need not be computed and asserted. The ability to generate hypotheses and make assumptions

- that implicit preconditions held in an effort to explain how an action led to an effect,
- that agents had certain desires in order to explain their emotional reactions or actions,
- that agents anticipated certain events in order to explain emotional reactions, and
- that actions are considered praiseworthy or blameworthy on the basis of emotional reactions to those actions

gives abduction advantages that are important for explanatory reasoning about emotions.

4.2 The Problem of Evaluating Explanations

The most important limitation of the method implemented here is that it does not address questions associated with evaluating the plausibility of multiple explanations. Such questions include the following:

How can one avoid a combinatorial explosion of explanations, many of which are completely implausible? Evaluation of plausibility cannot wait until all possible explanations have been produced. Sometimes infinitely many explanations are possible, so some evaluation must be done during explanation construction.

The machine-generated explanations we have presented here were generated using depth-first search. Most were the first acceptable explanations generated but, in some cases, the initial explanations were rejected by the user. Alternative explanations for a given example are often compatible but, in general, alternate explanations will include mutually inconsistent competitors. This raises the question of how one should decide what to believe and what to disbelieve when conflicts arise between alternate explanations? Methods for weighing the evidence would help decide which explanation is more plausible in many cases. But in other cases, it might be prudent to delay making a decision (Josephson, 1990). Or one could take some action aimed at acquiring new information that might help resolve the conflict.

Finally, how should one decide when to assume a hypothesis that would explain given observations? Currently, we rely on simple heuristics that specify inadmissible assumptions (see Section 2.4). After applying these heuristics, the abduction engine falls back on the user. The user is asked to validate each assumption and to accept each explanation. If the user rejects an assumption or explanation, backtracking occurs and the abduction engine seeks the next alternative.

4.3 Advantages of the Situation Calculus for Emotions

In this section, we list some desiderata for representations of theories of the cognitive structure of emotions. We show how the features of our situation calculus representation address goals relevant to representing and reasoning about emotion elicitation.

One of the most basic tenets of the theory of emotions sketched in Section 1.2 is the view that emotions are positively or negatively valenced reactions. Each emotion is paired with an emotion with a complementary valence. In addition, pairs of opposing extrinsic predicates play an important role in reasoning about emotions. For example, the opposing predicates *praiseworthy* and *blameworthy* play a crucial role in the emotion theory. Negations and opposites are important in the emotion-eliciting conditions. Our representation language provides support for these aspects of the theory by allowing for both positive and negative fluents and actions.

The situation calculus account for actions captures important causal information clearly needed in constructing explanations involving emotions. Situation calculus provides for a causal theory of actions that includes both positive and negative effects and preconditions. Emotion types (represented as fluents) are not caused directly by actions, in the sense that they do not appear as direct effects encoded in *causes* statements. Instead, they are caused indirectly; the theory specifies eliciting conditions that contain actions and other fluents. We saw several instances of actions causing effects that resulted in emotional reactions. In the first emotion example, Mary was happy to be at home. The fact that she was at home was caused by the fact that she went there. This is an instance of a positive effect of *ptrans*. We saw an example of a negative consequence of an action engendering an emotional reaction in the case of Mary's relief when T.C. vacated her home. Given that T.C. moved to another location, a negative effect was used to infer (and explain) the fact that T.C. was no longer at Mary's home.

Frame axioms capture the notion that fluents persist unless explicitly altered by actions. The emotion-eliciting conditions use this to advantage: They do not require fluents to be caused by the action most recently executed. Frame axioms are employed to propagate fluents caused by one action through successive actions into later states (provided they are not canceled by intervening actions). In the initial example, the reason Mary is at home is that she went there earlier and the fact that she ingested a meal did nothing to cancel this result.

Many examples of emotional reactions defined in the emotion-elicitation rules in terms of fluents are naturally expressed as responses to actions or other events.[14] Consider the example of *gloat*ing. John's unfortunate neighbor vomited. The ejection of contents of the neighbor's stomach through his mouth resulted in a relocation of said contents. But John's neighbor's distress was in reaction to his vomiting rather than its effect (the new location of the contents of his stomach). The function *did* provides us with fluents that enable us to refer to actions when we wish to focus on the action itself rather than on a specific effect of the action.

Chains of emotional reactions occur frequently. Our representation provides for such chains because emotions are represented as fluents that take fluents as arguments and because fluents appear in the eliciting conditions of emotions. Like other fortunes-of-others emotions, *happy_for* is an emotional chain reaction. Fortunes-of-others emotions are reactions to events, but instead of focusing exclusively on the event, they also focus on another's emotional reaction to that event. In the example of *happy_for*, the fact that Mary's roommate was going to Europe was not so important to Mary as her roommate's happiness. Mary's roommate's joy engenders Mary's joy.

Goals play a large role in emotion elicitation. Our representation supports reasoning about goals in constructing explanations involving emotions. (See Section 2.1.) A good example of reasoning about chains of desires occurs in a case involving John's gratification over a high score on the graduate record examination (GRE). In that example, the system is "told" that John wants to be enrolled in grad school. John's gratification is explained in terms of his desire to be admitted to graduate school. The rules for desires are used to infer that John wants to be admitted because this is a precondition of matriculation and matriculation results in achievement of enrollment.

---

[14] In fact (as discussed in Section 2.3) the original informal theory defined emotion types in terms of events rather than fluents.

Situation calculus provides support for temporal reasoning. A situation

$do\{a_n, do[a_{n-1}, \ldots, do(a_1, s_0)]\}$

defines a temporal sequence of situations, $s_0, s_1, \ldots s_n$ where for $i = 1$ to n, $s_i = do(a_i, s_{i-1})$ and s_{i-1} *precedes* s_i. Temporal precedence is used in eliciting conditions for the prospect-based emotions, *satisfied, fears_confirmed, relieved,* and *disappointed*. These emotions are reactions to the confirmation or disconfirmation of a hoped-for or feared fluent. The precedence constraints apply when the relevant fluents are hoped-for or feared in a situation prior to confirmation or disconfirmation. The compound emotions *grateful, angry_at, gratified,* and *remorseful* also employ temporal constraints. Each of these emotions is a reaction to an action and a fluent caused by the action. Two situations are relevant in these emotion types, the situation when the action causes the fluent, and the ensuing situation associated with the emotional reaction to the fluent. The eliciting conditions use the predicate *precedes* to ensure that the temporal precedence constraint between these situations is satisfied. We saw examples in the cases involving anger (Section 3.5), and relief (Section 3.4). In the situation prior to his leaving, Mary feared that T.C. would be at her home. She experienced relief after he left.[15].

Note that the situation calculus does not force a temporal ordering on all events. This is an advantage in the context of emotions. In the eliciting conditions of *happy_for,* there is no time constraint between the situations when the two agents are happy. In the case of *happy_for* (Section 3.2), Mary's roommate's emotional reaction and Mary's reaction are allowed to occur in different situations. Using separate situations is useful, for example, if some intervening action results in someone being informed of another's earlier good fortune. Avoiding temporal contraints on the situations is also useful because in some cases the usual temporal order is reversed. For example, one might be happy for another in anticipation of the other's happiness (e.g., upon learning that the other won a lottery even before the lucky winner knew it).

4.4 Limitations of the Situation Calculus of Emotion Elicitation

In this section, we discuss the main limitations of the situation calculus of emotion elicitation. These include limitations inherent in the situation calculus itself and limitations in the theory of emotion elicitation.

A major limitation of this study is that we did not attempt to represent or reason about intensities of emotions. It is important to extend the represen-

---

[15] As discussed in Section 2.3, in some cases of relief and disappointment, the attendant fears and hopes violate the constraint requiring them to occur prior to the relief and disappointment. Additional eliciting conditions allow for this, but because it is the exception rather than the rule, priority is given to the interpretation that includes the temporal constraint.

tation presented here to include intensities. Many emotion types are represented in natural language by a number of emotion tokens. Many tokens indicate specific relative intensities of a particular emotion type; for example, "annoyance," "exasperation," and "rage" denote increasingly intense subtypes of anger. Ortony et al., (1988) suggested that emotions only occur when their intensities are driven above thresholds. The approach taken here is to use predicates that are true or false in place of these continuous, real-valued variables. This approach may be viewed as a crude first approximation. We speculate that methods developed in AI research on qualitative reasoning about physical systems (e.g., Forbus, 1984; Kuipers, 1986; Weld & de Kleer, 1990) could be applied to the problem of representing and reasoning about intensities of emotions.

Another limitation of the emotion-eliciting conditions is that they are phrased in terms of facts rather than beliefs. This is because we wanted to avoid having to reason about beliefs and knowledge. But such reasoning is clearly relevant to emotions. Consider the eliciting condition for satisfaction. It states that a person is satisfied if the person hoped for a fluent earlier and now it holds. It seems clear that the eliciting condition is too simple. It should be phrased in terms of the person's epistemic state. For example, the rule for satisfaction might be rewritten: A person may be satisfied if that person *believes* that some hoped for fluent holds. Logics of belief and knowledge should help address such issues, but they are beset with their own complexities (e.g., referential opacity, computational intractability) and they might introduce more problems than solutions.[16]

The most important difficulty for the situation calculus underlying our representation is the famous qualification problem (McCarthy, 1977). The problem is that it is difficult, if not impossible, to specify all the preconditions relevant to the successful execution of an action. We do allow assumptions about the possibility of actions, which means that we can explain how an action occurred without knowing all the preconditions that might have made it possible. At present, we do not attempt explanations of inaction, so we do not have to deal with the difficult problem of inferring preconditions that failed, thus preventing an action from occurring. We do not claim to have solved the qualification problem but we believe our representation and reasoning methods are no more limited by it than are other approaches.

Some examples in the diary study are beyond the scope of our current methods because they require reasoning about actions not taken and the resulting negative effects. In one case, a woman's roommate fails to pay their phone bill. This triggers anger and fear. She is afraid that she will get a bad credit rating and that her phone will be disconnected. In another example, a student expresses anger because his mother failed to send his records

---

[16] For examples of AI approaches to the difficult problems of reasoning about belief, knowledge, and action, see Konolige (1985) and Moore (1985).

to a dentist and he can't get his teeth cleaned without them. Several cases involve students worrying about poor grades caused by not studying for examinations. The general principle in these examples is that the failure to do an action can often be said to be the reason that an effect does not hold. If we could capture this intuition in a "causal law of non-action" we would have something of direct importance for the attribution emotions because they often involve attributing praiseworthiness or blameworthiness to non-action. Such a law would also have indirect impacts, by combining with existing laws such as the rules for indirect goals, for example, to capture the idea that one may not want an action to be done if it causes an undesirable effect. The current situation calculus does not cover such cases because it says little about negative actions.

Time is important in reasoning about many emotions, especially the prospect-based emotions *hope, satisfaction, relief, fear, fears-confirmed,* and *disappointment.* Our situation calculus deals with temporal precedence but ignores all other temporal relationships such as simultaneity. In the situation calculus, information is stored in an initial state and then propagated to later states via frame axioms. This is a problem in reasoning about emotional reactions to ongoing events. For example, a woman can be grateful to her husband for giving her a massage while he is giving it to her, but limitations of the situation calculus prevent a formulation of the eliciting conditions for gratitude during an ongoing action. The use of representation and reasoning methods suggested by Allen (1981) might help address this limitation.

Actions are viewed as discrete, opaque transitions. Situation calculus provides no tools for describing what happens while an action is in progress; it provides no tools for describing continuous processes like the gradual dissipation of anger. Again, representation and reasoning techniques developed in research on qualitative physics (e.g., Bobrow, 1985; Hobbs & Moore, 1985) might help overcome this limitation.

Strict logical implication often fails to capture the reality of relationships among events, actions, and possible effects. Many contributions of actions toward the achievement of goals involved in examples drawn from the diary study are uncertain in the sense that the action is not guaranteed to achieve the goal. Often, actions facilitate or increase the probability that the goal will be achieved. Several examples in the case study data describe student's emotional reactions to the granting of extensions on due dates of assignments. Besides temporal reasoning, these examples require probabilistic reasoning, in the sense that the granting of extensions increases the likelihood of successful completion of projects. Instead of encoding these weak causal relationships as implications, qualitative representations of conditional probabilities could be associated with cause–effect relationships. Plausibility information such as probabilities could enter more directly into the eliciting conditions of several emotions. In particular, the *hopes* and *fears* emotion types depend on entertaining the possibility that the hoped-for or feared fluent will occur. The intensity of hope or fear (and its plausibility) depends in part on the subjective likelihood of the prospective event (Ortony et al., 1988).

5. RELATED AND FUTURE WORK

A similar approach to formalizing commonsense reasoning about emotions is presented in Sanders (1989). However, that work takes a deductive approach, using a deontic logic of emotions. The logic focuses on a cluster of emotions involving evaluations of actions—including what we have called admiration, reproach, remorse, and anger. The evaluation of actions is ethical, and involves reasoning about obligation, prohibition, and permission. The logic was used to solve problems involving actions associated with ownership and possession of property (e.g., giving, lending, buying, and stealing) by proving theorems. For example, the fact that Jack will be angry was proved given the following:

Jack went to the supermarket.
He parked his car in a legal parking place.
When he came out, it was gone.

It is not clear whetheer the theorems were proved automatically or by hand, so questions of complexity of inference and control of search in the deontic logic remain unanswered. We have argued that abduction offers advantages over deduction alone when applied to the task of constructing explanations involving emotions. Furthermore, our situation calculus of emotion elicitation is more comprehensive than the deontic logic for emotions in that it covers more emotion types. But our approach could benefit from the treatment of ethical evaluations. A detailed comparison and integration of the best parts of the two approaches would be worthwhile.

A number of theories of the cognitive determinants of emotions exist (e.g., Roseman, 1984). In principle, situation calculus could be used to codify these alternative theories. The abductive method we propose for explanatory reasoning about emotions does not depend on the particular emotion theory used.

Recent research on abduction addresses the issues of search control and plausibility mentioned earlier. Stickel (Hobbs, Stickel, Appelt, & Martin, 1993; Hobbs, Stickel, Martin, & Edwards, 1988) has suggested a heuristic approach for evaluating explanations in the context of natural language processing. Subsequent work by Charniak and Shimony (1990) gave Stickel's weighted abduction a probabilistic semantics. Still more recent work (Poole,

1991) incorporates probability directly into a logic-based approach to abduction. These methods promise to provide significantly improved abduction engines that could be used to construct explanations involving emotions, taking plausibility into account.

Related work on natural language comprehension (Dyer, 1983a, 1983b; Lehnert, 1981) argues that emotion words occur frequently in natural language discourse because emotions play a substantial role in our lives. Natural language systems encountering text involving emotions will need to identify and reason about emotions felt by characters in the text. Important subtasks involved in comprehension such as motivation analysis and plan recognition will often require reasoning about emotions (Charniak & McDermott, 1986).

This work focuses on explaining emotions in terms of eliciting situations. But, although situations give rise to emotional reactions, emotions in turn give rise to goals and actions that change the state of the world. Applications such as plan recognition will require the representation of knowledge for causal connections between emotions and subsequent actions. For a brief description of a system for recognizing plans involving emotions, see Cain, O'Rorke, and Ortony (1989). Cain et al. also described how explanation-based learning techniques can be used to learn to recognize such plans. For a fuller discussion of reasoning about emotion-induced actions, see Elliott and Ortony (1992).

This work is based upon a collection of 22 emotion types. In Ortony, Clore, and Foss (1987) about 270 English words are identified as referring to genuine emotions from an initial pool of 600 words that frequently appear in the emotion research literature. In another study, 130 of these emotion words were distributed among the 22 emotion types discussed here. Some emotion words map to several different types, for example, "upset" is compatible with at least *angry_at, distress,* and *shame.* In addition, many words map to the same type. Encoding the relationship between the affective lexicon and the emotion types is an important topic for future research aimed at automatically processing natural language text involving emotions.

Lehnert (1981) argued that it is important to embed a proposed method for representing and reasoning about affect into a larger information-processing system so that the method can be evaluated in terms of the effectiveness of the larger system. We have done this, in a limited sense, by embedding our situation calculus of emotion elicitation into an abductive reasoning system. However, it is still desirable to incorporate our method into a narrative understanding system comparable to BORIS (Dyer, 1983a). One way to do this would be to embed the system we describe here in TACITUS, a natural language processing system that uses abduction as the basis for comprehension (Hobbs et al., 1993).

6. CONCLUSION

We have presented a representation of a theory of the cognitive antecedents of emotions and we have described an abductive method for explaining emotional states. We sketched a computer program, an abduction engine implemented in a program called AbMaL, that uses the theory of emotion elicitation to construct explanations of emotions. We presented explanations of examples based on cases taken from a diary study of emotions. We examined the strengths and weaknesses of both the representation of knowledge of eliciting conditions and the method for constructing explanations.

The most important advantage of our approach to explanatory reasoning about emotions is that abduction allows us to construct explanations by generating hypotheses that fills gaps in the knowledge associated with cases where deduction fails. We found that the majority of the diary study examples could not be explained using deduction alone because they do not follow logically from the given facts. The abduction engine explained the emotions involved in these cases by making assumptions including valuable inferences about mental states such as desires, expectations, and the emotions of others.

REFERENCES

Allen, J.F. (1981). Maintaining knowledge about temporal intervals. *Proceedings of the International Joint Conference on Artificial Intelligence* (pp. 221-226). San Francisco, CA: Morgan Kaufman.

Birbaum, L. (1991). Rigor mortis: A response to Nilsson's "Logic and artificial intelligence." *Artificial Intelligence, 47,* 57-77.

Bobrow, D.G. (Ed.). (1985). *Qualitative reasoning about physical systems.* Cambridge, MA: MIT Press.

Cain, T., O'Rorke, P., & Ortony, A. (1989). Learning to recognize plans involving affect. In A.M. Segre (Ed.), *Proceedings of the Sixth International Workshop on Machine Learning* (pp. 209-211). San Francisco, CA: Morgan Kaufmann.

Charniak, E. (1987). Logic and explanation. *Computational Intelligence, 3,* 172-174.

Charniak, E., & McDermott, D.V. (1986). *Introduction to artificial intelligence.* Reading, MA: Addison-Wesley.

Charniak, E., & Shimony, S.E. (1990). Probabilistic semantics for cost-based abduction. *The Eighth National Conference on Artificial Intelligence* (pp. 106-111). Cambridge, MA: AAAI Press/MIT Press.

Davis, E. (1990). *Representations of commonsense knowledge.* San Francisco, CA: Morgan Kaufmann.

DeJong, G.F., & Mooney, R. (1986). Explanation-based learning: An alternative view. *Machine Learning, 1,* 145-176.

Dyer, M.G. (1983a). *In-depth understanding: A computer model of integrated processing for narrative comprehension.* Cambridge, MA: MIT Press.

Dyer, M.G. (1983b). The role of affect in narratives. *Cognitive Science, 7,* 211-242.

Elliott, C., & Ortony, A. (1992). *Point of view: Reasoning about the concerns of others.* In J. Kruschke (Ed.), *Proceedings of the Fourteenth Annual Conference of the Cognitive Science Society.* Hillsdale, NJ: Erlbaum.

Fikes, R.E., & Nilsson, N.J. (1971). STRIPS: A new approach to the application of theorem proving to problem solving. *Artificial Intelligence, 2,* 189–208.

Forbus, K.D. (1984). Qualitative process theory. *Artificial Intelligence, 24,* 85–168.

Geneserth, M.R., & Nilsson, N.J. (1987). *Logical foundations of artificial intelligence.* San Francisco, CA: Morgan Kaufmann.

Green, C.C. (1969). Applications of therem proving to problem solving. In D.E. Walker & L.M. Norton (Ed.), *International Joint Conference on Artificial Intelligence.* San Francisco, CA: Morgan Kaufmann.

Heider, F. (1958). *The psychology of interpersonal relations.* Hillsdale, NJ: Erlbaum.

Hobbs, J.R., & Moore, R.C. (Eds.). (1985). *Formal theories of the commonsense world.* Norwood, NJ: Ablex.

Hobbs, J.R., Stickel, M.E., Appelt, D.E., & Martin, P. (1993). Interpretation as abduction. *Artificial Intelligence, 63,* 69–142.

Hobbs, J.R., Stickel, M.E., Martin, P., & Edwards, D. (1988). Interpretation as abduction. *Proceedings of the 26th Annual Meeting of the Association for Computational Linguistics* (pp. 95–103).

Josephson, J.R. (1990). On the "logical form" of abduction. *Working notes of the AAAI spring symposium on automated abduction.* University of California, Irvine, Department of Information and Computer Science.

Konolige, K. (1985). Belief and incompleteness. In J.R. Hobbs, & R.C. Moore (Eds.), *Formal theories of the commonsense world.* Norwood, NJ: Ablex.

Konolige, K. (1992). Abduction vs. closure in causal theories. *Artificial Intelligence, 53,* 255–272.

Kowalski, R. (1979). *Logic for problem solving.* New York: Elsevier.

Kuipers, B.J. (1986). Qualitative simulation. *Artificial Intelligence, 29,* 289–338.

Lehnert, W.G. (1981). Affect and memory representation. *Proceedings of the Third Annual Conference of the Cognitive Science Society* (pp. 78–82).

McCarthy, J. (1968). Programs with common sense. In M. Minsky (Ed.), *Semantic information processing.* Cambridge, MA: MIT Press.

McCarthy, J. (1977). Epistemological problems of artificial intelligence. *Proceedings of the Fifth International Joint Conference on Artificial Intelligence* (pp. 1038–1044). San Francisco, CA: Morgan Kaufmann.

McCarthy, J., & Hayes, P. (1969). Some philosophical problems from the standpoint of artificial intelligence. In B. Meltzer & D. Michie (Eds.), *Machine intelligence* (Vol. 4). Edinburgh, Scotland: Edinburgh University Press.

McDermott, D. (1987). A critique of pure reason. *Computational Intelligence, 3,* 151–160.

Mitchell, T.M., Keller, R.M., & Kedar-Cabelli, S.T. (1986). Explanation-based generalization: A unifying view. *Machine Learning, 1,* 47–80.

Moore, R.C. (1985). A formal theory of knowledge and action. In J.R. Hobbs & R.C. Moore (Eds.), *Formal theories of the commonsense world.* Norwood, NJ: Ablex.

Nilsson, N.J. (1991). Logic and artificial intelligence. *Artificial Intelligence, 47,* 31–56.

O'Rorke, P. (in press). Abduction and explanation-based learning: Case studies in diverse domains. *Computational Intelligence, 10.*

Ortony, A., Clore, G.L., & Collins, A. (1988). *The cognitive structure of emotions.* New York: Cambridge University Press.

Ortony, A., Clore, G.L., & Foss, M.A. (1987). The referential structure of the affective lexicon. *Cognitive Science, 11,* 361–384.

Peirce, C.S.S. (1931–1958). *Collected papers of Charles Sanders Peirce (1839–1914).* Cambridge, MA: Harvard University Press.

Peng, Y., & Reggia, J.A. (1990). *Abductive inference models for diagnostic problem solving.* New York: Springer-Verlag.

Poole, D. (1991). Representing diagnostic knowledge for probabilistic horn abduction. *Proceedings of the 12th International Joint Conference on Artificial Intelligence* (pp. 1129-1135). San Francisco, CA: Morgan Kaufmann.

Poole, D.L., Goebel, R., & Aleliunas. R. (1987). Theorist: A logical reasoning system for defaults and diagnosis. In N. Cercone & G. McCalla (Eds.), *The knowledge frontier: Essays in the representation of knowledge.* New York: Springer-Verlag.

Pople, H.E. (1973). On the mechanization of abductive logic. *Proceedings of the Third International Joint Conference on Artificial Intelligence* (pp. 147-152). San Francisco, CA: Morgan Kaufmann.

Reiter, R. (1991). The frame problem in the situation calculus: A simple solution (sometimes) and a completeness result for goal regression. In V. Lifschitz (Eds.), *Artificial intelligence, and mathematical theory of computation: Papers in honor of John McCarthy.* San Diego, CA: Academic.

Roseman, I.J. (1984). Emotions, relationships, and health. In P. Shaver (Eds.), *Review of personality and social psychology* (Vol. 5). Beverly Hills, CA: Sage.

Sanders, K.E. (1989). A logic for emotions: A basic for reasoning about commonsense psychological knowledge. In E. Smith (Ed.), *Proceedings of the Annual Meeting of the Cognitive Science Society.* Hillsdale, NJ: Erlbaum.

Schank, R.C. (1972). Conceptual dependency: A theory of natural language understanding. *Cognitive Psychology, 3,* 552-631.

Turner, T.J. (1985). [Diary study of emotions: Qualitative data]. Unpublished raw data.

Weld, D.S., & de Kleer, J. (Ed.). (1990). *Readings in qualitative reasoning about physical systems.* San Francisco, CA: Morgan Kaufmann.

Woods, W.A. (1975). What's in a link: Foundations for semantic networks. In D.G. Bobrow & A.M. Collins (Eds.), *Representation and understanding: Studies in cognitive science.* New York: Academic. (Also appears in *Readings in Knowledge Representation* (1985), R.J. Brachman & H.J. Levesque (Eds.). San Francisco, CA: Morgan Kaufmann.

COGNITION AND EMOTION, 1987, *1* (3) 217–233

Motives, Mechanisms, and Emotions

Aaron Sloman

School of Cognitive Sciences, University of Sussex, U.K.

Intelligent animals are solutions to a design problem posed by: the varying requirements of individuals, the more permanent requirements of species and social groups, the constraints of the environment, and the available biological mechanisms. Analysis of this design problem, especially the implications of limited knowledge and a continuous flow of information in a rapidly changing environment, leads to a theory of how new motives are processed in an intelligent system. The need for speed leads to architectures and algorithms that are fallible in ways that explain why intelligent agents are susceptible to emotions and errors. This holds also for intelligent robots. A study of such mechanisms and processes is a step towards a computational theory of emotions, attitudes, moods, character traits, and other aspects of mind so far not studied in Artificial Intelligence. In particular, it turns out that no special emotional subsystem is required. This framework clarifies and refines ordinary concepts of mental processes, and suggests a computational approach to psychotherapy.

INTRODUCTION

Ordinary language makes rich and subtle distinctions between different sorts of mental states and processes such as mood, emotion, attitude, motive, character, personality, and so on. Our words and concepts have been honed for centuries against the intricacies of real life under pressure of real needs and therefore give deep hints about the human mind.

Requests for reprints should be made to Aaron Sloman, School of Cognitive Sciences, Arts Building D, University of Sussex, Brighton, BN1 9QN. This work is supported by a fellowship from the GEC Research Laboratories and a grant from the Rennaisance Trust. Some of the ideas derive from Simon's seminal paper. My ideas have benefited from discussions over several years with Monica Croucher and her thesis (Croucher, 1985) developed the ideas of this paper in greater depth. The view of ordinary language as a source of information about the human mind, sketched here and expanded in Chapter 4 of Sloman, 1978, owes much to Austin (1961). Keith Oatley's editorial comments have been most helpful.

Yet actual usage is inconsistent, and our ability to articulate the distinctions we grasp and use intuitively is as limited as our ability to recite rules of English syntax. Words like "motive" and "emotion" are used in ambiguous and inconsistent ways. The same person will tell you that love is an emotion, that she loves her children deeply, and that she is not in an emotional state. Many inconsistencies can be explained away if we rephrase the claims using carefully defined terms. As scientists we need to extend colloquial language with theoretically grounded terminology that can be used to mark distinctions and describe possibilities not normally discerned by the populace. For instance, it will be seen that love is an attitude, not an emotion, although deep love can easily trigger emotional states. In the jargon of philosophers (Ryle, 1949), attitudes are dispositions, emotions are episodes, although with dispositional elements.

For a full account of these episodes and dispositions we require a theory about how mental states are generated and controlled, and how they lead to action—a theory about the mechanisms of mind. The theory should explain how internal representations are built up, stored, compared, and used to make inferences, formulate plans or control actions. Outlines of a theory are given later. Design constraints for intelligent animals or machines are sketched, then design solutions are related to the structure of human motivation and to computational mechanisms underlying familiar emotional states.

Emotions are analysed as states in which powerful motives respond to relevant beliefs by triggering mechanisms required by resource-limited intelligent systems. New thoughts and motives get through various filters and tend to disturb other ongoing activities. The effects may interfere with, or modify the operation of other mental and physical processes, sometimes fruitfully sometimes not. These are states of being "moved". Physiological changes need not be involved. Emotions contrast subtly with related states and processes such as feeling, impulse, mood, attitude, temperament; but there is no space for a full discussion here.

On this view we need posit no special subsystem to account for emotions since mechanisms underlying intelligence suffice (cf. Oatley & Johnson-Laird, 1985). If emotional states arise from mechanisms required for coping intelligently in a complex and rapidly changing world, this challenges the common separation of emotion and cognition. This applies equally to human beings, other animals, or intelligent machines to come.

DESIGN CONSTRAINTS FOR A MIND

The enormous variety of animal behaviours indicates that there are different ways of designing agents that take in information from the environment and are able to act on it individually or co-operatively. Human beings merely occupy one corner of this "space of possible minds". Elsewhere I have sketched constraints determining the design solutions embodied in the human mind. I have space only to summarise the key results relevant to emotions.

Constraints include: a multiplicity of internal and external sources of motivation (often inconsistent), speed limitations, inevitable gaps and errors in beliefs about the environment and varying degrees of urgency associated with motives. Resource limits and urgency render inevitable the use of potentially unreliable "rule-of-thumb" strategies. Unpredictability of new information and new goals implies a need to be able to interrupt, modify, suspend, or abort ongoing activities, whether external or internal. This includes such things as hardware and software "reflex" actions, some of which should be modifiable in the light of experience.

Reflexes are inherently fast-acting but stupid. They may be partly controlled by context-sensitive filters using rules-of-thumb to assess priorities rapidly and allow extremely important, urgent, or dangerous ongoing activities to proceed without disturbance while allowing new, specially important, or urgent, motives to interrupt them. (The "insist-ence" of a motive is defined later in terms of its ability to get past such filters.) A major conclusion is that intelligent systems will have fast but stupid subsystems, including filters for new motives. Fast, dumb filters will sometimes let in undesirables.

Incomplete information and the need to cope with long-term change in the social or physical environment require higher order sources of action that provide learning: Not only generators and comparators of motives, but generators and comparators for the generators and comparators them-selves.

Although several independent subsystems can execute plans in parallel, like eating and walking, conflicts among requirements can generate incompatible goals, necessitating a decision-making mechanism. The two main options are a "democratic" voting scheme, and a centralised decision-maker. If subsystems do not all have access to the full store of available information or not all have equal reasoning powers, a "democra-tic" organisation may be dangerous. Instead, a specialised central mechan-ism is required for major decisions (Sloman, 1978, Chs. 6 and 10). This seems to be how normal human minds are organised.

Similar constraints determine the design of intelligent artefacts. Physical limitations of biological or artificial computing equipment necessitate major divisions of functions, including the allocation of the highest level control to a part with access to most information and the most powerful inference mechanisms. However, an occasional urgent need for drastic action requires overriding hardware or software reflexes that operate independently of higher level control—a mechanism enabling emotional processes described in the following sections.

Goal Generators

Many different sorts of motivators are pointed to by ordinary words and phrases such as:

aims, attitudes, desires, dislikes, goals, hates, hopes, ideals, impulses, likes, loves, preferences, principles, alluring, amusing, bitter, boring, charming, cheering, depressing, distressing.

There are many more. They mark subtle distinctions between different springs of action and the various ways things affect us. Conceptual analysis (Sloman, 1978, Ch. 4) brings out their presuppositions. A key concept is having a goal.

To a first approximation, to have a goal is to use a symbolic structure, represented in some formalism, to describe a state of affairs to be produced, preserved, or prevented. The symbols need not be physical structures: Virtual formalisms will do as well (Sloman, 1984). Goals can use exactly the same descriptive formalism as beliefs and hypotheses. The difference is solely in the roles they play.

A representation of a state of affairs functions as a goal if it tends (subject to many qualifications) to produce behaviour that changes reality to conform to the representation.

A representation functions as a belief if it is produced or modified by perceptual and reasoning processes which tend (subject also to many qualifications) to alter the representations to conform to reality.

(What "conform" means here cannot be explained without a lengthy digression.) The same representations may also be used in other roles, as instructions, hypothesised situations, rules, etc.

Some new goals subserve a prior goal and are generated by planning processes. Some are responses to new information, such as wanting to know what caused the loud noise around the corner. Goals are not triggered only by external events: A thought, inference, or recollection may have the same effect.

How can a goal be produced by a belief or thought? If goals involve symbolic structures, a computational explanation might be that a "goal-generating" condition-action rule is used. For example, a benevolence rule might be: "If X is in distress generate the goal [X is not distressed]". A retribution goal-generator underlying anger might be "If X harms me generate the goal [X suffers]". A full analysis would describe various "goal generators", "goal generator generators", and so on. A learning system would produce new goal generators in the light of experience, using generator generators.

Goal Comparators

Generators do not always produce consistent goals. Different design constraints lead to different co-existing goal generators. Social animals or machines need goal generators that produce goals for the benefit of others, and these can conflict with the individual's own goals and needs. Goal comparators are therefore needed for selection between different ends.

Some comparators apply constraint goals in planning, for instance using a "minimise cost" rule to select the cheaper of two subgoals. Others directly order ends, like a rule that saving life is always more important than any other goal, but not because of some common measure applicable to both. As there are different incommensurable sources of motivation and different bases for comparison, there need not be any *optimal* resolution of a conflict.

Higher Order Motivators

Despite possibly confusing colloquial connotations, the general term "motivator" is used to refer to mechanisms and representations that tend to produce, or modify, or select between actions, in the light of beliefs. Motivators recursively include generators and comparators of motivators. Some motivators are transient, like the goal of picking up a particular piece of cake, while others are long term, like an ambition to be slim.

Motivators should not be static—motivator generators are required for flexible production of new goals. Still higher intelligence involves the ability to learn from experience and modify the generators. Thus the requirement for generators is recursive. The same applies to comparators: If two generators, regularly come into conflict by generating conflicting goals, then it may be necessary to suppress or modify one of them. This requires a generator comparator. Comparator generators and comparator comparators are also needed. Higher order generators and comparators account for some personality differences. Their effects account for some of the subtleties of emotional states.

Theoretical research is needed to design generally useful higher level generators and comparators. Empirical research is needed to establish what the mechanisms are in people. Do we have a limit to levels of generators and comparators or can new levels be recursively generated indefinitely?

Varieties of Motivators

"Derivative" and "nonderivative" motivators can be distinguished. Roughly, a motivator is derivative if it is explicitly derived from another motivator by means–ends analysis and this origin is recorded and plays a role in subsequent processing. A desire to drink when thirsty would be nonderivative, whereas a desire for money to buy the drink would be derivative. A motive can be partly derivative, partly nonderivative, like a desire to quench one's thirst with whisky to impress others. An attempt will be made to show how nonderivative motivators are central to emotional states.

The distinction has behavioural implications. Derivative goals are more readily abandoned and their abandonment has fewer side-effects, e.g. if they appear to be unattainable or if the goals from which they are derived are satisfied or abandoned. An unpromising derivative goal can easily be replaced by another if it serves the supergoal as well. If a nonderivative goal is abandoned because it conflicts with something regarded as more important it may continue to demand attention—one source of emotions. Abandonment will produce regret and a disposition to revive the goal if the inconsistency can be removed.

Human nonderivative goals include bodily needs, desire for approval, curiosity, aesthetic wishes, and the desire to succeed in tasks undertaken. Because these goals serve more general biological purposes some theorists regard them as derivative. However, the mechanisms that create a goal need not explicitly associate it with higher level goals, but simply give it the causal power to produce planning and action, for instance, by simply inserting a representation of the goal in a database whose contents constantly drive the system. Despite its implicit function such a goal is nonderivative for the individual.

Quantitative Dimensions of Variation

Motives can be compared on different dimensions, definable in terms of the mechanism sketched above. *Insistence* of a motive is its interrupt priority level. Insistent desires, pains, fears, etc., are those that more easily get through interrupt filters, depending on the threshold set in relation to current activities.

Goals that get through filters need to be compared to assess their relative *importance*. This (sometimes partial) ordering is determined by beliefs and comparators, and may change if they do. Complex inferences may be required. Importance of a derivative goal is linked to beliefs about effects of achieving or not achieving it. Insistence concerns how likely a goal is to get through the interrupt filter in order to be considered, whereas importance concerns how likely it is to be adopted as something to be achieved if considered. Insistence has to be assessed very quickly, and should correlate with importance but sometimes will not. A bad filter will assign low priorities to important goals, and vice versa. A desire to sneeze does not go away just because silence is essential for survival. (Not all animals have such complex motivational systems.)

Urgency is a measure of how much time is left before it is too late. This is not the same as insistence or importance: Something not wanted very much may be urgent, and vice versa.

Intensity of a goal determines how actively or vigorously it is pursued if adopted. It is partly related to urgency and importance, and partly independent of them. Obstacles to an intense goal tend to be treated as a challenge rather than a reason to abandon the goal. Often a long-term important goal will lose out to something much less important but more intense—the age-old conflict between desire and duty. Ideally insistence, intensity, and importance should be correlated, but the relationship can be upset by interactions with urgency and the way reflex assignments of insistence or intensity derive from prior experience or evolutionary origins.

Another measure of a motive is how distressing or disruptive failure to achieve it is. Different again is how much pleasure is derived from fulfilment. This can be assessed by how much effort tends to go into preserving the state of fulfilment, or achieving a repetition at a later time. Both are normally expressed as how much someone "cares", and relate to the potential to generate emotional states as described later.

These different kinds of "strength" of motives all play a role in cognitive functioning and may be needed in sophisticated robots. They can have subjective correlates in a system with self-monitoring, although self-monitoring is not always totally reliable. Objectively they are defined in terms of dispositions to produce effects or resist changes of various sorts, internal or external. Different combinations of strengths will affect what happens at various stages in the evolution of a goal, from initial conception to achievement, abandonment, or failure.

SUMMARY OF PROCESSES INVOLVING MOTIVES

So far the following intermediate processes through which motives may go have been sketched:

1. Initiation—by a body monitor, motive generator, or planner creating a new subgoal.
2. Reflex prioritisation of a new goal—assigning insistence.
3. Suppression or transmission by the interrupt filter.

4. Triggering a reflex action (internal or external, hardware or software).

5. Evaluation of relative importance, using comparators.

6. Adoption, rejection, or deferred consideration—adopted motives are generally called "intentions". Desires may persist as desires, although not adopted for action.

7. Planning—"intrinsic" planning is concerned with how to achieve the goal, "extrinsic" planning with when, and how to relate it to other activities, e.g. should it be postponed?

8. Activation—starting to achieve the motive, or re-activating temporarily suspended motives.

9. Plan execution.

10. Interruptions—abandonment or suspension.

11. Comparison with new goals.

12. Plan or action modification in the light of new information or goals, including changes of speed, style, or subgoals.

13. Satisfaction (complete or partial).

14. Frustration or violation.

15. Internal monitoring (self-awareness).

16. Learning—modification of generators and comparators in the light of experience.

These are all computational processes, capable of being expressed in terms of rule-governed manipulation of representations of various sorts, although filling out the details is not a trivial task. I will now try to indicate how they relate to emotions. The full story is very complicated.

An Example: Anger.   What is meant by "X is angry with Y". This implies that X believes that there is something Y did or failed to do and as a result one of X's motives has been violated. This combination of belief and motive does not suffice for anger, since X might merely regret what happened or be disappointed in Y, without being angry. Anger also requires a new motive in X: A desire to hurt or harm Y. Most people and many animals seem to have retributive motive generators that react like this, alas. The new motive is not necessarily selected for action however intense it may be: Fear of consequences and appropriate comparators may keep it inoperative.

Production of the new desire is still not sufficient for anger. X may have the desire, yet put it out of mind and calmly get on with something else: In that case he is not angry. Alternatively, X's desire to do something unpleasant to Y may be entirely derivative: Purely a practical measure to reduce the likelihood of future occurrence, without any ill-will felt towards

Y. Then if X can be assured somehow that there will be no recurrence, the motive will be dropped. That is not anger.

Anger involves an *insistent* and *intense* nonderivative desire to do something to make Y suffer. High insistence means the desire frequently gets through X's filters to "request attention" from X's decision-making processes. So even after rejection by comparators, the desire frequently comes back into X's thoughts, making it hard for him to concentrate on other activities. Filters designed for speed can be too stupid to reject motives already ruled out by higher levels. Moreover, the desire must not be derivative, that is a subgoal that will disappear if a supergoal is removed. In socially sophisticated agents, anger may include a belief that Y's action had no social or ethical justification.

So emotions are states produced by motivators, and involve production of new motivators.

The violation of the original motive, and the insistence of the new motive, may be associated with additional secondary effects. For example, if X becomes aware of his anger this can make him annoyed with himself. If other people perceive his state, this can also affect the nature of the emotion. The episode can revive memories of other situations which enhance the anger.

Sometimes, in human beings, emotional states produce physiological disturbances too, probably as a result of the operation of physical and chemical reflexes driven by "rule-of-thumb" strategies, as suggested earlier. However, if X satisfied enough of the other conditions he could rightly be described as very angry, even without any physical symptoms. Strong anger can exist without any physical side-effects in so far as it constantly intrudes into X's thoughts and decisions, and in so far as he strongly desires to make Y suffer, and suffer a great deal. Although nonphysical anger might be called "cold", it would still have all the socially significant aspects of anger.

Anger is partly dispositional in that it need not *actually* interfere with other motives: For instance, if the new motive to punish Y is acted on, there need be no further disturbance. However, the anger has the *potential* to disturb other activities if the new motive has high insistence.

Anger is sometimes *felt*, as a result of self-monitoring. However, it is possible to be angry, or be in other emotional states, without being aware of the fact. For example, I suspect that dogs and very young children are unaware of their anger (although very much aware of whatever provoked it).

Emotions like anger can vary along different quantitative and qualitative dimensions, such as: How certain X is about what Y has done, how much X cares about it (i.e. how important and intense the violated motive is); how much harm X wishes to do to Y; how important this new desire is, how intense it is, how insistent it is, how long lasting it is; how much mental disturbance is produced in X; how much physiological disturbance there is; which aspects of the state X is aware of; how many secondary motives and actions are generated. Different dimensions will be appropriate to different emotions.

Variations at different stages of the scenario correspond to different states, some not emotional. When there is no desire to cause harm to Y, the emotion is more like exasperation than anger. If there is no attribution of responsibility, then the emotion is simply some form of annoyance, and if the motive that is violated is very important, and cannot readily be satisfied by some alternative, then the emotion involves dismay. Because arbitrarily many motives, beliefs, and motive generators can be involved, with new reactions triggered by the effects of old ones, the range of variation covered by this theory is bound to be richer than the set of labels in ordinary language. It will also be richer than the range of physiological responses.

TOWARDS A GENERATIVE GRAMMAR FOR EMOTIONS

Analysing anger and other emotions in the light of the mechanisms sketched above, suggests the following components of emotional states:

1. There is at least one initiating motive, M1, with a high level of importance and intensity.

2. A belief, B1, about real or imagined or expected satisfaction or violation of M1 triggers generators of various kinds, often producing new motives.

3. Different sorts of cases depend on; (a) whether M1 is concerned with something desired or disliked; (b) whether B1 is a belief about M1 being satisfied or violated; (c) whether B1 concerns past, present, or future; (d) whether B1 involves uncertainty or not; (e) whether the agent is aware of his emotion or not; (f) whether other agents are thought to be involved or not; (g) whether M1 is concerned with how other agents view one (cf. Roseman, 1979).

4. In more complex situations several motives simultaneously interact with beliefs, e.g. a situation where B1 implies that important motives M1(a) and M1(b) are inconsistent, e.g. in dilemmas.

5. Sometimes M1 and B1 trigger a generator that produces a secondary motive, M2, for instance a desire to put things right, preserve a delight, punish a perpetrator, or inform others. This in turn can interact with other

beliefs, to disturb, interrupt, or otherwise affect cognitive processes. This would be a "two-level" emotional state. Several levels are possible.

6. Sometimes M1 and B1 trigger several motive generators simultaneously. The resulting interactions can be very complex especially when new motives are in conflict, e.g. a desire to undo the damage and to catch the culprit.

7. Sometimes the newly generated motives conflict with previously existing motives.

8. New motives with high insistence get through interrupt filters and tend to produce (although they need not actually produce) a *disturbance*, i.e. continually interrupting thinking and deciding, and influencing decision-making criteria and perceptions.

9. Thoughts as well as motives can interrupt. Even with no new motive there may simply be a constant dwelling on M1 and B1. This is especially true of emotions like grief, involving what cannot be undone. Such compulsive dwelling might derive from triggering of automatic learning mechanisms concerned with re-programming generators.

10. New motives need not be selected for action. M2 may be considered and rejected as unimportant, yet continue to get through interrupt filters if its insistence is high.

11. In some emotional states, like fright, M2 triggers reflex action, by-passing deliberation and planning, and interrupting other actions (Sloman, 1978, Ch. 6). "Software reflexes" are called "impulsive" actions. Reflexes make it possible to take very rapid remedial action or grasp sudden opportunities. Sometimes they are disastrous, however. Some reflexes are purely mental: A whole barrage of thoughts and feelings may be triggered.

12. Some emotional states arise out of the individual's own thoughts or actions, for instance, fear generated by contemplating possible errors. Secondary motives may be generated to take extra care, etc. These secondary motives may generate so much disturbance that they lead to disaster.

13. Some emotions involve interrupting and re-directing many ongoing processes, for instance, processes controlling different parts of the body in restoring balance. If sensory detectors record local changes, the system's perception of its own state will be changed.

14. Self-monitoring processes may or may not detect the new internal state. If not, X will not be conscious of, or feel, the emotion. Internal monitoring need not produce recognition, e.g. relevant schemata might not have been learnt (Sloman, 1978, Ch. 10). People have to learn to discriminate and recognise complex internal states, using perceptual processes no less complex than recognising a face or a typewriter.

15. Recognition of an emotion can produce further effects, e.g. if the internal state fulfils or violates some motive. It may activate dormant motives or motive-generators and possibly lead to successively higher order emotions (recursive escalation).

The interruptions, disturbances, and departures from rationality that characterise some emotions are a natural consequence of the sorts of mechanisms arising from constraints on the design of intelligent systems, especially the inevitable stupidity of resource-limited interrupt filters that have to act quickly. A robot with an infinitely fast computer and perfect knowledge and predictive power would not need such mechanisms. However, not all emotions are dysfunctional: When walking on a narrow ledge it is important that you do not forget the risks.

These mechanisms allow so many different subprocesses in different situations that no simple table of types of emotions can do justice to the variety. The same rich variation could characterise the detailed phenomenology of emotions in clever robots with self-monitoring abilities.

A full account of how people typically *feel* anger, elation, fear, etc. would have to include bodily awareness. Yet what makes many emotions important in our lives is not this sort of detail, but the more global cognitive structure. Fury matters because it can produce actions causing harm to the hater and hated, not because there is physical tension and sweating. Grief matters because the beloved child is lost, not because there is a new feeling in the belly. So it would be reasonable for us to use terms like "afraid", "disappointed", "ecstatic", "furious", or "grief-stricken" to describe the state of mind of an alien being, or even a sufficiently sophisticated robot, without the physiological responses (contrast Lyons, 1980).

MOODS, ATTITUDES, AND PERSONALITY

A *mood* is partly like an emotion: It involves some kind of global disturbance of, or disposition to disturb, mental processes. However, it need not include any specific beliefs, desires, inclinations to act, etc. In humans, moods can be induced by chemical or by cognitive factors, for instance, drinking or hearing good or bad news. A mood can colour the way one perceives things, interprets the actions of others, predicts the outcome of actions, makes plans, etc. As with an emotional state, a mood may or may not be perceived and classified by the individual concerned. A more detailed theory would have to distinguish different mechanisms, for instance, global "hardware-induced" speed changes of certain subprocesses and global "software-induced" changes in relative priorities of motives or inference strategies.

An *attitude*. such as love or admiration, is a collection of beliefs, motives, motive generators, and comparators focused on some individual, object, or idea. People who love their children will acquire new goals when they detect dangers or opportunities that might affect their well-being. The strength of the love determines the importance and interrupt priority levels assigned to such goals. Selfishness is a similar attitude to oneself. In communities of intelligent systems, able to think and care about the mental states of others, the richness and variety of attitudes makes them an inexhaustible topic for study by poets, novelists, and social scientists. Attitudes are often confused with emotions. It is possible to love, pity, admire, or hate someone without being at all emotional about it. Attitudes are expressed in tendencies to make certain choices *when the opportunity arises*, but need not include continual disturbance of thoughts and decisions. One can love one's children without having them constantly in mind, although news of danger to loved ones may trigger emotions.

Character and personality include long-term attitudes. Generosity, for example, is not a goal but a cluster of goal generators that produce new goals in response to information about another's needs and comparators that select them over more self-centred goals. Hypocrites produce similar goals but never adopt them for action. A personality or character is a vast collection of unfocused general dispositions to produce certain goals in specific situations. The set of such collections is too rich for ordinary adjectives. A whole novel may be required to portray a complex personality. More generally, the space of possible mental states and processes is too rich and complex for colloquial labels like "attitude", "emotion", "mood" to survive in an adequate scientific theory.

There are many kinds of deep and moving experiences that we describe as emotions, for lack of a richer, more fine-grained vocabulary: For instance, delight in a landscape, reading poetry, hearing music, being absorbed in a film or a problem. These involve powerful interactions between perception and a large number of additional processes, some physical as well as mental. Listening to music can produce a tendency to move physically and also a great deal of mental "movement": Memories, perceptions, ripples of association, all controlled by the music. Such processes might be accounted for in terms of aspects of the design of intelligent systems not discussed here, such as the need for associative memories and subtle forms of integration and synchronisation in controlling physical movements. The synchronisation is needed both within an individual and between individuals engaged in co-operative tasks. Music seems to take control of some such processes.

I conjecture that the mechanisms sketched here are capable of generating states we ordinarily describe as emotional—fear, anger, frustration, excitement, dismay, grief, joy, etc. The mechanisms are generative in the sense that the relevant motives, beliefs, plans, and social contexts can be indefinitely complex and the emotional processes they generate can be correspondingly complex and varied (Abelson, 1973; Dyer, 1981; Lehnert, Black, & Reiser, 1981). This means that no simple bounded taxonomy of emotional states can begin to capture the variety, any more than a taxonomy can capture the variety of sentences of English (cf. Roseman, 1979).

Does a Scientific Theory of Mind need Such Concepts?

It is sometimes suggested that although concepts like "belief", "desire", "emotion" play an important role in individual thoughts about other people, they are not required for a fully developed scientific theory of the mind. In its extreme form this is materialist reductionism, but that is as implausible for psychology as the suggestion that concepts of software design can be replaced by concepts relating only to computing hardware.

A more subtle suggestion (S. Rosenschein, SRI, pers. commun.) is that an entirely new collection of "intermediate level" concepts, unrelated to beliefs, desires, intentions, etc., will suffice for a predictively and explanatorily successful scientific theory of how people and other intelligent organisms work. Because it is unlikely that ordinary concepts can be dispensed with entirely in expressing significant generalisations about human behaviour (Pylyshyn, 1986) I have taken a weaker stand: Instead of totally replacing ordinary concepts we need to extend and refine them, showing how they relate to a working design specification.

Even if this sort of theory is wrong, it may be deeply implicated in semantics of natural language concepts concerning human mental states and actions. If so, a machine able to understand ordinary language and simulate human communication will require at least an implicit grasp of the theory.

Implications

Not all these mechanisms can be found in all animals. In some less intelligent creatures, selection of a motive might be inseparable from the process of initiating action: Operative motives could not be dormant. In such animals or machines lacking the mechanisms required for flexibility in a complex environment, emotions in the sense described here would be impossible.

It is also unlikely that all of this richness exists in young children. By investigating the development of the cognitive and computational mechanisms in children, including the motivational mechanisms, we can hope to understand more about their emotional states. In particular, it seems that many higher order generators and comparators are not available to infants, and that interrupt filters are far less selective than in most adults, which is not surprising if software filters are the result of learning.

The very complexity of the mechanisms described reveals enormous scope for "bugs". Motive generators and comparators could produce unfortunate desires and preferences. Interrupt priorities may be assigned in a way that doesn't correlate well with reflective judgements of importance. Thresholds for interrupts may be set too high or too low. Learning processes that modify generators and comparators may be too quick to change things on flimsy evidence. Given the inevitable stupidity of some of the faster reflexes and filters, we can expect some kinds of malfunctions of generators and comparators to lead to intense emotions that interfere with normal cognitive or social functioning. Reactions to unfulfilled motives may be too strong, or too weak for the long-term good of the individual or his associates. The relative importance assigned to different sorts of motives by the goal assessment procedures may produce a tendency to select goals that are unachievable or achievable only at enormous cost. Dormant, temporarily suspended motives may too often go unattended because the monitoring process fails to detect opportunities, perhaps because of inadequate indexing. The pervasiveness of "rules-of-thumb" for coping with inadequate information, limited resources, and the need for speed, provides enormous scope for systematic malfunction. Recursive escalation of emotions might account for some catatonic states.

The inevitability of familiar types of fallibility should be a matter of concern to those who hope that important decisions can be taken very rapidly by machines in the not too distant future.

In fact, if people are as complex and intricate as we have suggested, it is amazing that so many are stable and civilised. Perhaps this theory will reveal types of disturbance we previously could not recognise.

The theory implies that processes of learning and cognitive development, occur in a framework of a complex and frequently changing collection of motivators. These and the processes they generate must have a profound influence on what is learnt when, and it is to be expected that there will be enormous variation between individuals. The implications for educators have yet to be explored.

CONCLUSION

A theory of this general sort is a *computational* theory of mind. The computations may occur in a *virtual* machine implemented in lower level machines, brain-like or computer-like: They need not be implemented directly in physical processes. Thus, the theory is neutral between physically explicit representations as found at low levels in conventional computers and implicit or distributed representations studied in neural-net models.

The test of this approach will be the explanatory power of the theories based on it. We need both a systematic explanation of the whole range of possibilities we find in human behaviour and an account of how people differ from one another and from other actual and possible behaving systems. (Concerning explanations of possibilities, see Sloman 1978, Ch. 2).

Understanding computational mechanisms behind familiar mental processes may enable us to reduce suffering from emotional disturbances, learning disabilities, and a range of social inadequacies. Some problems may be due to brain damage or neural malfunction. Other problems seem more like software faults in a computer. I conjecture that many emotionally disturbed people are experiencing such software "bugs".

The analysis still has many gaps. In particular, an account of pleasure and pain is missing, and I am not yet able to give an acceptable analysis of what it is to find something funny. There are states like being thrilled by rapid motion, spellbound by a sunset, moved to tears by reading a book or watching a play, that require more detailed analysis. I have not discussed the many aspects of human emotional life that arise contingently from our evolutionary history and would not necessarily be found in well-designed robots. So there is much yet to be done. Nevertheless, the theory provides a framework for thinking about a range of possible types of intelligent systems, natural and artificial—part of our general study of the space of possible minds. Attempting to test the ideas in working computer simulations will surely reveal gaps and weaknesses.

Manuscript received 5 June 1986
Revised manuscript received 1 September 1986

REFERENCES AND BIBLIOGRAPHY

Abelson, R. A. (1973). The structure of belief systems. In R. C. Schank and K. M. Colby (Eds), *Computer models of thought and language*. San Francisco: Freeman.

Austin, J. L. (1961). A plea for excuses. In *Philosophical papers*. Oxford University Press. Reprinted 1968. In A. R. White (Ed.), *Philosophy of action*. Oxford University Press.

Boden, M. (1978a). *Purposive explanation in psychology*. Harvester Press.

Boden, M. (1978b). *Artificial intelligence and natural man*, Harvester Press.

Croucher, M. (1985). *A computational approach to emotions*. Unpublished Thesis, University of Sussex.

Dennett, D. C. (1979). *Brainstorms*. Harvester Press.

Dyer, M. G. (1981). The role of TAUs in narratives. *Proceedings Cognitive Science Conference*. Berkeley.

Edelson, T. (1986). Can a system be intelligent if it never gives a damn? In *Proceedings of the Fifth National Conference on Artificial Intelligence* (AAAI-86). Philadelphia, Pennsylvania.

Lehnert, W. G., Black, J. B., & Reiser, B. J. (1981). Summarising narratives. In *Proceedings of the Seventh International Joint Conference on Artificial Intelligence*. Vancouver, British Columbia.

Lyons, W. (1980). *Emotion*. Cambridge University Press.

Oatley, K. & Johnson-Laird, P. N. (1985). Sketch for a cognitive theory of the emotions. *Cognitive Science Research Paper No. CSRP.045*. University of Sussex, Cognitive Studies.

Pylyshyn, Zenon, W. (1986). *Computation and cognition: Toward a foundation for cognitive science*. Cambridge, Mass.: MIT Press.

Roseman, Ira (1979). Cognitive aspects of emotion and emotional behaviour. Paper presented to the *87th Annual Convention of the American Psychological Association*.

Ryle, Gilbert (1949). *The concept of mind*. London: Hutchinson.

Simon, H. A. (1979). Motivational and emotional controls of cognition. In *Models of thought*. Yale University Press.

Sloman, A. (1965). How to derive "Better" from "Is". *American Philosophical Quarterly*.

Sloman, A. (1978). *The computer revolution in philosophy: Philosophy, science and models of mind*. Harvester Press.

Sloman, A. (1981). Skills learning and parallelism. *Proceedings Cognitive Science Conference*. Berkeley.

Sloman, A. (1984). Why we need many knowledge representation formalisms. In M. Bramer (Ed.), *Research and development in expert systems*. Cambridge University Press.

Sloman, A. (1985). Real-time multiple-motive expert systems. In Martin Merry (Ed.), *Expert systems 85*. Cambridge University Press.

Sloman, A. & Croucher, M. (1981). Why robots will have emotions. In *Proceedings of the Seventh International Joint Conference on Artificial Intelligence*. Vancouver, British Columbia.

COGNITION AND EMOTION, 1992, 6 (3/4), 201–223

Basic Emotions, Rationality, and Folk Theory

P.N. Johnson-Laird

Department of Psychology, Princeton University, U.S.A.

Keith Oatley

Centre for Applied Cognitive Science, Ontario Institute for Studies in Education, Toronto, Canada

Answering the question of whether there are basic emotions requires considering the functions of emotions. We propose that just a few emotions are basic and that they have functions in managing action. When no fully rational solution is available for a problem of action, a basic emotion functions to prompt us in a direction that is better than a random choice. We contrast this kind of theory with a componential approach which we argue is either a version of the theory of basic emotions or else leads to the doctrine that emotions are mistaken tenets of folk psychology. We defend the psychological reality of the folk theory of emotions, and we argue that universal basic emotions make it possible to understand people from distant cultures, and to translate emotional terminology from one language to another. Finally, we show how theories of basic emotions can be tested, and indicate the kinds of empirical result that can bear on the issue.

INTRODUCTION

How many emotions are there? There are several possible replies to this question. One is that the question is meaningless, but this response amounts to rejecting the ordinary concept of emotion. It is akin to answering that emotions do not exist—that they are false tenets of folk theories, i.e. of the common sense theories that lay people have about mind and behaviour, which contrast with scientific and other kinds of specialist theories (D'Andrade, 1987). Another answer is that although individuals experience only a finite number of emotions in their lifetimes, there are indefinitely many possible emotions that they might experience. This view can be coupled with the claim that one never experiences the

Requests for reprints should be sent to Dr P.N. Johnson-Laird, Department of Psychology, Princeton University, Princeton, NJ 08544, U.S.A.

same emotion twice. If there is an indefinite number of emotions, then we can ask how they relate, if at all, to one another. Each emotion might be unique and unrelated to any other. Such a conception is metaphysically defensible but removes emotions from the domain of scientific investigation. Each wave on the seashore is unique, but the science of hydrodynamics idealises waves, and from this abstract standpoint treats waves that are, in fact, distinguishable as the same.

In this paper we will put a different view, that folk psychology and scientific psychology both have something in common. Folk psychology treats different experiences as instances of the same emotion, so in daily life, we talk of different occasions of fear, anger, happiness, and so on. Likewise, a science of emotions is bound to treat different emotional experiences as members of the same class. Hence, we can ask our question again, but slightly differently. How many sorts of emotion are there?

One answer is that there is a small finite set of distinguishable emotions that are the bases of all emotional experiences. This is the hypothesis of basic emotions as primitive unanalysable elements at the psychological level of the system. Another answer is that there are many sorts of emotion, but every distinct sort is generated from among the same finite set of components, much as, say, each chemical molecule is constructed from atoms. Hence, all emotions would be on a par—there would be no sense in which some emotions were basic whereas others were complex.

To be more specific, a theory of basic emotions might analyse embarrassment as founded on the basic emotion of fear, with some other non-emotion component such as a cognition of a particular sort, i.e. knowing that one is the object of unwelcome attention. A componential theory in which no emotions are basic might distinguish embarrassment from fear in terms of different appraisal-response elements that make up each kind of emotion. Such a theory then holds that because of its components each type of emotion has its own unique psychology and physiology, although perhaps with some components in common. Both the theory of basic emotions and the componential theory postulate that an emotional experience depends on various elements. The crucial distinction is that no components can be an emotion *per se* in the componential approach, whereas one component of any emotional experience is always a basic emotion according to the basic theory. Both theories assume that a science of emotions is possible only if there is a finite basis for emotional experience. Both assume that emotions can be taken to pieces analytically. The question is: Is there always a piece that consists in an irreducible basic emotion or are there sub-emotional elements, such as appraisal-response components, into which all emotions can be analysed?

There are various ways in principle of establishing the nature of the finite basis of emotions. In this paper, we will examine the lesson that the rationality. Next, we will enquire into the set of basic emotions. Once we have outlined a theory of them, we will contrast it with a componential theory, and we will show that the componential approach turns out to be a version of the theory of basic emotions or else leads to the doctrine that emotions are a false tenet of folk psychology. We will defend the psychological reality of the folk theory of the emotions, and we will counter arguments based on the difficulty of translating emotional terminology from one language to another. Finally, we will show how the theory of basic emotions can be corroborated, and describe some of these results.

THE NEGLECT OF FUNCTION

What function, if any, do emotions serve? The question has been somewhat neglected by theorists, and one can read much on the cognitive underpinnings of emotions that does not address this issue. Yet, in our view, it is the key to whether or not there is a small set of basic emotions: The hypothesis of basic emotions makes sense only if it elucidates problems faced by the cognitive system. Although elements of current theories of emotions can be traced back at least to the nineteenth century, the intellectual history of the topic has lacked cumulative coherence. Without any clear sense of the psychological function of emotions, it has been difficult to generate more than a patchwork of ideas and observations. With such a sense, perhaps the scientific understanding of emotions will become cumulative, analogously, say, to the understanding of visual perception.

One reason for the neglect of function is probably the pervasive influence of William James. Like Descartes (1649/1911) he classed emotions with perceptions (e.g. James, 1890): They are perceptions of events inside the body. Beyond his claim that not all emotions are accidental, he had little to say about what purposes they might have. Because, according to James, emotions are percepts of bodily feedback from physiological changes, or from actions that have already taken place, emotions occur too late to affect either the control of these actions or the decisions that led to them. Just as there are indefinitely many percepts of the outer world there are, in James's view, indefinitely many feelings, each reflecting a particular pattern of proprioception and physiological perturbation.

James's influence may account for the importance that many theorists place on emotions as primarily pleasant or unpleasant, that is, as "valenced" (e.g. Frijda, 1987; Ortony & Clore, 1989). If emotions are not a system for the immediate control of actions, then they are important endpoints giving colour to experience. Incidentally they can then have motivational consequences. People strive to attain them if pleasant, and to avoid them if unpleasant (Hammond, 1990).

In short, James's theory of emotion contributes to the development of a powerful tradition. His conclusions, perhaps serendipitously, resonate with the high valuation of "experience" as such in Western culture. Pleasantness and unpleasantness have become the crucial characteristics of emotions over and above their own strict individuality. Within this tradition, the possible existence of a set of basic emotions seems both unattractive theoretically and intractable empirically.

RATIONALITY AND THE FUNCTION OF EMOTIONS

Analyses of mental processes in cognitive science assume that each process has functions independently of its particular embodiment. From this perspective, emotions should have a function that could be embodied in a system based either on carbon-like humans or on silicon-like computers. As many people within cognitive science have argued, function is accordingly best thought of in terms of the design of the system. During natural selection systems are fitted to functions, although as a means of design, evolution is notoriously a "tinkerer" not a grand architect. A priori there are many possible designs to enable organisms to cope with their environment.

The simplest possible design relies on "fixed action patterns" and is found in insects. Consider, for instance, the common tick, which is a parasite of mammals. According to von Uexkull (1957), the female tick lacks eyes, but at one stage in her life cycle the photosensitivity of her skin triggers the action of climbing a bush from which she then hangs. She lets go only when a second trigger occurs: The smell of butyric acid, which is secreted by the sweat glands of all mammals. If she happens to land on a passing animal, a third trigger comes into operation: The warmth of her host's body. Propelled by this taxis, she burrows through the hair to the skin, and there she punctures the skin and fills herself with blood. Once full, she drops off the animal to lay her eggs on the earth. And the cycle continues anew.

This sort of design works well when classes of events can be mapped one-to-one on to appropriate responses. Perfect performance is impossible, e.g. a tick may drop but miss the passing mammal. Yet, the solution is rational in the following sense: All that is necessary for a reasonable chance for individuals to survive and to reproduce is built into the species-specific procedures for action. In principle, there are no uncertainties about what to do: The stimulus either unlocks the fixed action pattern, or not. Of course, this certainty can be the undoing of a species if there is a significant change in its environment.

At the other extreme in the theoretical series of designs are those that are impeccably rational. They are maximally flexible because they enable the organism to determine which goals to pursue at any point in time, and to decide at each choice point the best course of action in pursuit of those goals. No contingency is unanticipated, and performance is invariably optimal. Creators of artificial intelligences have aspired to such designs; philosophers have argued that they are realised in human thought (e.g. Dennett, 1978, p. 20; Cohen, 1981); and psychologists have claimed that apparent errors are merely failures in performance that do not impeach the underlying rationality of the system (e.g. Henle, 1978).

In designs based on fixed action patterns or on impeccable rationality, there is no occasion for anything corresponding to an emotion. There are no surprises, no misunderstandings, no irresolvable conflicts. Human beings are neither equipped with a set of responses each matched to an important stimulus, nor do they possess impeccable rationality. A fully rational system of thought is a paragon that cannot be realised by any finite device. Any set of observations is compatible with an infinitude of different valid conclusions, and so no finite organism can follow up all of them (Johnson-Laird, 1983; Cherniak, 1986; Stich, 1990). Moreover, human reasoners make genuine mistakes in reasoning—mistakes that they even acknowledge in some cases. They make invalid inferences that should not occur if their thinking were guided by valid formal rules of inference (Johnson-Laird & Byrne, 1991). In short, to paraphrase de Sousa (1987): Human beings are neither insects nor omniscient, omnipotent gods.

If impeccable rationality is impossible, what design is embodied in human beings? Johnson-Laird and Byrne (in press) argue for a significant modification of the competence-performance distinction. The original distinction hinged on the idea that rational competence is based on valid rules of inference, which, like the rules of grammar, might sometimes be inadequately reflected in actual performance. The new notion of rational competence depends instead on a meta-principle: An inference is valid provided that there is no model of the premises in which its conclusion is false. Individuals have a tacit grasp of this meta-principle, and they put it into practice by building mental models of premises, drawing useful conclusions from them, and then searching for alternative models that might refute such conclusion. But they have no grasp of any specific logical rules, and they have no comprehensive algorithm for valid thinking, i.e. for searching for models that refute conclusions. The meta-principle is defensible as a rational requirement for any system for deductive inference, although it alone does not guarantee the validity of inferences. To argue that errors arise as result of performance factors is misleading, however, because it suggests a failure to put into practice correct rules, whereas there are no rules to put into practice, only the higher-order meta-principle. This principle is compatible with the observations of deductive failure, and with the arguments against impeccable rationality.

Granted that reasoning is fallible and time-consuming, Oatley and Johnson-Laird (1987), following Simon (1967), and a tradition of cognitive theorists, proposed that the function of emotions is to fill the gap between fixed action patterns and impeccable rationality. For many species, including *homo sapiens*, the world is too complex to form perfect mental models, so events and the outcomes of actions are often unanticipated. The problem that Simon identified is that complex systems acting in the natural world, as opposed to a simplified microworld, need something equivalent to interrupt signals in computation. Such signals are necessary in systems that have limited resources, and that need to be influenced by unforeseen events demanding urgent attention. Emotions, as Simon noted, seem to be co-extensive with the occurrence of such problems. They arise particularly when individuals have many concurrent goals, including mutually incompatible ones, and their resources of time, ability, and processing power, are too limited to make a fully rational choice. Moreover, social mammals often cannot achieve their more valuable objectives alone, and so they need to interact with others. Co-operation calls for mutual plans, but it is impossible to guarantee that copies of the plan kept by each partner are identical. Competition calls for antagonistic plans, and it is impossible to determine their outcome. The biological system of emotions offers a solution to these problems, particularly those that arise from the limits of rational principles to govern or to predict complex social interactions. Emotions enable social species to co-ordinate their behaviour, to respond to emergencies, to prioritise goals, to prepare for appropriate actions, and to make progress towards goals. They do so even though individuals have only limited abilities to cogitate.

Emotions guide individual and group behaviour. Social mammals are unable to determine the best course of action at many of the junctures in their lives. Even in humans, the resources for rational thought are often too slow and too error-prone to solve this problem. The function of emotions is accordingly to bridge the gaps of rationality. We argue that this bridge is possible only if many specific junctures can be mapped into a few broad classes of reaction.

We have proposed (Oatley & Johnson-Laird, 1987) that the cognitive evaluation of a juncture in action calls into readiness a small and distinctive suite of action plans that has been selected as appropriate to it. Each basic emotion thus prompts both the individual and the group in a way that in the course of evolution has been more successful than alternative kinds of prompting in broadly defined, recurring circumstances that are relevant to goals. Thus, when the broad class of event occurs that indicates achievement of a subgoal that increases the probability of attaining a goal, then its cognitive evaluation initiates an internal emotional signal. We propose that emotion signals of this kind have no propositional content or syntactic structure: They have a control function rather than an informational function. The signal that is sent when subgoals are achieved acts to prompt the individual to continue the same line of action. When a goal is lost, a different emotion signal is sent. It prompts the individual to disengage from that goal. The internal emotional signals have causal effects within the organism, preparing it physiologically for each general class of actions. In the case of human beings, the signals can in addition be experienced subjectively as emotions. The signal caused by a successful achievement is experienced as happiness, and the signal caused by the loss of a goal as sadness. An important consequence of the ensuing actions is the communication of the individual's emotional state to others in the same social group—an example is the distinctive type of alarm signal sent by certain social mammals and birds. The receipt of such external signals has emotional consequences for these other individuals too.

If the emotional guidance of action is to be rapid, successful, and independent of reasoning which is too time-consuming, then the cognitive evaluations must be coarse and the resulting suites of actions must be broad and flexible. There are two key issues here. First, many events in the world must be mapped on to a relatively small number of categories, which each elicit a distinct set of bodily, behavioural, and (at least in the case of humans) phenomenological consequences. If there were very many categories, then the problem of deciding amongst them would re-emerge as a time-consuming matter. Secondly, the small repertoire of actions triggered by a particular emotion must be useful to a wide class of specific triggering events. For example, if there is a conflict in goals because an event threatens an individual's safety during the course of another action, then the emotion of fear prepares a small repertoire of actions, which includes stopping the current action, checking everything that has been done recently, monitoring the environment, fleeing, being prepared for fighting, physical exertion, or bodily harm. In the case of human beings, the repertoire can be supplemented with action sequences that have been practised. The purpose of fire drills, for instance, is to enable people to learn how to leave a building in the event of a fire without having to think about what to do.

Although we have not yet developed a computer simulation of this theory, it is based on computational considerations. Such considerations are called "computational" because they are at a particular level of analysis (Marr, 1982), in which knowledge of aspects of the social and physical environment is mapped on to a design for the kinds of operations that could cope with these aspects. Emotions function in real time to redistribute cognitive resources and to manage goal priorities. When an event has been detected that requires re-computing these priorities, an emotion occurs and it helps to manage either the continuation of the current course of action or the transition to another sequence of action. Emotions help to specify which goals will be actively pursued, and which abandoned, or assigned to a subsidiary or dormant status (see also Stein & Levine, 1990). Emotions have further consequences by way of external signals that co-ordinate group behaviour.

We can summarise the argument so far in three propositions:

Proposition 1. Events and their significance for goals are often unforeseen because: (a) finite organisms cannot be impeccably rational, and they have imperfect models of the world; (b) individuals with several goals are often unable to satisfy all of them simultaneously; and (c) social animals interact together in ways that cannot always be anticipated.

Proposition 2. It follows that junctures in action will occur at which an individual needs to act, but for which there is no fully rational method to select the next action.

Proposition 3. Emotions function to redistribute cognitive resources at junctures in action, particularly where neither cogitation nor reflexes (the residue of fixed action patterns) determine an appropriate course of action. Because some action is probably better than becoming lost in thought, a biologically based system makes ready a small repertoire of actions appropriate to a recognisable type of goal-relevant event. The mechanism tends to constrain the individual to choose the next action sequence from this repertoire. Such a mechanism is a result of natural selection, and the repertoires of actions include both species-specific patterns and individually acquired habits.

WHICH EMOTIONS ARE BASIC?

Many theorists have proposed sets of basic emotions. There are differences among the theories and among the sets of basic emotions that have been proposed. These differences prompt sceptics to argue that it is no longer clear what is meant by the claim that some emotions are basic, and that it has no testable content (see, for example, Ortony & Turner, 1990). Most previous theories, however, have not been based on a functional analysis. Their principal motivation has been to bring order to the disparate set of human emotions by seeking to derive them from a set of basic emotions, e.g. by postulating a set of opposites, by analogy to chemistry or to the mixing of colours (e.g. McDougall, 1926; Plutchik, 1962). Pride, for instance, has been proposed to be a combination of joy and anger; and love a combination of joy and acceptance. Some of the postulates of such theories, however, have no empirical support either subjectively or physiologically, and this again has been noted by sceptics. It is a common experience to have "mixed" feelings, but this state is characterised by an awareness of alternative and conflicting emotions (see also Ellworth & Smith, 1988; Stein & Levine, 1989). Indeed, our own research (Oatley & Duncan, in press) shows that in more than a third of episodes of happiness, sadness, anger, and fear, a person experiences simultaneously two basic emotions. The most common such mixture is sadness and anger—caused, for instance, by a loss which also frustrates some plan. Our method does not discriminate between true simultaneity and rapid alternation of under-lying states. What individuals do not report, however, is the existence of a single emotion made up from phenomenally remote constituents.

According to our theory, emotions are a result of coarse cognitive evaluations that elicit internal and external signals and corresponding suites of action plans. They are emotions because they have cognitive rather than physiological causes. From an analysis of the ontology of simple social mammals, we have proposed the following set of basic emotions: happiness, sadness, anger, fear, disgust (Oatley & Johnson-Laird, 1987), and perhaps desire (Oatley & Johnson-Laird, 1990). Hence, specific emotions are typically caused by the perceptions of general cate-gories of event: (1) happiness with perception of improving progress towards a goal; (2) sadness when a goal is lost; (3) anger when a plan is blocked; (4) fear when a goal conflict or a threat to self-preservation occurs; (5) disgust with a perception of something to reject; and (6) desire with a perception of something to approach. These emotions are indeed basic—however, depending on how the evidence points, other emotions may be basic too. The names of the basic emotions have misleading enthno- and anthropocentric connotations, but in English they come close to suggesting the emotional behaviours of social mammals.

We argue that the status of the basic emotions is corroborated in five ways. First, each of them is an emotion that appears to be universal, and to have universal concomitants, such as a corresponding facial expression (see Ekman, 1973, and Ekman, this issue). Second, each has either a bodily or phenomenological component that can be experienced without the individual knowing the cause of the emotion. Third, the semantics of the large emotional vocabulary of English can be explicated without having to appeal to any other emotions (see Johnson-Laird & Oatley, 1989). Fourth, each term denoting a basic emotion is primitive in the sense that it is semantically unanalysable. It refers to a phenomenological primitive that one needs to have experienced in order to grasp the meaning of the terms. If Mr Spock (of *Star Trek*) does not experience emotions, then it is impossible to explain to him what happiness or sadness are. We could explain what kinds of events are likely to cause these states; we could explain what physiological changes they are likely to bring about and what actions they are likely to elicit. With some perceptual training, such as

297

experience with Ekman's Facial Action Coding System (Ekman & Friesen, 1978), he would be able to discriminate amongst facial displays of emotions. But we could not explain to him what it was like to feel happy or sad, any more than we could explain what red was like to a person who was completely colour blind. Fifth, the apparent complexity of human emotional experience comes from the diverse cognitive evaluations that can elicit and accompany the basic emotions, and that can differ from one culture to another. The accompanying cognitions are also reflected in the vocabulary of emotions. An emotion term accordingly refers to a subset of the basic emotions, typically just a single basic emotion, perhaps with an indication of the intensity of the emotion, as in the series: "contentedness", "happiness", "joy", "ecstasy". A term can also convey that the state has a known cause or object. For example, to use the term, "glad", properly, is to imply a conscious propositional knowledge of what caused the happiness: That one is glad that something has, or has not, happened.

Basic Emotions vs. Components of Emotions

Our theory of basic emotions contrasts with a recent componential proposal made by Ortony and Turner (1990), which is also computationally motivated (see Ortony, Clore, & Collins, 1988). Ortony and Turner reject the hypothesis of basic emotions, and instead they consider it more profitable to analyse emotional expressions and responses in terms of dissociable components that are innate. Their theory is akin to the notion that the underlying components of facial expressions, and other emotional responses, are governed by a system of production rules of the form:

If an Event E1 occurs, then do Action A1
If an Event E2 occurs, then do Action A2
. . . and so on.

As an example, Ortony and Turner (p. 332) consider the apparently universal facial expression of anger, which they analyse in terms of separate and dissociable components. We can capture the essence of their claims in the following production rules:

If you become conscious of being unable to attain a goal, then furrow your brow.
If you desire to be aggressive towards the agent responsible for the blockage, then form an open, square mouth that shows your teeth.
If you are determined to remove the source of the goal blockage, then compress your lips.
If you devote considerable attention to the visual environment, then raise your upper eyelids.

These dissociable elements of the prototypical facial expression of anger are invoked by an event in relation to a goal: a goal blockage. This event, however, has attributes such as the "existence of an identifiable agent responsible for the blockage", which may, or may not, be present in any given episode. Ortony and Turner go on to make a case for dissociable physiological components underlying emotional experiences. "Our view", they write (p. 322) "is that such differences in physiological responses are usually better interpreted as indicating not so much the presence of specific emotions as the presence of dissociable components of emotions, namely specific appraisals and their corresponding responses".

We see several problems with a componential analysis of this sort. First, Ortony and Turner allow only "external causes of co-occurrences of sub-components" (p. 323). So, although their system is like a set of production rules, because the independence of sub-components is so fundamental to them, anything that might bind rules together internally is excluded. In contrast to computational production systems which have the power of universal Turing machines, no logical operations between rules are descri-bed. Moreover, although alternative environmental events may trigger a single rule, there is no indication that the same event might trigger alternative rules. Hence, there is no indication that their mechanism might generate default operations for states of uncertainty. The system they discuss, therefore, does not address the functional issue of filling the gap between fixed action patterns and impeccable rationality that we have discussed in the previous section. It is hard, indeed, to see how the system differs in principle from the fixed action patterns of insects. Secondly, their account takes a critical step towards treating emotions as a myth of folk psychology. As Ortony and Turner make clear, the dissociable actions that they propose are not caused or linked by anger. On the contrary, the theory dismantles anger into a set of components, which can differ from one case to another. There need be nothing in common to all occasions of anger. How is it possible then for individuals to refer to anger in so many diverse situations? One possible answer is that there is a prototypical set of components underlying all experiences of anger (cf. Fehr & Russell, 1984). Granted a certain number of the characteristic components of anger, then individuals experience the emotion. Such an approach has some plausibil-ity as an account of the concept of an emotion such as anger. However, Ortony and Turner (p. 323) specifically argue against it as an analysis of the emotion itself.

In our view anger typically corresponds to the following sequence:

- an individual's goal is frustrated, often but not necessarily by another agent;
- the individual perceives the blockage;

- a basic anger signal propagates through the cognitive system which the individual experiences as feeling angry;
- as a result of this signal, physiological mechanisms prepare the body for aggression, the face assumes an expression in which the brows are furrowed, etc; and
- plans are made ready for removing the blockage.

The important component in this sequence is the third one: a specific signal propagates which the individual experiences as feeling angry. This component alone is sufficient for an individual to be angry. One can be angry for no known reason, that is, without any awareness of a goal-blockage and without betraying one's feelings by facial expressions or bodily behaviours. If an individual feels angry in such circumstances, then, according to our theory of basic emotions, the feeling is mediated by a primitive unanalysable signal of anger that impinges on consciousness, but without knowing anything else about the state. For a componential analysis, however, a feeling of anger must be mediated by a set of dissociable components that can differ from one such experience to another. Ortony and Turner make this point with great clarity in their discussion of fear (p. 327): "There are various kinds of fear, each consisting of somewhat different components."

It is hard to see what these components of a feeling could be. One putative view is that emotions are valenced experiences, anger is a negative experience, and so the subjective experience is composed of the following components:

$$\text{Emotion} + \text{Negative valence} + X$$

where X is a set of subjective components that distinguish anger from other negative emotions, such as fear. There is a striking dilemma for such a view, however. Either X includes a set of components common to all subjective experiences of anger, or else it does not. If it does contain a common set of components, then we are back once again at a theory of basic emotions: underlying any experience of anger is a common set of components. Hence, on this side of the dilemma, the componential theory is entirely compatible with the theory of basic emotions.

We suspect that Ortony and Turner prefer the other side of the dilemma in which subjective experiences of anger do not contain common elements. In this case, there really is nothing in common to all occasions of anger other than:

$$\text{Emotion} + \text{Negative valence}$$

But these two components fail to distinguish anger from other negative emotions. Hence, this view leads ineluctably to the conclusion that emotions as distinct subjective experiences, such as anger, fear, and sadness, have no real existence. To reject common components is to reject, not just basic emotions, but all the everyday categories of emotion: One is indeed forced to treat them as a myth of folk psychology.

FOLK THEORIES AND SCIENTIFIC THEORIES OF EMOTIONS

We argued in the previous section that the rejection of basic emotions leads to a rejection of the naïve everyday categories of emotion too. This sceptical view of emotion has always attracted adherents, who regard folk psychology as based on errors that are as egregious as those that underlie naïve physics. Ultimately, according to this form of reductionism, the ideas and terms of folk theory will be replaced by proper scientific explanations. Once again, William James anticipated the critique of folk psychology. Here he is on the pointlessness of studying emotion terms of ordinary language, and of trying to sort them into categories, such as basic and non-basic (James, 1890, p. 485):

> If one should seek to name each particular one of them [emotions] of which the human heart is the seat, it is plain that the limit to their number would lie in the introspective vocabulary of the seeker, each race of men having found names for some shade of feeling which other races have left undiscriminated. If we should seek to break the emotions, thus enumerated, into groups, according to their affinities, it is again plain that all sorts of groupings would be possible, according as we chose this character or that as a basis, and that all groupings would be equally real and true.

In more recent times, the argument from this side has gone somewhat as follows: Accounts which include intentional terms, such as "desiring that something", or "believing that something", are folk theories that seek to explain and predict individual's actions. Just as naïve physics depends on the misleading idea of impetus, so folk psychology depends on the misleading idea that beliefs and desires cause behaviour. Newton replaced impetus by coherent laws of motion; so, too, the psychology of belief and desire will be replaced by a proper scientific account of behaviour that will be based, not on such "intentional" concepts, but on the neurophysiology of the nervous system (see, for example, Stich, 1983; Churchland, 1984).

The view implicit in our theory of emotions is that folk psychology is not a myth. It embodies important truths: that individuals have beliefs and desires and needs, that they use their beliefs to decide what to do to attain their goals and then try to carry out these actions—and that emotions have effects on behaviour. An achievement of cognitive science is to rehabilitate mental terms following their banishment during "Behaviourism", and to show how the psychology of "belief and desire" can be modelled computa-

301

tionally. There is no warrant for the generalisation from naïve physics to the conclusion that all folk theories are mistaken. In particular, psychological phenomena and physical phenomena are different. A putative account of, say, physical motion is corrigible. But the subjective experience of an emotion is incorrigible in the sense that it is not a hypothesis that could be falsified by evidence in the way that hypotheses about the physical world may be. If you feel definitely happy, you will not be mistaken that you are happy, because, according to us, the feeling of happiness is a direct phenomenological result of a certain kind of signal in the cognitive system. The (folk theoretical) concept which in English is called happiness indicates just such a feeling. It indicates something real. Like sleepiness, or pain, or thirst, it is subjective, not open to consensual validation or evidential refutation.

When an emotion signal does impinge on consciousness it does not have to be interpreted to determine which emotion it represents. It does not represent an emotional state. A conscious emotion is the experience of an emotion signal. Such an experience leaves room for various kinds of doubt, for instance about its cause, about the interpretation of the emotion-eliciting event, about whether the feeling is strong enough to be sure that an emotion really is occurring, or about what kind of emotion it is—particularly if for some reason the emotion is suppressed, or if two emotions occur as a mixture. But, we argue, in straightforward cases where the emotion is felt strongly, e.g. feeling happy at seeing a good friend, feeling angry if someone lets you down, feeling afraid at a traffic accident, there is no doubt about the nature of the emotion itself. So, in the structured diaries of 30 patients attending a gastrointestinal clinic, each asked to record four episodes of emotion of any kind, half of their emotion episodes were experienced in this way. For each episode subjects were asked: "Would you call it a type of any of the following?—happiness/joy, sadness/grief, anger/irritation, fear/anxiety or disgust/hatred." They were then asked to rate how sure they were about this choice on an 11 point scale from 0 (not at all sure) to 10 (completely sure), (Duncan & Oatley, in prep.). All episodes of emotion were rated as one of the five types. Subjects rated 50.4% of their categorisations as "completely sure", and only 14% of episodes at 5 or below on the scale of certainty of categorisation. Our claim is also supported by the ready ability of children to learn and to understand the causal sequence of events underlying emotions—the chain from the perception of a goal-related event, to the emotion, and then to a change in action (Stein & Levine, 1989). The cause of the emotion is typically obvious; and this defence of folk psychology is consistent with the existence of basic emotions.

Subjective experiences, of having beliefs, desires, emotions, lie at the heart of folk psychology. As a theory, however, the folk theory of emotions provides little account of psychological mechanisms, or their physiological bases. The goal of a cognitive science of emotions is thus to spell out a mechanism that is at least consistent with common observations of the causes and consequences of emotions. The persistence over time of these observations does not indicate a stagnation of explanation as a result of isolation from evidence. The evidence is the set of observations of the causes and consequences of emotions to which people are continuously open.

Sceptics might imagine that this hypothesis of a convergence between folk theory and scientific theory is a quirk peculiar to us. But other researchers too, with quite different theories from ours, have come to the same conclusion. Ortony and his colleagues argue that individuals can be usefully consulted about what terms refer to emotions, and that these everyday intuitions map on to the scientific theory of emotions (Ortony et al., 1988). Similarly, Fehr and Russell (1984) and Shaver, Schwartz, Kirson, & O'Connor (1987), have consulted people in a range of ways about their categorisations of emotion terms. These investigators also assume that people know that emotions are caused by certain types of events related to goals. They postulate a correspondence between the results of their studies and scientific categories, and, in the case of Shaver et al., their results support basic categories of emotions, which correspond to some degree to those that we have postulated.

Our hypothesis of a convergence between folk theories of emotions and scientific theories of emotions is, like any other scientific claim, open to refutation. There are indeed several ways to challenge it. One is to argue that self-reports are neither reliable nor valid, and only objective reports of behaviour or physiology should have any part in science. Evidence for such assertions can be derived from the work of Nisbett and Wilson (1977) and Nisbett and Ross (1980), which shows that people are often poor judges of the causes of their judgements and behaviour. The true causes include social conformity, compliance to subtle conditions of experimental designs, and attributional biases. Individuals are not conscious of these factors, and their explanations of their own behaviour ignore them. Instead, they focus on events that are salient, without weighing in any statistically appropriate manner relevant causal factors. They also display pervasive mental shortcomings. They can make gross errors of judgement about the causal effects of their own or others' actions (Jenkins & Ward, 1965), they overlook falsifying evidence (Wason, 1960), and they are biased by information that is more immediately available or that appears to be more representative of the case in hand (Kahneman, Slovic, & Tversky, 1982).

Thus, the argument goes, people do not know the causes of either their behaviour or of their mental states in any way that resembles a scientific account. Not only do they lack a privileged introspective access to how events cause behaviour, but they are regularly misled by their introspections. They are subject to inbuilt mental deficits in reasoning that will necessarily lead them astray. Hence, folk psychology is not merely irrelevant to scientific theories, but to attend to it is positively misleading.

This kind of argument has encouraged many to eschew evidence based on self-reports, but we believe this is mistaken, for three inter-related reasons.

First, as many of the psychologists studying the shortcomings of the human inferential system have themselves pointed out, their studies deliberately focus on cognitive illusions much as perceptual psychologists seek visual illusions with the goal of revealing the workings of the cognitive system. No psychologist argues from the existence of visual illusions to the claim that all vision is illusory and non-veridical. Likewise, the failures of inference in the psychological laboratory hardly justify the claim that human reasoning is intrinsically irrational (see Johnson-Laird & Byrne, in press).

Secondly, as Craik (1943) proposed, the brain models important entities, attributes, and relations in the world. If it had not converged on successful models of important sequences, we would not be able to operate in the world. Thought, behaviour, and communication, are successful more often than not—the central postulates of folk psychology are based on essentially correct, though radically incomplete, mental models. Actions are caused by goals in conjunction with beliefs. The reason, for example, that the engineers in charge at Chernobyl did not report the destruction of the nuclear reactor to the authorities in Moscow is because they did not believe that the reactor had been destroyed. They persisted in the view that the reactor was intact, despite much evidence to the contrary, including the reports of two young probationary engineers whom they had sent to examine it and who paid with their lives for their observations (Medvedev, 1990). Work on inferential failure may reveal causes of such pathological disbelief, but what is clear is that the belief led to a failure to report the scale of the disaster, and that this failure contributed to the appalling delay in evacuating the area.

Thirdly, emotions usually follow immediately after the events that cause them. Therefore, people will not ordinarily suffer the kinds of illusions of thinking just indicated. Such errors occur easily, for instance when causes are probabilistic and temporally distant from effects, as in the studies of Jenkins and Ward. The mechanisms of human learning have been successfully tuned by evolution to sequences in which a causal event is regularly and closely followed by a caused event, as routinely demonstrated in both classical and instrumental conditioning experiments. People are indeed bad in intuitions and judgements made outside this range, but their judgements about emotions derive from many experiences within it. Even if, as we agree, people do not have introspective access to many kinds of mental process, they can introspect the distinctive phenomenal occurrence of an emotion and they can connect such an occurrence with a putative causal event, which in the typical case is obvious rather than hidden or subtle. Both emotions and their usual causes fall within focal attention. As Ericsson and Simon (1980) argue, it is precisely such data that can be verbalised. Data that are outside attention require inferences of the kind that are subject to the errors pointed out by Nisbett, Wilson, Ross, and others.

LANGUAGE AND THE UNIVERSALITY OF BASIC EMOTIONS

A different argument against the existence of basic emotions concerns language and cross-cultural studies. Wierzbicka (this issue), argues that theorists have assumed the universality of categories and facial expressions that correspond to English terms. This ethnocentricity is immediately revealed, the argument goes, if one takes an emotional term from some other culture and tries to apply it to an English-speaking culture. For example, the Ifaluk emotion of *fago* (translated by Lutz, 1982, as "compassion-love-sadness") seems natural and basic to the culture, but it also seems to have no counterpart in English. Likewise, Lutz describes the Ifaluk emotion called *song*, which she translates roughly as "justified anger". Wierzbicka argues that this emotion does not correspond to any basic notion of anger, and that it should not be referred to by the English word, "anger". This argument is important; and we would like to clarify our position.

When a theorist proposes that the emotion or facial expression of, say, "happiness" is a basic and thus universal emotion, the claim is that among the basic emotions, which have evolved in social mammals and which are experienced and communicated among humans, is one that in English is most closely referred to as "happiness". If we have seemed to imply that the English "happiness" is the basic emotion, we apologise. What we mean is that there is a basic emotion, for which in English "happiness" or perhaps "enjoyment" or, to use Wierzbicka's phrase "something like happy" are the nearest indicators. The underlying emotion can be communicated between people nonverbally, and its communication can be effective despite deep gulfs of language and culture. In another language, the emotional terminology will be different, and whatever term corresponds most closely to "happiness" is likely to differ in its connotations. Thus, on Ifaluk, Lutz describes a concept *ker*, which she translates as "happiness/excitement". Cultural attitudes differ: People on Ifaluk do not believe that they have a Jeffersonian right to the pursuit of *ker*. Although pleasant, it has a negative social connotation, and people are distrustful of it because it can lead to showing off, and neglect of concern for others which is highly valued on Ifaluk. Nuances of this kind thus reflect different conscious attitudes to each emotion, cultural differences in its causation, and differences in the forms of morally acceptable behaviour to which it may lead. Moreover, most emotion terms in a language have a meaning that combines reference to a basic emotion with other semantic information, such as the cause of the emotion. Thus, for example, "embarrassment" refers to a state corresponding to fear (a basic emotion) caused by finding oneself an object of unwelcome social attention—a common experience in the English-speaking world. Different languages are therefore likely to focus on different causes and objects of emotion, and so emotional terms may be difficult to translate from one language to another. As many philosophers from Quine (1960) onwards have pointed out, when you seek a translation of a word or expression from one language to another, then you must attribute a certain degree of common rationality to the other culture. You are likely to be sceptical about the accuracy of the translation if it implies total irrationality and that you should cease to treat the individuals of the alien culture as having any meaningful beliefs. Indeed, some philosophers go further and argue that complete rationality is a prerequisite if an individual is to be said to hold any meaningful beliefs (Davidson, 1975; Dennett, 1978, p. 20). If emotion terms were fundamentally untranslatable, as Wierzbicka sometimes seems to imply, then it should be impossible for native speakers of incommensurable languages ever to learn one another's terminology. The emotional life of the Ifaluk should remain forever beyond Lutz's empathic grasp. The emotion of the inhabitants of some alien planet may truly be beyond our comprehension, but no such individuals have ever been found on earth. It may be difficult to translate words denoting emotions, but it is not impossible to empathise with a culture and to learn to experience the corresponding emotions.

In short, our general theory of the semantics of emotion terms, which was applied to English terms in the first instance, should be equally applicable to other languages. It preserves the notions—common to both folk theories and scientific theories—that emotions are distinctive states, that they are caused by recognisable events of which people can be consciously aware, and that they can be directed to objects or to other people.

IS THE THEORY TESTABLE?

In this final section, we will counter the criticism that the theory of basic emotions is too vaguely defined to be susceptible of empirical test (Mandler, 1984; Ortony & Turner, 1990). We believe that this criticism is prompted not by any conceptual difficulty in testing the existence of basic emotions, but by the practical difficulty of such investigations. Indeed, few investigations have been performed that fulfil the conditions to make a compelling case.

One way of falsifying the hypothesis of basic emotions would be to show that the apparent diversity of emotions cannot be reduced to a small basic set because different varieties of, say, fear, have no underlying components in common. What is needed is a set of cumulative studies that test for the universal existence of a small set of basic emotions corresponding to folk theoretical categories. These studies should investigate whether such emotions are experienced, communicated, and recognised universally; and they should investigate whether they have common components in their underlying neurophysiology. Hence, the studies need to examine different cultures, infants on whom culture has yet to impinge, and the physiological systems of animals and human beings (see Panksepp, 1982). It is even possible that certain eliciting conditions for basic emotions will prove to be universal, or at least common to diverse cultures, although the theory does not strictly call for this condition to hold.

Studies of basic emotions are complex, difficult, and time-consuming. Yet, various researchers have begun to undertake them. Ekman and his colleagues have carried out a paradigmatic set of studies that meet the necessary conditions (see Ekman's paper in this issue). They have shown in particular that facial expressions of a basic set of emotions are common across diverse cultural groups, and that basic emotions have distinctive physiological accompaniments.

A stringent hypothesis is that basic emotions should be perceived categorically, just as, for example, the contrast between certain English consonants, has been tested by Etcoff (1990). What distinguishes "bit" from "pit" is a few milliseconds of onset in voicing, i.e. the vibrations of the larynx in the articulation of the phonemes /b/ and /p/. For equal physical differences in voicing onset time, it is difficult to discriminate between two sounds lying on one side or the other of the boundary between /b/ and /p/, but easy to discriminate between two sounds that straddle this boundary. Etcoff argued that if there are basic emotions, then the perception of facial expressions should also be categorical in the same way. Happy faces should be sorted into one category, sad ones into another, and so on. She argued that if she could create equal physical increments in scales ranging between different basic emotions, then there would be categorical boundaries. On one side of them people would see one emotion, on the other a different one, but on either side discrimination should be poorer than across the boundary. She created equal increments using Brennan's (1985) computer program for drawing faces in a way that includes details of eyebrows, eyelids, and mouth. She traced 21 photographs from Ekman and Friesen's

(1976) pictures of facial affect from three models, who each posed expressions of a putative set of basic emotions: sadness, anger, fear, disgust, surprise, and a neutral state. She then used the program to create an incremental series of 11 faces that changed in equal physical increments from one emotion to another. For instance, in one series faces number 1 and number 11 were respectively drawings from the digitised photographs of happy and sad faces of one of the models. Face number 2 derived from the average positions of 10 sets of points from the happy face +1 set from the sad face; face number 3 derived from 9 sets from the happy face +2 sets from the sad face, and so on. With standard psychophysical methods, she then tested the hypothesis of categorical perception of these faces. She observed an abrupt shift in discriminability between the faces in all the series except the one from surprise to fear. She also observed the same effect between the emotion faces and the neutral faces, although the gradations of the neutral faces were more discriminable than those between the emotion faces.

A further corroboration of basic emotions has been obtained by Conway and Bekerian (1987). They found in studies of similarity judgements that emotion terms fell into groups corresponding to basic emotions: happiness/love/joy—misery/grief/sadness—fear/panic/terror—and anger/jealousy/hate. They then used lexical decision tasks to investigate the representation of these concepts in memory. In one experiment, the subjects read two sentences that had previously been judged appropriate to a particular emotion, such as love. They then immediately carried out a lexical decision task in which they were shown a string of letters and had to decide whether or not it was a word. It was either another emotion word from the same basic group, e.g. "joy", an emotion word from another basic group, e.g. "sadness", or a nonword. Interspersed with emotion trials were trials with emotionally neutral filler sentences and words and nonwords. The subjects' reaction times were faster for words from the same basic emotion group than for words from a different emotion group.

These experiments corroborate basic emotions within a single culture. Because the theory postulates an innate and universal foundation for basic emotions, it predicts that the phenomena observed by Etcoff and by Conway and Bekerian should generalise in the same way across different cultures.

CONCLUSION

We have made a case for the psychological reality of emotions and for their foundation on a small set of basic emotions: happiness, sadness, anger, fear, desire, and disgust. Each basic emotion depends on an innate and universal internal mental signal, which can be elicited by rapid and coarse cognitive evaluations that may be common to diverse cultures. These evaluations concern progress towards goals. The internal signals are causal precursors of subjective experience, somatic change, and plans for action. They are also precursors to external signals, such as facial expressions, that communicate the emotion to others. The theory can be contrasted with the rival hypothesis that there are no basic emotions, but instead more fundamental components, out of which all emotional experiences are constructed (Ortony & Turner, 1990). On the one hand, if there are supposed to be components in common to all subjective experiences of, say, fear, including cases where individuals have no knowledge of the cause of the emotion and react in no outward way to it, then the theory is entirely compatible with basic emotions. On the other hand, if there are not supposed to be any components in common to all subjective experiences of an emotion such as fear, then the theory amounts to a rejection of the folk categories of emotion. Emotions are nothing more than naïve illusions. Once dispelled, they will cease to exist as useful pre-theoretical categories for cognitive science. We have argued that there are no strong grounds for rejecting folk psychology; Ortony and his colleagues have defended a similar position (Ortony et al., 1988). Yet, Ortony and Turner (1990) have questioned both the concept of the basic emotions, and what would count as empirical evidence for or against them. They say that "current uses of the notion do not permit coherent answers to be given to such questions" (p. 329). Their own componential theory, however, seems to be either a variant of the basic emotion hypothesis or else a repudiation of the folk theory. The case for basic emotions has not convinced everybody, but the tests that have been carried out appear to corroborate it.

Manuscript received 4 April 1991
Revised manuscript received 22 November 1991

REFERENCES

Brennan, S. (1985). The caricature generator. *Leonardo*, *18*, 170–178.
Cherniak, C. (1986). *Minimal rationality*. Cambridge, MA: MIT Press.
Cohen, L.J. (1981). Can human irrationality be experimentally demonstrated? *Behavioral and Brain Sciences*, *4*, 317–370.
Churchland, P.M. (1984). *Matter and consciousness*. Cambridge, MA: MIT Press.
Conway, M.A. & Bekerian, D.A. (1987). Situational knowledge and emotions. *Cognition and Emotion*, *1*, 145–191.
Craik, K.J.W. (1943). *The nature of explanation*. Cambridge University Press.
Davidson, D. (1975). *Thought and talk*. In Guttenplan, S. (Ed.), *Mind and language*. Oxford University Press.
Dennett, D.C. (1978). *Brainstorms*. Cambridge, MA: MIT Press.

Descartes, R. (1911). *Passions de l'âme*. In E.L. Haldane & G.R. Ross (Eds and Trans.), *The philosophical works of Descartes*. [Originally published 1649.] Cambridge University Press.

D'Andrade, R. (1987). A folk model of the mind. In D. Holland & N. Quinn (Eds), *Cultural models in language and thought*. Cambridge University Press.

de Sousa, R. (1987). *The rationality of emotions*. Cambridge, MA: MIT Press.

Ekman, P. (1973). Cross-cultural studies of facial expression. In P. Ekman (Ed.), *Darwin and facial expression: A century of research in review*. New York: Academic Press.

Ekman, P. & Friesen, W.V. (1976). *Pictures of facial affect*. Palo Alto, CA: Consulting Psychologists Press.

Ekman, P. & Friesen, W.V. (1978). *Facial Action Coding System (FACS): A technique of the measurement of facial action*. Palo Alto, CA: Consulting Psychologists Press.

Ellsworth, P.C. & Smith, C.A. (1988). From appraisal to emotion: Differences among unpleasant feelings. *Motivation and Emotion, 12*, 271–302.

Ericsson, K.A. & Simon, H.A. (1980). Verbal reports as data. *Psychological Review, 87*, 215–251.

Etcoff, N. (1990). *Categorical perception of facial expressions*. Paper presented to the Fifth Annual Meeting of the International Society for Research on Emotions, Rutgers University NJ, 25–28 July.

Fehr, B. & Russell, J.A. (1984). Concept of emotion viewed from a prototype perspective. *Journal of Experimental Psychology: General, 113*, 464–486.

Frijda, N.H. (1987). Comment on Oatley and Johnson-Laird's 'Towards a cognitive theory of emotions'. *Cognition and Emotion, 1*, 51–59.

Hammond, M. (1990). Affective maximization: A new macro-theory in the sociology of emotion. In T.D. Kemper (Ed.), *Research agendas in the sociology of emotions*. Albany, NY: State University of New York Press.

Henle, M. (1978). Foreword to R. Revlin & R.E. Mayer (Eds), *Human reasoning*. Washington: Winston.

James, W. (1890). *The principles of psychology*. New York: Holt.

Jenkins, H. & Ward, W. (1965). Judgements of contingency between responses and outcomes. *Psychological Monographs, 79*, No. 594.

Johnson-Laird, P.N. (1983). *Mental models*. Cambridge, MA: Harvard University Press/Cambridge University Press.

Johnson-Laird, P.N. & Byrne, R.M.J. (1991). *Deduction*. Hove: Lawrence Erlbaum Associates Ltd.

Johnson-Laird, P.N. & Byrne, R.M.J. (In press). Models and deductive rationality. In K. Manktelow & D. Over (Eds), *Rationality*. London: Routledge.

Johnson-Laird, P.N. & Oatley, K. (1989). The meaning of emotions: Analysis of a semantic field. *Cognition and Emotion, 3*, 81–123.

Kahneman, D., Slovic, P., & Tversky, A. (1982). *Judgement under uncertainty: Heuristics and biases*. Cambridge University Press.

Lutz, C. (1982). The domain of emotion words on Ifaluk. *American Ethnologist, 9*, 113–128.

Mandler, G. (1984). *Mind and body: Psychology and emotions and stress*. New York: Norton.

Marr, D. (1982). *Vision*. San Francisco, CA: Freeman.

McDougall, W. (1926). *An Introduction to social psychology*. Boston: Luce.

Medvedev, Z.A. (1990). *The legacy of Chernobyl*. New York: Norton.

Nisbett, R.E. & Ross, L. (1980). *Human inference: Strategies and shortcomings of social judgement*. Englewood Cliffs, NJ: Prentice Hall.

Nisbett, R.E. & Wilson, T.D. (1977). Telling more than we can know: Verbal reports on mental processes. *Psychological Review, 84*, 231–259.

Oatley, K. & Duncan, E. (In press). Structured diaries for emotions in daily life. In K. Strongman (Ed.), *International review of studies of emotion*, Vol. 2. Chichester: Wiley.

Oatley, K. & Johnson-Laird, P.N. (1987). Towards a cognitive theory of emotions. *Cognition and Emotion, 1*, 29–50.

Oatley, K. & Johnson-Laird, P.N. (1990). Semantic primitives for emotions. *Cognition and Emotion, 4*, 129–143.

Ortony, A. & Clore, G.L. (1989). Emotions, moods, and conscious awareness: Comment on Johnson-Laird & Oatley's "The language of emotions: An analysis of a semantic field". *Cognition and Emotion, 3*, 125–137.

Ortony, A., Clore, G.L., & Collins, A. (1988). *The cognitive structure of emotions*. Cambridge University Press.

Ortony, A. & Turner, T.J. (1990). What's basic about basic emotions? *Psychological Review, 97*, 313–331.

Panksepp, J. (1982). Towards a general psychobiological theory of emotions. *Behavioral and Brain Sciences, 5*, 407–467.

Plutchik, R. (1962). *The emotions: Facts, theories, and a new model*. New York: Random House.

Quine, W.V.O. (1960). *Word and object*. Cambridge, MA: MIT Press.

Shaver, P., Schwartz, J., Kirson, D., & O'Connor, C. (1987). Emotion knowledge: Further exploration of a prototype approach. *Journal of Personality and Social Psychology, 52*, 1061–1086.

Simon, H.A. (1967). Motivational and emotional controls of cognition. *Psychological Review, 74*, 29–39.

Stein, N.L. & Levine, L.J. (1989). The causal organisation of emotional knowledge: A developmental study. *Cognition and Emotion, 3*, 343–378.

Stein, N.L. & Levine, L.J. (1990). Making sense out of emotion: The representation and use of goal-structured knowledge. In N.L. Stein, T. Trabasso, & B. Leventhal (Eds), *Psychological and biological approaches to emotion*. Hillsdale, NJ: Lawrence Erlbaum Associates Inc.

Stich, S. (1983). *From folk psychology to cognitive science*. Cambridge, MA: MIT Press.

Stich, S. (1990). Rationality. In D.N. Osherson & E.E. Smith (Eds), *Thinking: An invitation to cognitive science*, Vol. 3. Cambridge, MA: MIT Press.

von Uexkull, J. (1957). A stroll through the worlds of animals and men. In C.H. Schiller (Ed. and Trans.), *Instinctive behavior: The development of the modern concept*. London: Methuen.

Wason, P. (1960). On the failure to eliminate hypotheses in a conceptual task. *Quarterly Journal of Experimental Psychology, 12*, 129–140.

Acknowledgments

Rosenthal, David M. "Two Concepts of Consciousness." *Philosophical Studies* 49 (1986): 329–59. Reprinted with the permission of Kluwer Academic Publishers.

Nelkin, Norton. "What Is Consciousness?" *Philosophy of Science* 60 (1993): 419–34. Reprinted with the permission of the University of Chicago Press.

Lycan, William G. "Consciousness as Internal Monitoring, I." In *Philosophical Perspectives, 9, AI, Connectionism and Philosophical Psychology, 1995,* edited by James E. Tomberlin (Atascadero, Calif.: Ridgeview Publishing, 1995): 1–14. Reprinted by permission of Ridgeview Publishing Company.

Crick, Francis and Christof Koch. "The Problem of Consciousness," *Scientific American* 267 (September 1992): 153–59. Reprinted with the permission of Scientific American, Inc.

Damasio, Antonio R. and Hanna Damasio. "Images and Subjectivity: Neurobiological Trials and Tribulations." In *The Churchlands and Their Critics,* edited by Robert N. McCauley (Cambridge: Blackwell, 1996): 163–75. Reprinted with the permission of Blackwell Publishers.

Flanagan, Owen. "Consciousness and the Natural Method." *Neuropsychologia* 33 (1995): 1103–15. Reprinted with the permission of Elsevier Science Ltd.

Dennett, Daniel C. "The Evolution of Consciousness." In *Speculations: The Reality Club,* edited by John Brockman (New York: Prentice Hall, 1988): 87–108. Reprinted with the permission of Prentice Hall.

Churchland, Paul M. "The Rediscovery of Light." *Journal of Philosophy* 93 (1996): 211–28. Reprinted with the permission of the Journal of Philosophy, Inc., Columbia University, and the author.

Searle, John R. "Consciousness, Explanatory Inversion, and Cognitive Science." *Behavioral and Brain Sciences* 13 (1990): 585–642. Reprinted with the permission of Cambridge University Press.

Jackson, Frank. "Epiphenomenal Qualia." *Philosophical Quarterly* 32 (1982): 127–36. Reprinted with the permission of Basil Blackwell Ltd.

Chalmers, David J. "Facing Up to the Problem of Consciousness." In *Explaining Consciousness: The "Hard Problem,"* edited by Jonathan Shear (Cambridge:

MIT Press, 1995): 9–30. Reprinted with the permission of Cambridge University Press.

O'Rorke, Paul and Andrew Ortony. "Explaining Emotions." *Cognitive Science* 18 (1994): 283–323. Reprinted with the permission of Ablex Publishing.

Sloman, Aaron. "Motives, Mechanisms, and Emotions." *Cognition and Emotion* 1 (1987): 217–33. Reprinted with the permission of Lawrence Erlbaum Associates Ltd.

Johnson-Laird, P.N. and Keith Oatley. "Basic Emotions, Rationality, and Folk Theory." *Cognition and Emotion* 6 (1992): 201–23. Reprinted with the permission of Lawrence Erlbaum Associates Ltd.